OCEAN WASTE DISPOSAL PRACTICES

OCEAN WASTE DISPOSAL PRACTICES

Alexander W. Reed

NOYES DATA CORPORATION
Park Ridge, New Jersey London, England
1975

Library of Congress Catalog Card Number: 75-15205
ISBN: 0-8155-0591-4
Printed in the United States

Published in the United States of America by
Noyes Data Corporation
Noyes Building, Park Ridge, New Jersey 07656

FOREWORD

This book is an evaluation of waste discharges and practices as they affect the littoral regions of the United States, with many projections of the ultimate influence upon the world's oceans.

The problems of rapidly increasing waste loads arising from the exponential increase in populations make a constant review of the assimilation capacity of coastal waters a pressing necessity. The overall problem is not a static one. The level of research and study must keep pace in order to meet the growing demand for transforming the wastes into an assimilable condition before dumping them into the ocean.

Waste generation is inevitable—most wastes eventually reach the ocean whether they are buried on land, discharged into the air or run into streams. But there is now a growing public awareness that the ocean must be kept reasonably intact not only as a means for marine transport, but also to develop heretofore untapped resources such as unfamiliar but abundant fish species, offshore sand and gravel, deep water manganese deposits, thermal gradients and currents and artificially induced upwellings to produce electrical energy. Clean seawater, beaches, and shorelines are valuable resources and indispensable to mankind for their recreational and aesthetic values.

The major purpose of the government-sponsored reports from which this book is constructed is to provide scientific findings, evaluations, and conclusions with respect to possible long-range effects of oceanic pollution and man-induced changes of ocean ecosystems. This book should help to weigh these effects against the need for and benefits to be derived from ocean dumping and its alternatives.

A list of references is included to provide easy access to the information contained in this book.

CONTENTS AND SUBJECT INDEX

STATING THE CASE FOR OCEAN DUMPING

The following sources were used for the information provided in this chapter.

PB 198 032
PB 233 019
PB 240 917
COM 74 50632

The scientists and engineers concerned with the general field of environmental quality control have, in the past 50 to 100 years, accomplished a considerable amount towards the understanding and alleviation of certain of the waste problems facing the world society. However, because realization of the magnitude of the total problem has been difficult by both the professional and by the general public, there has been a tendency for most efforts to be concentrated in "putting out fires" as they became recognized. An overall viewpoint has been slow in arriving. The history of discharges into Santa Monica Bay on the Southern California coastline furnishes a case history to point this up.

In earlier and simpler times all raw wastes, though relatively small quantities, were simply dumped into the nearshore waters of the Bay on the basis that the discharge point and area were so far removed from the population that they constituted a bottomless pit where the problem could be simply forgotten. As population and quantity of waste discharges grew and beach use increased, the area of biological hazard expanded and it became necessary to take measures to avoid endangering the beach users by waterborne diseases. Continued population growth and a better understanding of the overall waste problem prompted design for both biological and aesthetic control.

The continuing increase in population, pressure from an affluent society for recreational use of the Bay and its beaches including fishing, and concern for the ecology of the Bay and coastline led to its recognition as a separate system. A research project is now proposed to study the whole Southern California nearshore water system as an ecological unit including interfaces with wastes versus air and land. Ultimately, effects upon the world ocean must be considered.

Throughout this time there was an implication that the ocean would remain the final wastes sink. It is felt that this is still true. Most waste materials which reach waters at any place throughout the world will eventually be deposited in the oceans as the ultimate sink. But this fact does not necessarily imply disaster, however, for Santa Monica Bay even with the volume of liquid wastes now being received, all of the beneficial uses described in the regulations of the State of California Water Quality Control Board are being met.

The problems of waste discharges into the coastal waters from other relatively uncontrolled sources and of the rapidly increasing waste loads arising from the exponential increase in population make a constant review of treatment requirements and of coastal water assimilation capacity necessary. It is not sufficient to simply stand on past practices and past successes; the overall problem is not a static one so the level of research and study must keep pace.

In terms of the evaluation of sources and characteristics of waste discharges, as they affect the littoral region of the coasts of the United States, the operational agencies have expended a great deal of effort in obtaining information and in controlling sources of discharge under their jurisdiction. In a general way a considerable amount of information on these is available. However, there are a number of relatively uncontrolled sources of wastes which reach the marine environment about which little is known. The following waste quantities have been estimated for the United States.

Agricultural wastes	1.3 billion tons/yr
Mining: (reserve mining, Lake Superior; uranium in Colorado River; coal mining discharges in Pennsylvania, Virginia and Kentucky)	1.0 billion tons/yr
Air pollution discharges	142 million tons/yr
Solid and liquid wastes from municipalities	350 million tons/yr
Automobile bodies, etc	15 million tons/yr

Assuming these estimates to be reliable, solid and liquid wastes from municipalities listed above, makes up perhaps 12.5% of the total. For better understanding of the overall problem, the contribution by these other sources must be better defined.

In considering a subject as wide and as diffuse as the evaluation of the sources and characteristics of waste discharge, it is difficult to make a neat and concise outline of all facets of the subject. The following is an attempt to list a number of the items of concern in terms of degree of knowledge.

High Level of Knowledge

(1) Naturally occurring organic wastes: their composition, decomposition and treatment; for instance, the nitrogen cycle, carbon cycle, sulfur cycle, etc., are well-known.

  (a) There are limited data on subjects such as virus, pathogens, parasites, and the effect of some metabolic products.

(2) The quantity of wastewaters from sanitary sewers, storm drains, rivers, streams, etc.

(3) Industrial wastes quantity and quality from well established processes (except trace elements, etc.).

(4) The relatively standard methods of sampling, monitoring, and analysis.

Incomplete Levels of Knowledge

(1) Wastes from newer industries and those making exotic products. (Half-life determinations, impact on ecology, etc., are factors to be explored.)

(2) Fallout of material entering atmosphere either by recognized avenues of air pollution or by methods such as evaporation after agricultural application.

(a) Information has been developed although incomplete, covering the following to some degree:
1. DDT, chlorinated hydrocarbons, chlorinated biphenyls.
2. The metabolic products of the materials of Item 1 above.
3. Oil and oxidized oil products.
4. Lead alkyls and some other trace metals.
5. Radioactive materials.

(3) Materials received into the ocean from storm drainage systems.

(a) Pesticides, herbicides, and other poisons.

(b) Material from industries which are presumably too small in quantity or too mild to require passage through treatment systems.

(c) Oil field wastewaters, particularly the composition of organic substrates.

(d) Urban runoff consisting of wash from the land and building surfaces. This includes fallout from air pollution, street litter, lawn fertilizers, etc., and is an increasingly important problem.

(e) Material reaching the storm sewer system as a result of washings from or losses from transportation carriers.
1. Trucks and trains carrying toxic and/or dangerous materials.
2. Losses from pipelines (understream crossings, loading stations, etc.).
3. Radioactive substances in aircraft washings.

(4) Effects of solid waste disposal; what are effects on adjacent marsh lands; effect of proposed refuse islands.

(5) Effects of many common chemicals in general use in industry, commercial establishments and the home.

The Need for Integration of Present "Knowns"

A need exists for systematic summation of findings and collation of known facts to various disciplines. For example, the quantity of surface runoff entering the coastal waters has been determined accurately;

a total inventory to include quality parameters of the water is needed as a corollary to the above need.

Speculative Unknowns

(1) The many and interrelated results of breaking a more or less closed ecological chain are not understood.

(2) Many strange metabolic products are entering the environment. In addition chemicals which may encourage mutations are being used and discharged. What are the threats to the environment, if any, from these materials? If these materials are having an effect, what is the biological "half-life" of each of them? Little information is available on this subject, except in the field of radioactivity and perhaps to a small degree for DDT.

(3) The whole question of synoptic testing and/or measurements is pertinent. There is very little understanding of the total effect on the environment as a result of the society's waste products. The phenomena beggar description and the few samples and tentative conclusions now possible may miss significant points such as the mass effect. There is a need to develop new techniques to adequately measure complex interactions over large areas.

Sampling and Testing

Test procedures must be developed which will permit more complete evaluation of the biological effects on receiving waters and their biota from waste discharges. It is fruitless to measure wastewater constituents if their relevance is not better understood.

Ocean disposal of wastes will continue to increase in the future unless technological breakthroughs occur in the economic recycling of these misplaced natural resources. Although it may be temporarily necessary to apply restrictions and sanctions to reduce the degradation of the marine environment by pollution, the long term cure must come from the recovery and reuse of these wasted sources of metals, fuels, fibers, chemicals, soil conditioners and fresh water.

Thus, a positive policy of supporting and encouraging research in the technology of resource reclamation while minimizing the negative "thou shalt not" approach of control by edict and decree should be emphasized. Alternate methods of disposal for virtually all waste except polluted dredge spoil exist. However, the economic factors of transportation costs by rail or truck, capital expenditures for landfill areas or incinerators and energy requirements to use these alternates will result in pressures to use the oceans as disposal sites. Most of these alternates are also objectionable from an aesthetic standpoint and may cause greater air, land, and water pollution problems than ocean disposal.

The following projections of increases in volumes of the various waste types disposed in the ocean are based upon current trends and do not reflect any new reclamation processes or major changes in present disposal techniques. By comparison, coastal population is increasing at an annual rate of 1.5%, and industrial production is increasing 4.5% annually.

Solid Wastes – Because of the reduction in availability of landfill sites and restrictions to protect air and groundwater quality, ocean dumping of garbage, trash and construction debris is expected to increase 17% annually.

Sewage Sludge – Because of increased use of metropolitan collection and treatment facilities by both industry and septic tank users and the greater efficiency of solids removal by advanced treatment processes, ocean disposal of this waste is estimated to increase 5% annually.

Industrial Wastes – Because of increased recycling of acids and caustics, ocean dumping of this class of waste is declining in volume. However, the remaining chemical wastes are both more complex and more concentrated. Thus the volumes of seawater required for dilution and biodegradation to subtoxic levels are estimated to remain static.

Conventional Munitions – Alternate methods of disposal are now used for explosives and no further ocean dumping of significant volume is expected. This waste category was the most innocuous and had the least environmental impact.

Dredge Spoil – Because of the economic advantages of larger cargo vessels, there is pressure to deepen ports and channels. Maintenance dredging at one-fourth of the present intermediate depth ports will increase by 40% annually. The national annual increase in ocean dumping of spoil is estimated at only 8% because more of this material will be used for beach replenishment and development of wildlife habitat. Spoil quality may improve because of stricter regulation of waste discharges into harbors and navigable waterways.

Radioactive Wastes – The present moratorium on ocean dumping of high-level wastes is expected to remain in effect for the next fifteen years. In the same period, there will be a great increase in the number of nuclear power generating plants, nuclear powered ships, and small isotope powered thermoelectric sources for remote weather stations, lighthouses, buoys, etc. Beginning about 1990 there will be a sudden increase in the number of obsolete devices and unless recycling or conversion techniques are developed, the pressure to use the ocean deeps for disposal will be immense because of the cost, and both real and imaginary hazards of land disposal.

Total volumes of material being dumped in the ocean are increasing five times more rapidly than the general population. Pressures will be exerted in the future to relax certain restrictions on ocean disposal. An adequate body of knowledge should be assembled to provide for immediate control of present waste disposal problems and for evaluation of future disposal needs.

In October, 1972, Congress provided legislation for Federal control of water pollution: the Federal Water Pollution Control Act Amendments of 1972 (Public Law 92-500) and the Marine Protection, Research and Sanctuaries Act of 1972 (Public Law 92-532).

Under Public Law 92-500, discharges of pollutants into navigable waters, which includes the territorial sea and the contiguous zone, are unlawful without a permit. Section 402 establishes the National Pollution Discharge Elimination System (NPDES) for all public and private dischargers to be administered by the Environmental Protection Agency (EPA). Individual states may issue permits under their own program if it is authorized and approved by EPA.

Rules and regulations governing the NPDES were published in the *Federal Register* on May 22, 1973. Permits for transporting and dumping wastes anywhere in ocean waters are required under Public Law 92-532, effective April 23, 1973. Interim regulations and criteria governing the issuance of dumping permits were published in the *Federal Register* on April 5 and May 16, 1973, respectively. Final regulations and criteria were issued October 15, 1973 (38 FR 28610-28621).

On the international level, a "Convention on the Prevention of Marine Pollution by Dumping of Wastes and Other Matter" was approved at an International Conference in London, October 3 to November 13, 1972. The Convention was ratified by the United States on August 3, 1973 (S. 15800) and will become effective following ratification by fifteen Contracting Parties (countries). In March 1974 Public Law 92-532 was amended (Public Law 93-254) to incorporate those provisions of the international convention not included in the original legislation.

The London Convention prohibits the dumping of some materials, requires special care for the dumping of other identified substances, and provides for a general permit for others. The prohibited substances (except as trace contaminants) include organohalogen compounds, mercury and cadmium and compounds of these metals, persistent plastics and other persistent synthetic materials, crude oil and certain other petroleum products.

Absolutely prohibited are high level radioactive wastes and materials produced for biological and chemical warfare. The prohibition does not include substances which are rapidly rendered harmless by physical, chemical, or biological processes in the sea, unless they make edible marine organisms unpalatable or endanger human health.

The list of materials which require special care includes arsenic, lead, copper, zinc, and their compounds, organosilicone compounds, cyanides, fluorides, and those pesticides not contained in the list of prohibited substances. In addition to these, the disposal of industrial wastes containing large quantities of acids and alkalis are to be assayed for beryllium, chromium, nickel, vanadium, and their compounds. The regulations adopted by EPA under Public Law 92-532 have been designed to be consistent with the International Convention.

Public Law 92-532 gives responsibility for issuing permits for ocean disposal of dredged material to the U.S. Army Corps of Engineers, subject to the criteria developed by EPA. Permits for dumping all other materials, e.g., municipal sewage sludge, industrial waste, are issued by EPA. Radiological, chemical, or biological warfare agents, and high-level radioactive waste may not be dumped under any circumstances.

The provisions of the Marine Protection, Research, and Sanctuaries Act of 1972 reflect the concern and public awareness that coastal and ocean resources important to man can be threatened by unregulated ocean dumping. Clean seawater, beaches, and shorelines are valuable resources, as are bountiful and disease-free finfish and shellfish stocks, and the natural areas that support them. Other truly unique areas that will be designated as marine sanctuaries under other provisions of this Act have also been formally recognized as resources of recreational, scientific, and aesthetic values. The Marine Protection, Research, and Sanctuaries Act of 1972 addresses the potential conflicts between the impacts of man-induced changes in the ocean ecosystem, and certain nonconsumptive uses as well as the integrity of primary resources themselves. The Act directly controls dumping by requiring, prior to issuing permits, the EPA, and the Corps of Engineers in the case of dredged spoils, to formally and publicly consider the effects of such dumping on human health and welfare, including economic, aesthetic and recreational values, fisheries resources, wildlife, shorelines and beaches, marine ecosystems, and on nonliving resource exploitation, etc., and to weigh these effects against the need for and benefits to be derived from ocean dumping and its alternatives.

The Act also directs the Department of Commerce to work with the Coast Guard and to initiate a comprehensive and continuing program of monitoring and research regarding the effects of ocean dumping, and, in consultation with other agencies, to initiate a comprehensive and continuing research program with respect to possible long-range effects of pollution, overfishing, and man-induced changes of ocean ecosystems. A major purpose of these research and monitoring activities is clearly intended to provide scientific findings, evaluations, and conclusions that are relevant to defining use conflicts that could influence the decisions to grant or deny dumping permits.

Waste generation is inevitable—most wastes will eventually reach the waters of the world ocean whether they are buried on land, discharged into the air, or discharged directly to receiving waters. Initially, this load will affect the littoral or nearshore waters. This makes the problems more acute for with the exception of agricultural valleys these waters represent the most productive areas in the world.

The technology of waste disposal has not kept pace with production. The threat is ominous, however, action is commencing toward a better understanding of where the major effort should be placed. Once constraints and guidelines have been developed compatible techniques can be adopted.

DEFINING OCEAN WASTES AND ASSESSING THE EFFECTS

The following sources were used for the information provided in this chapter.

PB 204 868
PB 213 473
PB 224 793
PB 240 917
AD 757 599
COM 74 50632

PARAMETERS OF COASTAL WASTE CHARACTERIZATION

Physical Description

Wastes transported aboard barges and ships for disposal at sea from the U.S. coastal cities surveyed can be summarized into several major categories and ranked according to total volume: (1) dredge spoil; (2) industrial wastes, (3) sewage sludge, (4) refuse, (5) radioactive wastes, (6) construction and demolition debris, (7) military explosives and chemical wastes, and (8) miscellaneous.

Dredge Spoil: Dredge spoil consists of sediments containing various concentrations of alluvial sand, silt, clay, and municipal or industrial waste sludges dredged to improve and maintain navigation channels. It is these sludges which put the Corps of Engineers into the municipal and industrial waste disposal business giving the Corps a responsibility neither within its charter nor desired by that agency. Most of the dredging operations are conducted by the Corps with seagoing hopper dredges. These spoils are normally disposed of in open coastal waters generally not more than three to four miles from the dredging site.

Dredge spoils currently account for 80% of the ocean waste disposal by volume. Land disposal of dredge spoil is becoming an increasing problem in many areas. To compound the situation, the energy crisis is calling for extension of harbor facilities to handle the more modern vessels which have drafts from 35 to 100 feet. Not only does initial dredging of harbors, for super port capability, result in tremendous initial volumes of ocean disposal, but maintenance dredging will be increased proportionally. Along with the routine problem of disposing of non-polluted dredge spoil as the harbors are enlarged to meet the new draft requirements, it will be necessary to remove highly polluted dredge spoil from the inner harbor areas. At present, 34% of dredge material is considered by the Corps of Engineers to be polluted. Since dredge spoil disposal represents such a large portion of ocean wastes, it will be treated more thoroughly later in this chapter.

Industrial Wastes: Industrial wastes originate from a variety of manufacturing and processing operations including petroleum refining, steel and paper production, pigment processing, insecticide-herbicide-fungicide manufacturing, chemical manufacturing, oil-well drilling operations, and metal finishing, cleaning, and plating processes, as well as many others. Industrial wastes vary greatly in both physical characteristics and toxicity. The severe effects that may result from the disposal of these wastes present the greatest potential hazard to the total marine environment of any of the wastes discussed.

The rate of disposal (by volume) of industrial waste in the ocean is decreasing. Reasons given for the decreases are: new methods of recycling waste material; higher costs of barge disposal of industrial waste; increasingly stringent state and Federal control regulations; new industrial processes which have resulted in lesser volume of waste, but with higher concentrations of toxic materials; adverse public opinion; and hazards associated with disposal in the ocean environment, and increasing use of municipal facilities.

Materials for which ocean disposal is likely to remain necessary are: (1) outdated drug stocks from pharmacies and warehouses which, because of the danger of falling into illegal channels of use, are disposed of at sea, and (2) highly hazardous materials such as metallic sodium, the transportation of which presents a major problem. It should be noted that about 40% of the nation's industrial activity is concentrated in the estuarine zone; therefore, disposal of industrial waste in the ocean frequently represents the shortest and safest transportation route for disposal of wastes.

Refinery Wastes — Wastes from petroleum refining operations are produced during the chemical refining processing used to extract various products from the crude oil. These include spent caustic solutions, sulfuric acid sludges, dilute process water solutions, spent catalysts, petrochemical wastes, and chemical cleaning wastes. In many cases solid wastes, such as spent clay, catalysts, and sludges are slurried with liquid chemical wastes for disposal at sea. The spent caustics and acid sludges also contain varying amounts of contaminants including: sulfides, sulfates, phenolates, naphthenates, sulfonates, cyanides, heavy metals, mercaptides, chlorinated or brominated hydrocarbons, and other organic and inorganic compounds.

Spent Acids — Large quantities of spent sulfuric acid and ferrous sulfate are produced by steel mills. The acid, or pickle liquor, is usually withdrawn from the operation and disposed of when one-half or two-thirds of the acid is converted to ferrous sulfate. Typical wastes contain up to 7% free acid and up to 30% ferrous sulfate. Some steel mills have converted a portion of their pickling operations to hydrochloric acid which has resulted in reduced volumes containing spent hydrochloric and sulfuric acids.

Another major waste results from titanium pigment manufacturing in which the ore, with iron as a principle impurity, is digested with sulfuric acid. The wastes include the liquor with 7 to 9% free sulfuric acid and 8 to 10% ferrous sulfate and a mud slurry with 15 to 20% inert solids.

Pulp and Paper Mill Wastes – Pulp and paper mill wastes include the organic constituents of wood being pulped, dissolved salts from the digestion process, undigested pulp, and fibers that escaped processing. Depending on the process, the wastes may contain a sulfate cooking solution (calcium, magnesium or ammonia), black liquor (sodium hydroxide, sulfide, sulfate) and organic constituents including lignin, hexose and pentose sugars.

Chemical Wastes – Chemical manufacturing, chemical laboratories, metal cleaning finishing and electroplating processes, and a variety of other industrial operations all produce a host of complex waste chemicals. These include mercuric or arsenical compounds, organic acids, pesticide chlorinated hydrocarbons, alkalis, anilines, cyanides, and other highly toxic substances. Chemical industry wastes are particularly complex in their composition and behavior.

Oil Drilling Wastes – Oil-well drilling wastes disposed of at sea consist primarily of drilling muds containing barite and diatomaceous clays and cuttings (rock chips) from the drill hole. This type of waste is physically similar to dredging spoils.

Waste Oil – These wastes consist principally of oil sludge that remains after any rerefining process. In the case of oil tankers, the wastes are oil residues resulting from tank cleaning operations (butterworthing). Sources of waste oil are numerous, diverse, and include service stations, ship tanks, tank cars, etc. (the Oil Pollution Control Act of 1961 restricts the disposal of these wastes at sea to areas beyond the 50-mile limit).

Sewage Sludge: Indications are that point source disposal of sewage sludge will increase. Barge disposal costs of sewer sludge range from $6.90 to $18.40 per ton. The cost per ton/mile range is from $.04 per ton/mile to $.30 per ton/mile. Piping of digested sewage sludge runs approximately $.22 per ton/mile to $.44 per ton/mile. Although there are no apparent cost advantages, the operational advantages are significant. One of the major advantages is that pipe disposal is an automated, year-round system that is not dependent upon such variables as labor disputes or weather.

Pipeline disposal of treated sewage sludge is used by many large metropolitan areas, such as Los Angeles, whose Hyperion plant disposes of approximately five MGD of sludge per week from a seven mile long outfall. The Orange County Sanitation District Plant, which processes all municipal sewer waste for the western part of Orange County, California, also uses a pipe outfall.

Increasing population in the coastal zone and the higher level of treatments required to meet water quality standards will result in even more sludge. Environmental problems resulting from sludge disposal in the ocean are significant, both in terms of volume and the toxic, and sometimes pathogenic, materials involved. Communication with sanitation plant officials indicates that industrial disposal into municipal sewer systems is on the increase because of elimination of industrial outfalls into rivers and harbors. Most municipal waste treatment plants are not properly equipped to handle this industrial load; therefore, the extent of heavy metals and other substances toxic to marine biota in the sewer sludge is increasing. The Council on Environmental Quality (CEQ) report on ocean disposal estimates that there will be a 50% increase in sewer sludge disposal within the next 30 years.

Total solids in sewage include large and small suspended particles as well as matter in true solution. The weight of these solids is usually 0.03% or less of the total sewage solution by weight. Solids are removed and treated by screening, grit removal, primary sedimentation, final sedimentation, anaerobic digestion and thickening. Thickened sludges currently being disposed of at sea from barges generally range between 3 to 5% solids by weight.

Sludge properties needed to predict its fate are the same as those of dredge spoils. Bulk specific gravity can be controlled by watering processes or use of additives. Primary sludges have solids contents of 2 to 5% with 70 to 80% volatile matter. A well-digested sludge contains 40 to 50% volatiles and 5% solids which can be increased to 10% by dewatering. Sludge properties for aerobic and anaerobic processes are given in Table 2.1. Results of studies of settling velocity distribution of digested sludges in sea water indicate a high proportion of solids have low settling velocities (90% at Santa Monica and 100% at San Diego are <1 cm/sec).

TABLE 2.1: TYPICAL DIGESTED SLUDGE CHARACTERISTICS

Characteristics	Anaerobic		Aerobic	
Total solids	34,000	68,000	–	–
Percent volatiles	48	48.6	47	48.9
Specific gravity	1.011	–	–	–
pH	7.7	7.2	6.7	5.6
Maximum particle size, microns	2,000	–	–	–

Refuse: Limited disposal of refuse and garbage at sea occurs. These wastes originate primarily from canneries and from commercial and naval vessels. Cannery wastes consist of ground fruit pits, skins, etc., and are disposed of on a seasonal basis. The composition of vessel refuse depends on the type of vessel, geographic location, and season, and consists of varying amounts of food, paper, plastics, metal, glass, and similar wastes.

However, the disposal of processed refuse and garbage to the ocean is receiving more attention as it approaches an economically feasible status relative to land disposal. One process being considered involves sinking of properly compressed bales and depends on the increased pressure with depth to maintain a density sufficient to keep the material on the bottom.

Radioactive Wastes: The nuclear energy industry produces high activity, intermediate activity, low activity, and so-called nonactive wastes. Activities range from hundreds of thousands of curies per gallon for high activity wastes to microcuries per gallon for low activity wastes. Solid wastes include contaminated laboratory or process equipment, clothing, and other items. Between 1946 and 1967 the U.S. Atomic Energy Commission (AEC) disposed of limited quantities of solid packaged radioactive waste materials at designated disposal sites in the Atlantic and Pacific Oceans. In most cases these wastes were packaged in weighted 55-

gallon drums. Most of these wastes are now disposed of on land.

Although ocean disposal of radioactive waste is virtually nonexistent, this is an area which must be closely followed by the Ocean Disposal Program because, in the near future, the volume of nuclear power plants on-line is going to increase by a factor of 20. A significant number of nuclear powered submarines and surface ships (approximately 100) are in service. Nuclear vessels regularly discharge low-level liquid coolant wastes at sea under strict AEC regulations.

With the lifetime of 1 to 30 years for the power plants and reactor vessels, EPA must be prepared to meet the pressure for disposal of the used reactor vessels in the ocean. Alternative methods for disposal of these vessels are by entombment on site which involves expensive industrial work, or dismantling and storage at an inland Atomic Energy Commission site.

Construction and Demolition Debris: Typically, these wastes consist of masonry, tile, stone, plastic, wiring, piping, wood, and excavation dirt. (At this time the only sea disposal of this type of waste is from New York City, and is covered in the study of the New York Bight.)

Explosives and Munitions: This category includes unserviceable or obsolete ammunition such as shells, mines, solid rocket fuels, propellents, agents. Until 1964, the disposal of explosives was conducted primarily from barges and ships. Since then, gutted World War II Liberty ships have been utilized, being sunk with the wastes aboard in water depths greater than 6,000 feet.

If properly carried out, the ocean disposal of conventional munitions (nonradioactive) has little or no effect on a marine environment. Other methods of disposal of such munitions, burning, exploding, or dismantling, are hazardous and, as such, are not attractive alternatives. Poorly regulated operations could result in an explosive device being brought to the surface by fishing nets, or become snagged by a boat's lines, or by a manned submersible.

Miscellaneous: Included in this category are various types of rejected or contaminated products, such as foodstuffs, appliances, small batches of toxic wastes (barreled) such as pesticides and complex chemical solutions. Wastes in this category are usually disposed of in small lots, and records regarding their disposal are difficult to obtain as the disposal operations are not often sanctioned by any regulatory agency.

Location of Disposal Areas

Two-hundred eighty-one ocean areas designated for the disposal of wastes were identified and used to summarize the regional distribution of disposal areas for the principal waste categories (Table 2.2). It is evident that the number of disposal sites for dredge, spoil, industrial wastes, and explosives was not equal for the three coastal areas. Most radioactive waste disposal sites are located off the Atlantic Coast. Disposal sites for refuse were found on the Pacific Coast and those for sewage sludge, only on the Atlantic Coast. The Atlantic Coast has 48% of all disposal sites. Differences of location are the result of specific locations of some wastes, whereas other sites are shown as center points of large dump areas.

TABLE 2.2: NUMBER OF MARINE WASTE DISPOSAL AREAS (BY REGION AND WASTE TYPE)*

	Disposal Areas			
Waste Type	**Pacific**	**Atlantic**	**Gulf**	**Total**
Dredge spoil	15	83**	63	161
Industrial waste	9**	15**	16	40
Sewage sludge	0	2	0	2
Refuse	3**	0	0	3
Radioactive waste	10**	25**	2	37
Explosive and chemical ammunition	19**	19**	11	49
TOTAL (no duplicates)	54	135	92	281

*Revised and updated by James L. Verber, Food and Drug Administration, U.S. Department of Health, Education, and Welfare.

**Areas used for two or more types of wastes.

Sixty percent of the dredge spoil areas are located within three miles of the coast in water depths less than 100 feet. Most, but not all, of the radioactive and explosive waste disposal areas are situated at depths of 6,000 feet or more. Ocean depths of this magnitude are located 10 or more miles from the Pacific coast and 100 or miles from the Atlantic and Gulf coasts. The explosive disposal areas are identified on U.S. Coast and Geodetic Survey charts, and in some cases these also serve as disposal areas for toxic industrial waste material.

Industrial and municipal waste disposal areas are situated at distances from 15 to 100 miles offshore depending on the type of waste and established regulatory procedures. For example, acid wastes are disposed of in a designated area 15 miles from the coast of New York City; by contrast, the disposal site for certain toxic chemical wastes is nearly 125 miles offshore. Most industrial waste disposal operations in the Gulf of Mexico are conducted beyond the 400-fathom line (2,400 feet), which off Galveston requires disposal about 125 miles off the coast.

Past determinations of whether or not a given offshore area was suitable for waste disposal have been based primarily on the following considerations: (1) the disposal area should be far enough away from the coast and/or have an ocean depth deep enough to ensure that no identifiable wastes return to public beaches or interfere with fishing or recreation in coastal areas; (2) the disposal areas should be within a reasonable distance from the coast so that the costs of tug and barge operations are minimized.

In a small number of cases, the potential harm to the environment and the restrictions that the waste disposal operations might place on future exploitation of natural resources (i.e., fishing, minerals, oil research) were taken into consideration. Some studies have been conducted of various environmental aspects of individual disposal operations; these, however, have been limited in both scope and duration and in most cases were conducted after the disposal operations were in progress.

Disposal Methods and Costs

Methods employed for sea disposal of the various wastes consist primarily of transporting the wastes in bulk or containers aboard towed or self-propelled barges. Dredge spoil is handled routinely by U.S. Corps of Engineers aboard their oceangoing hopper dredges. Bulk wastes are normally discharged from tank barges while underway. Containerized wastes might, depending on local practices, be weighted and sunk, or ruptured at the sea surface by axes or rifle fire and allowed to sink. In a few cases highly toxic chemical wastes such as arsenic and cyanide are carried to sea regularly as deck cargo on merchant ships.

The containers are then discharged overboard in undetermined areas once the ship is at least 300 miles from shore. In addition, an accepted method of disposal of spent caustics from oil refineries is as ballast on outgoing crude oil tankers that is discharged far at sea.

Dredge Spoil: Corps of Engineers hopper dredges resemble small tankers in general appearance except for a large amount of deck equipment. The Corps presently owns and operates a fleet of 15 hopper dredges in U.S. coastal and inland waterways. Hopper dredges load by sucking up the bottom sediments through drags, or underwater pipes, while underway in the dredging area. In general, the hauling distance to the spoil disposal areas averages 3 to 4 miles, although in the New York area the trip to the designated offshore disposal area is more than 30 miles.

The 1968 costs per ton for dredge spoil disposal, ranged from 20 to 55 cents per ton, representing the reported total cost of the dredging operations and converting cubic yards to tons by utilizing a factor of 1.2 tons per cubic yard for average sediments consisting of a mixture of sand, silt, and clay (Table 2.3). Tonnages were generally based on average annual discharges, reflecting the years 1967 through 1969.

Industrial Wastes: Bulk industrial wastes are most commonly transported to sea in tank barges ranging in capacity from 1,000 to 5,000 short tons. These barges must be of double-skinned bottom construction and certified for ocean waters by the U.S. Coast Guard, which also promulgates regulations regarding the bulk shipment of chemicals by water. The five-thousand ton capacity barge *Sparkling Waters*, for example, was designed to handle insecticides and other toxic chemical wastes and operates out of New York. *Sparkling Waters* reportedly can be emptied in 30 minutes. Discharge rates of other waste disposal barges vary between 4 and 20 tons per minute.

The depth at which the wastes are discharged is typically between 6 and 15 feet, and towing speeds from 3 to 6 knots are normally used during the discharge operations. Oceangoing tugs used to haul the barges to the disposal areas are generally of the 1,000 to 2,000-hp class. Average and reported range of costs for disposal of bulk industrial wastes were computed on the basis of the weight of the volume of water in which the wastes were contained. Costs varied according to geographical area, type of waste, distance to the disposal area, and annual volume of wastes handled.

Estimated disposal costs for industrial wastes represent a synthesis of the data supplied from three sources: (1) published, (2) waste producers, and (3) barging firms. These figures represent actual barging costs and do not include other costs incurred by the waste producers for treatment, storage, loading, etc., of the wastes.

TABLE 2.3: AVERAGE AND REPORTED RANGE OF COSTS PER TON FOR MARINE DISPOSAL OF WASTES IN U.S. COASTAL WATERS 1968

Type of Waste	Total U. S.		Pacific Coast		Atlantic Coast		Gulf Coast	
	Average cost/ton	Reported Range $/ton	Average cost/ton	Reported Range $/ton	Average cost/ton	Reported Range $/ton	Average cost/ton	Reported Range $/ton
Dredging spoils	$0.40/ton	$.20 - .55	$0.43/ton	None	$0.54/ton	$.40 - .55	$0.25/ton	$0.20 -.25
Industrial wastes								
Bulk	$1.70/ton	$0.60-9.50	$1.00/ton	$0.60-9.50	$1.80/ton	$.60-7.00	$2.30/ton	$.75-3.50
Containerized	$24/ton	$5-130	$53/ton	$50 - $130	$7.73/ton	$5 - $17	$28/ton	$10 - $40
Refuse and Garbage	$15/ton	$5 - $60	$15/ton	$5-$60				
Sewage sludge	$1.00/ton	$.80 - 1.20			$1.00/ton	$.80-1.20		
Construction and demolition debris	$0.75/ton	$.70 - 1.35			$0.75/ton	$.70 -1.35		
Explosives	$15/ton	$15 - $90						
Miscellaneous	$15/ton	$5 - $600	$15/ton	$5 - $600				

Toxic industrial liquids and sludges are also disposed of at sea in containers, usually 55-gallon drums, either routinely or on demand. These wastes are transported to sea as deck cargo on either merchant ships or contract disposal vessels. Once at sea, the barrels are jettisoned. Depending on local practices the barrels may be weighted for sinking or ruptured at the sea surface. The wide range in the reported costs for disposal of containerized industrial wastes reflects the differences between regular ongoing disposal operations and those where occasional disposal of toxic or hazardous wastes is required by a waste producer.

Refuse and Garbage: Refuse and garbage originating from naval and commercial vessels are transported to sea in various versions of the U.S. Navy's YG class scows. Load capacity varies between 25 and 136 tons. About one trip per week of a 136 ton scow is the average for these disposal operations at San Diego, California. Disposal areas are located at least 20 miles from the coast as prescribed by California law. Once at sea, flood gates are opened, and the refuse is discharged on the sea surface. Upon completion of the unloading, the tanks are hosed out, and the vessel returns to port. The average total round trip time is six to eight hours depending on sea conditions.

Cannery wastes are disposed of at sea on a weekly basis from the San Francisco area from June to October each year. The wastes are taken to sea aboard a 1,000-ton capacity barge and discharged at the sea-surface as a slurry consisting of seawater and wastes. Discharge operations are conducted at least 20 miles from the coast while the barge is under tow. The average costs for the disposal of refuse and garbage are about twice the figure usually found quoted for various proposed schemes involving the disposal of municipal wastes at sea.

Sewage Sludge: About one half of the total tonnage of sewage sludge disposed of at sea from the New York area is handled by a fleet of five self-propelled barges operated by the city of New York at a reported annual cost of $1.25 million. These barges have capacities of between 1,200 and 3,200 tons.

In addition, contract firms regularly transport sludge to sea from various outlying communities in the New York-New Jersey area. Equipment owned by these independent firms ranges from 3,000 to 6,300 tons capacity. The largest is the *Ocean Disposal No. 1* which is 226 feet long, 56 feet wide and has a 20-foot draft. The unloading and tank cleaning operations are radio controlled from the towing barge, and the 6,300-ton load reportedly can be discharged in 30 minutes.

The round-trip distance to the offshore sludge disposal area from the New York City area averages about 30 miles, although it varies depending on the location of a particular treatment plant. In contrast to the large-scale sludge disposal operations from New York City, sewage sludge from Philadelphia is handled by a single converted tank barge having a capacity of 2,700 tons. Total annual cost of this barging is $336,000. About 150 trips a year are made to the disposal area situated about 10 miles off Cape May, New Jersey. The round-trip barge-haul to this area is 227 miles and takes about 30 hours.

Average and reported ranges of costs for sludge disposal at sea reflect the variations in hauling distances, annual amount of wastes handled by private contractors and the capacity of individual barges.

Construction and Demolition Debris: Construction and demolition debris from the New York City area is disposed both in landfill and at sea. Sea disposal is conducted with 3,000 to 5,000-ton-capacity hopper barges that are towed to the offshore disposal area situated nine miles seaward of the Sandy Hook Light or some 15 to 20 miles off New York City.

Costs for the disposal of these wastes is from $.70 to $1.35 per ton and averaging $.75 per ton. Variations in cost are due to differences in quantities and the length of barge haul.

Munitions: Deep-water disposal of outdated munitions and other dangerous military explosives was practiced for many years. Prior to 1964, the wastes were loaded aboard various types of barges and towed to sea for disposal. It has been estimated that it costs $78 per ton to dispose of explosives by this method. The method was hazardous because of repeated handling of the wastes both onshore and at sea.

In 1963 the U.S. Navy conceived the idea of the CHASE (Cut Holes and Sink Em) disposal program. The inspiration for this method came from the U.S. Army, which in 1958 scuttled the obsolete World War II vessel, *S.S. Wm. Ralston*, with 8,000 tons of mustard and lewisite gas aboard. The CHASE program obtained obsolete, surplus World War II cargo ships, which were stripped of any useable equipment or machinery, filled with explosive wastes and towed to sea to be sunk by flooding in the designated deepwater disposal sites. Through 1968, 12 such disposal operations were conducted.

These disposal operations were not conducted without problems. Two disposal ships unexpectedly exploded upon contact with the sea floor. In addition, as part of the AEC's Vela Uniform seismic test series, the *S.S. Robert L. Stevenson,* loaded with 2,400 tons of explosives, was scuttled in 1967 in 4,000 feet of water off Amchitka, Alaska, with her cargo set for detonation at middepth. Instead of sinking as planned, she drifted and eventually sank in water shallower than that required to trigger the hydrostatic detonator. Although subsequent search located the vessel, attempts to detonate her cargo were fruitless.

The Navy terminated the operation on the basis that the detonation attempts indicated that the hulk was no longer a threat. Hydrographic notices and charts now inform interested mariners of the exact location of the *Stevenson*.

Radioactive Wastes: The disposal of low-level radioactive wastes utilized weighted containers carried to sea on the decks of barges or ships. The most common methods of packaging were to encase the wastes in concrete contained in 55-gallon steel drums. Atomic Energy Commission regulations required a minimum weight of 550 pounds to ensure sinking.

Pressure tests on the various types of containers were conducted both in the laboratory and at sea to determine if the containers would rupture under increased water pressure as a result of air voids inside the container. The results of these tests have shown that although physical deformation of the containers does occur, the containers maintain their integrity in containing the wastes.

Low-level radioactive wastes resulting from the operation of U.S. Navy nuclear submarines are regularly discharged at sea from the submarines themselves and from storage tanks aboard submarine repair ships. For example, the *U.S.S. Sperry* has three 15,000-gallon tanks for this purpose; the wastes in these tanks must be disposed of at sea in accordance with specific requirements such as depth and rate of pumping.

Miscellaneous Wastes: Various types of wastes including barreled chemicals and sludges, defective industrial products, etc., are transported to sea on the decks of barges and ships. In some areas these wastes are added to a regular load of wastes whose disposal has been sanctioned by a regulatory agency. As might be expected, no records are kept regarding these operations. The costs to handle disposal of wastes in this manner are strictly based on individual cases.

Tonnage

During 1968 some 62 million tons of wastes from the civilian and military sector were disposed of at sea at an estimated cost of $37 million (Table 2.4). If dredging spoils are excluded, this estimate is reduced to 9.8 million tons disposed of at a cost of roughly $13.7 million. Excluding spoils, the tonnage from Atlantic coastal cities is more than eight times that from either the Pacific or Gulf coastal cities.

Dredge Spoil: Although dredge spoil is by far the largest of the waste categories being discharged at sea, it does not rank low on the scale with respect to producing adverse effects on the marine environment. Dredge spoil discharged at sea off the Pacific Coast is about one-quarter that disposed off the Atlantic and more than one-half off the Gulf Coasts. The reason for this is the practice in the State of California of returning sand and other sediments dredged from harbor entrances to the land for beach restoration or landfill operations. By estimates from the Corps of Engineers nearly 34% of the dredge spoils are polluted (Table 2.5).

TABLE 2.4: SUMMARY OF TYPE, AMOUNT AND ESTIMATED COSTS OF WASTES DISPOSED OF IN PACIFIC, ATLANTIC, AND GULF COAST WATERS FOR THE YEAR 1968*

Waste type	Pacific Coast		Atlantic Coast		Gulf Coast		Total		Total	
	Annual tonnage	Estimated cost$	Annual tonnage	Estimated cost $	Annual tonnage	Estimated cost $	Annual tonnage	Estimated cost $[e]	% Tonnage	% Cost
Dredging spoils	8,320,000	3,608,000	30,880,000[a]	16,810,000	13,000,000	3,228,000	52,200,000	23,646,000	84	63.5
Industrial wastes										
bulk	981,000	991,000	3,011,000	5,406,000	690,000	1,592,000	4,682,000	7,989,000	8	21.7
containerized	300	16,000	2,200	17,000	6,000	171,000	8,500	204,000	<1	<1
Refuse, garbage[b]	26,000	392,000					26,000	392,000	<1	1
Sewage sludge[c]			4,477,000	4,433,000			4,477,000	4,433,000	7	12
Miscellaneous	200	3,000					200	3,000	<1	<1
Construction and demolition debris			574,000	430,000			574,000	430,000	1	1
Explosives			15,200	235,000			15,200	235,000	<1	<1
Total, all wastes[d]	9,327,500	5,010,000	38,959,400	27,331,000	13,696,000	4,991,000	61,982,900	37,332,000	100	100

(a) Includes 200,000 tons of fly ash.
(b) At San Diego 4700 tons vessel garbage at $280,000 per year were discontinued in November 1968.
(c) Tonnage on wet basis. Assuming average 4.5 percent dry solids, this amounts to approximately 200,000 tons dry solids per year being barged to sea.
(d) Radioactive wastes omitted. There were no dumps during 1968. Average Annual disposal in 1969–1970 was 4.2 tons.
(e) Estimated costs were increased proportionately for each area from the original Tonnage/cost data.
* Revised and updated by James L. Verber, FDA.

TABLE 2.5: ESTIMATED POLLUTED DREDGE SPOILS*

Area	Total spoils (tons)	Estimated percent of total polluted spoils	Total polluted spoils (tons)
Altantic	30,880,000	45	13,896,000
Gulf	13,000,000	31	4,030,000
Pacific	8,320,000	19	1,580,800
Totals	52,200,000	34	19,506,800

* Table revised and updated by James L. Verber, Food and Drug Administration.

Industrial Wastes: After dredge spoil, industrial waste tonnage and costs are the second largest waste category. The 4.7 million tons of bulk industrial wastes (10% of the total tonnage) discharged at a cost of nearly $8 million (27% of the total cost) by 50 separate industrial waste producers was distributed as follows:

Major Types of Industrial Waste	Percent of total industrial wastes (Tons)	$
Waste acid	58	59
Refinery wastes	12	7
Pesticide wastes	7	15
Paper mill wastes	3	7
Others	20	12

Most of the pesticide wastes are from the Mississippi Valley, principally the Memphis area. An estimated 14,000 tons per month are loaded onto barges and shipped, not through New Orleans but via the intracoastal canal to Beaumont. From here they are shipped to approximately 100 miles offshore and discharged into 200 fathoms of water.

In a similar vein, proposals have been made for barging down the Ohio and Mississippi Rivers and ocean disposal of industrial wastes originating in West Virginia. Digested and dried sewage sludge is being barged via the Mississippi and Gulf of Mexico to users in southeastern states. These events are significant because they demonstrate that marine disposal of wastes is, in the minds of authorities, within the reach of most major urban and industrial areas of the United States.

When coastal cities disposing of more than 12,000 tons of industrial waste per year at sea are ranked according to tonnage, New York's operations are more than four times as great as Los Angeles, and nearly one-third more than the total tonnage of the other eight areas considered.

	Industrial Waste, tons/year
New York City	2,700,000
Los Angeles	660,000
Houston-Galveston-Beaumont, etc.	320,000
New Orleans	300,000
Philadelphia	290,000
San Francisco	200,000
Seattle	120,000
Norfolk	60,000
Mobile	12,000

Waste acid makes up over 90% of the total industrial wastes going to sea from New York City. In contrast, the bulk of the tonnage discharged from Los Angeles consists of sediment-like oil drilling muds and cuttings. Both of these have been considered to be relatively harmless in terms of environmental pollution.

Refuse and Garbage: The total volume of refuse and garbage being disposed of at sea is small (less than 1% of tonnage and cost) compared to most other wastes, and is confined to the Pacific coast. Major sources are from naval and commercial vessels and cannery operations.

Sewage Sludge: Sewage sludge from New York and Philadelphia ranks third in importance with 7% of the tonnage and 12% of cost. The total amount of sludge from the New York area is about ten times as great as that discharged by the city of Philadelphia. Both these operations are projected to increase significantly during the next ten years. In addition, the city of Baltimore is planning to dispose of some 400,000 tons of sewage sludge per year in 1970 in the disposal area now used by the city of Philadelphia.

Construction and Demolition Debris: The disposal of these wastes at sea occurs only from the city of New York and represents an alternate solution to offset the lack of available landfill. In 1968, 574,000 tons were disposed, however, yearly quantities of these wastes vary considerably in accord with construction operations.

Explosives and Munitions: The total tonnage of military explosives disposed at sea in 1968 represents the total cargo that was sunk during CHASE operations of that year (Table 2.4). The disposal of these wastes at sea is intermittent and depends on the backlog of outdated munitions that develops at individual military ordnance depots.

Radioactive Wastes: These wastes were not considered for 1968 since for all practical purposes the disposal of radioactive wastes at sea from the United States was almost nonexistent in that year. As a basis for comparison, however, the sea disposal of U.S. radioactive wastes from its inception in 1946 to 1967 (prepared from data supplied by the AEC) has been summarized (Table 2.6).

TABLE 2.6: TOTAL U.S. SEA DISPOSAL OF RADIOACTIVE WASTES (1946 through 1967)

Disposal area	Number of containers	Byproduct material (curies)	Source material (pounds)
Pacific Ocean	52,538	14,708	1,895
Atlantic Ocean	33,998	79,484	Unknown
Gulf of Mexico	1	10	Unknown
TOTALS	86,537	94,202	1,895

Miscellaneous Wastes: The data available for this type of waste are only crude approximations (Table 2.4). No clear picture as to tonnage or cost is available.

ENVIRONMENTAL EFFECTS OF COASTAL WASTE DISPOSAL

Little is known about the immediate effects of solid and liquid wastes on the marine environment, even though a number of pioneer studies have been carried out. Even less information is available on long-term or genetic effects.

An adequate understanding of the response of the marine environment to solid and liquid wastes barged to sea requires continuing investigation programs. Attention has been paid to the disposal of sewage and sewage sludges, both raw and treated, because of the well-known ability of shellfish to concentrate pathogenic bacteria for human consumption. Sludges barged and discharged to the New York Bight have resulted in high levels of fecal coliform contamination in surf clams and the Food and Drug Administration has closed affected areas to further harvesting.

In contrast, some 400 tons per day of sewage solids, of which approximately one-fourth is organic, have been discharged through submarine outfalls into Santa Monica Bay, California, with no apparent effects on fish abundance. Although the fish were able to avoid possibly harmful concentrations, studies of other Southern California receiving waters indicated severe effects on giant kelp (Macrocystis). The discharge attracted scavengers (sea urchins) that grazed on the mature plants and increased turbidity so that the young kelp did not develop. The great differences in effects of waste disposal at sea underscore the need for broad and critical research.

Observed Effects

For ease of discussion, the waste categories are presented in the same order as in the preceding section: Dredge Spoils, Industrial Wastes, Sewage Sludge, Refuse, Radioactive Wastes, and Explosives.

Dredge Spoils: The discharge of dredge spoils at sea results in anomalously rapid sediment buildup at the disposal site, in addition to temporary discoloration of the water by turbidity. The sediment buildup would be expected to have significant effects on bottom-dwelling organisms and the life forms dependent upon them in the food chain. Turbidity may be expected to affect various aspects of the water column and the contained biologic assemblage.

Bottom Sediment Buildup — The effects on fish and shellfish of rapid local buildup of sediment as the result of dredge spoil disposal can include destruction of spawning areas, reduction in food supplies and vegetational cover, trapping of organic matter (with resultant development of anaerobic bottom conditions), and the absorption or adsorption of organic matter (including oil).

Turbidity — The effects of turbidity on fish and crustaceans can be direct or indirect. Direct effects cause an immediate response or even mortality by suffocation; turbidity can also result in reduced growth and decreased survival of larval stages of fish and shellfish. The possible indirect effects of turbidity (and siltation) are: (1) reduction in light penetration resulting in reduced photosynthesis; (2) reduction of visibility to some feeding organisms; (3) destruction of spawning areas; (4) reduction of food supplies; (5) reduction of vegetational cover; (6) trapping of organic matter, resulting in anaerobic bottom conditions; (7) flocculation of planktonic algae; (8) absorption or adsorption of organic matter (including oil) or inorganic ions. The effect of dredge spoils depends in large part on the location and characteristics of the disposal site. For example, if dredge spoils are discharged in inshore waters normally quite turbid because of wave action or the tidal flushing of turbid waters out to sea, then the effects of the associated turbidity (but not necessarily the sedimentation) would almost certainly be negligible.

The situation would be significantly different where spoils polluted with oil, sewage, chemicals, etc. are dredged from channels and the confines of harbors and transported to sea for disposal. Once these sediments are discharged in other locations, the resulting turbidity (and associated pollution) can be spread extensively by coastal currents. In this regard, it was estimated that disposal of spoils from the approaches of Chesapeake Bay increases the turbidity of the water over an area of about 1 to 1.5 square nautical miles around the disposal site, and that discharged solids are dispersed over a bottom area at least five times that of the defined disposal area.

The effects on local fishing resources (particularly the northern lobster) of increased turbidity and siltation in relation to the disposal of dredge spoils in 100 feet of water off Narragansett Bay were also investigated. The results of biological assays and field experiments showed that adult lobsters were able to tolerate, with no adverse effects, turbidities as high or higher than those actually encountered in the disposal area for periods of time equivalent to those during which the spoils seem to remain in suspension. The one case of high mortality observed in the laboratory bioassay tests was believed to be the result of an unidentified toxic substance contained in the dredging spoil, rather than the concentration of suspended particles.

Interpretation of turbidity measurements at the point of discharge revealed that a fraction of the total turbidity cloud consisting of oils and/or light organic material remained near the surface. If present in sufficient quantites for an adequate length of time, this material could pose a possible pollution threat to the adjacent shore area, particularly during the summer season when a drift from the disposal site is predominately shoreward. The dispersion rate of the remaining turbidity cloud was sufficiently rapid that no direct adverse effects to marine resources of economic importance would be expected.

Comparison of the bottom topography before and after the disposal operation indicated that the volume of sediment added to the volume of excess material present in the site was about 2,004,000 cubic yards, which compares favorably with 1,960,000 cubic yards dredged by the Corps of Engineers; on the basis of this comparison, most of the dredging spoils remained within the designated disposal area. On the basis of the biological and physical data obtained prior to and during the December 1967 to July 1968 Narragansett disposal operation, it was concluded that the spoil disposal site was acceptable from the point of view of minimizing the damage to locally important marine resources.

Another aspect of dredge spoil disposal showed that the volumes of dredge spoils and other sediment-like wastes (construction and demolition debris) disposed of in Long Island Sound and the New York Bight disposal areas represent the largest single source of sediment entering directly into the Atlantic Ocean from North America. From Corps of Engineers records, it was estimated that the mass of waste solids from the New York City area disposed at sea increased from an average of 6.8 million tons per year during 1960 to 1963 to 8.6 million tons per year for the period 1964 to 1968.

Most of the disposal sites referred to are in areas where there is little natural deposition of sediment to dilute or bury the man-moved solids that in many cases consist of highly polluted harbor sediments. The effect of these sediments on the biota is not well understood at this time. Disposal of spoil from barges or dredges is significantly different from natural processes whereby river-borne sediment is deposited on the ocean floor. Disposal operations are highly localized and involve the instantaneous release of several thousand cubic yards of spoil, which produces a thick layer of sediment that may or may not subsequently be dispersed over a wider area by current action.

There is a lack of knowledge regarding the deposits formed by the present disposal operations and an absence of comparative data on existing conditions before disposal operations began. Along these lines it may be pointed out that the present disposal operations are, in effect, gigantic, unplanned experiments and, if treated as such, much could be learned from them.

Industrial Wastes: Industrial wastes include waste acids, paper mill wastes, chemical wastes, oil drilling wastes, and waste oil. These wastes are typically dissolved or suspended in a water media for bulk discharge at sea from tank barges. In some cases, the material is barreled and dropped over the side.

For industrial wastes, the solid concentration, bulk specific gravity and solubility in sea water are the minimum characteristics necessary for the evaluation of the local physical fate of the material. Characteristics of some industrial waste components may be determined from standard references.

The majority of studies of the environmental effects of industrial wastes in the sea have been carried out in Gulf of Mexico waters, because the Galveston District of the Corps of Engineers requires filing of laboratory and field studies of the wastes in support of disposal applications. In order to facilitate comparison of the environmental studies of the various types of industrial waste disposal operations, key information from the study results has been summarized (Table 2.7).

Waste Acid — Studies of the dispersion and environmental effects of acid-iron water discharged from barges in the New York Bight disposal area show that these wastes consist of ferrous sulfate and spent sulfuric acid in a freshwater solution, and are the byproducts of a titanium paint pigment manufacturing process.

The studies showed that the toxic effects of these wastes are minimal. Laboratory toxicity tests to determine if the acid-iron wastes had any effect on the growth of green plankton algae showed concentrations of wastes which severely limited growth are found only close to the discharge barge and are diluted rapidly at sea. Zooplankton from samples immediately behind the barge were temporarily immobilized, but recovered rapidly on dilution with unpolluted water a short distance astern of the disposal barge.

Additional findings showed that the accumulation of iron hydroxide, resulting from the chemical reaction of the wastes with seawater, was about twice as great in 1957 as that observed in 1948, and 50% more than that observed in 1950. This was less than might have been expected, however, from the increase in the discharge rate that had occurred over the same period. No evidence was found to indicate that the general turbidity of the waters caused by the iron precipitate has increased outside the general disposal area. Since establishment of the acid-iron disposal operation in 1948, a concentration of bluefish has developed on the outer boundaries of the disposal area, thus creating a highly popular sport fishery that did not previously exist.

Paper Mill Wastes — This waste is the so-called black liquor produced by the sulfate process used in paper manufacture; it includes sodium carbonate, sodium sulfate, sodium hydroxide, and other chemicals. The wastes were discharged off the Texas coast in 1955 and contained about 45% solids, were water-soluble but quite viscous, and had a pH greater than 13. Laboratory studies using four species of phytoplankton showed the concentration of black liquor affecting photosynthesis (i.e., threshold toxicity) was about 0.05 grams per liter, and the lethal concentration was on the order of 1 gram per liter. Tests on zooplankton produced no acute toxicities in concentrations up to 0.4 gram per liter; greater concentrations were not tested because it was impossible to observe the zooplankton as a result of the opaque characteristics of the wastewater mixture.

Visual observations at sea confirmed the nontoxic character of the wastes, as zooplankton in samples taken immediately astern of the discharge barge were observed to be swimming about normally. Field evidence also showed that the the absorbent component of the waste sank almost immediately, leaving only a slight discoloration at the surface that was visible for only 30 to 40 minutes after discharge. A minimum dilution ratio of waste to water immediately behind the barge was calculated to be on the order of 1 to 300,000 or a concentration of 0.03 gram of waste per liter, which is well below the values given previously for threshold and acute toxicity on phytoplankton.

Bacterial action should result in the ultimate disintegration of the organic components of the black liquor wastes while the inorganic fraction, principally sodium sulfate, would rapidly lose its identity when mixed with seawater.

Chemical Wastes — Chlorinated Hydrocarbons: The chlorinated hydrocarbon waste category includes beta-chloropropylene, trichloropropane, isopropyl chloride, allyl chloride, miscellaneous chlorides, and other constituents discharged by chemical and petrochemical plants. The disposal operations discussed here are confined to the Gulf of Mexico. Although wastes containing various percentages of chlorinated hydrocarbons are discharged off the Atlantic and Pacific coasts, their environmental effects have not been investigated primarily because in those areas such studies are not a prerequisite for obtaining a disposal permit.

Four separate chlorinated hydrocarbon disposal operations in Gulf waters have been studied. The work included field and laboratory investigations of the effects of these wastes on various marine organisms, with emphasis on phytoplankton, zooplankton, and fish. Observed effects at the disposal site included sinking of the bulk of the waste, discoloration of the water, and the death of fish and plankton on direct contact with the undiluted waste. After two to four hours, the surface waste field was found to be sufficiently diluted so that no effect on marine life was observed.

TABLE 2.7: SUMMARY OF ENVIRONMENTAL STUDIES ON INDUSTRIAL WASTES DISCHARGED AT SEA

Waste Type	Description of Disposal Area				Barging Characteristics				Waste Characteristics	Laboratory Toxicity Studies		
Industrial	Distance from shore (miles)	Water depth (feet)	Latitude	Longitude	Amount of waste/trip	Depth of discharge	Rates of discharge tons/min.	Towing speed (knots)	Given description	Type of test organisms	TLM (range)	Remarks
Spent sulfuric acid	15 from New Jersey coast	80	40°20'N	73°40'W	3200 tons 5000 tons	15 feet	18 70	6 7	Fe SO_4 (10%) H_2SO_4 (8.5%)	Marine phytoplankton.	2.7-35.5 mg/l 6-16 day test duration.	Investigated effects of various concentrations of iron on growth of algae.
Chlorinated hydrocarbons	125 SE of Galveston, Texas	2400	27°36'N	94°36'W	1200 tons	12 feet	5	6	Beta-chloropropylene (22%), trichloropropane (5%), isopropylchloride (38%), allyl chloride (11%), misc. chlorides (33%), heavy ends (3%), pH - 9.8, specific gravity 0.9 - 1.34.	Marine phyto-, and zoo-plankton, anemones, brine shrimp, bacteria, fish.	0.02-2.5% saturated 0.0005-1 g/l, 2-24 hr test duration.	Investigated acute toxicity and inhibition to photosynthesis.
Paper mill wastes (black liquor)	125 SE of Galveston, Texas	2400	27°36'N	94°36'W	1300 tons	10 feet	5	6	Paper mill wastes, 47% solids, BOD_5-100,000 ppm Na_2CO_3, Na_2SO_4, NaOH etc., pH - 13. Specific gravity 1.27 at 60°C.	Marine phytoplankton zooplankton.	0.001-1.0g/l, 22 hr test duration.	Investigated acute toxicity and inhibition to photosynthesis.
Ammonium sulfate (mother liquor)	100 S of Freeport, Texas	2760	27°35'N	95°20'W	1700 tons		7	4	Ammonium sulfate (23%), nitrogen (8%), carbon (12%), organics (29%) (alcohols, esters, amides), IOD-90 MG/L, BOD_5 57,000 ppm, pH 4.3, S.G. 1.23.	Brineshrimp, top-water minnows.	1.25% by volume 200 ppm, 24 hr test duration.	Acute toxicity.
Waste liquor	70 S of Freeport, Texas	960	27°48'N	94°54'W	2400 tons		5-24	6	Na_2S (Na_2S_2) (6%) (Na_2S_3) NaHS 1% S (total) (6%), NaCl (21%), organic 2%, solids (dissolved and suspended), specific gravity 1.24 at 60°C.	Marine phytoplankton, top-water minnows, brine shrimp.	0.0005-0.18% by volume, 24 hr test duration	Acute toxicity.
Chlorinated waste liquor	125 SE of Galveston, Texas	2400	27°36'N	94°36'W	1400 tons (proposed)		13-25 (recommended)	5 (recommended)	(Organic waste) chlorinated organics (10-15%), inorganic salts (Na_2SO_4) (5-6%); (acid) chlorinated organics (1%), sulfuric acid (10-15%), nitric acid (0.1%).	Marine phytoplankton, top-water minnows, brine shrimp.	0.02-0.64% by volume, 24 hr test duration.	Acute toxicity, waste acid more toxic to fish than brine shrimp.
Sodium sludge (containerized)	110 S of Galveston, Texas	2400	27°36'N	94°36'W	15 55-gal. drums (500-570 pounds per drum)	Surface	Variable	10	Metallic sodium (75%), calcium (24%), barium, magnesium, potassium (1%).			
Pesticides	95 SSE of Galveston, Texas	720	27°49'N	94°30'W	50 55-gal. drums per trip	1200 ft	1 barrel/2 mins. (600' intervals on bottom	3	Anilines (chloroaniline, monochlorobenzene), liquid organics (methanol, p-xylene, chlorobenzene), dry chemicals–insoluble (*thiram*, *thiram-E*, *thionex*, *zineb*, *ferbam*, *monuron*, carbon disulfide).	Minnows, largemouth bass, salmon, bluegills, mosquito fish, channel catfish, bluegill sunfish.	1-135 mg/l, 48-96 hr test duration.	Toxicity values obtained from literature.
Caprolactam wastes	35 S of Sabine Bank Light	60	29°10'N	93°40'W								

(continued)

TABLE 2.7: (continued)

Waste Type	Reported Field Observations			Mixing Characteristics		
Industrial	Physical-Chemical	Biological	Observed effects	Initial dilution	Diffusion coefficient cm^2/sec.	General study conclusions
Spent sulfuric acid	pH, iron concentration (0-60'), wind, weather, sea state.	Plankton (0-60'), benthic organisms, pelagic fish (30').	Water discoloration, plankton temporarily immobile, iron settled rapidly from surface layer, no appreciable accumulation of iron found in bottom sediments.	1:5,000	$2\text{-}9 \times 10^3$	Mixing and diffusion of wastes occurs rapidly in the wake of the barge. No evidence to indicate adverse effects. Each new proposed waste disposal operation needs careful study prior to allowing ocean disposal.
Chlorinated hydrocarbons	Temp. (0-900'), salinity oxygen, waste concentration (0-500'), surface currents, wind, weather, sea state, bottom mud.	Plankton, O_2 evolution, C_{14} uptake, chlorophyll-A, visual inspection of *sargassum* weed.	Water discoloration: fish, plankton killed on direct contact of waste, no harmful effects seen after 2-4 hrs. Bulk of waste sank. Low diffusion of waste at depth.		2.5×10^3 (average)	Disposal of toxic wastes at sea can be accomplished with only a slight effect on organisms in the biomass within a limited oceanic area. Each new waste disposal operation needs careful study prior to sanctioning it.
Paper mill wastes (black liquor)	Temp. (0-900'), salinity (0-600'), waste concentration surface samples, pH, oxygen (0-600'), wind, weather, sea state.	Plankton (0-100'), O_2 evolution, C_{14} uptake, chlorophyll-A.	Slight water discoloration. No mortality to marine life. Bulk of waste sank.	1:300,000	2.5×10^3 (average)	Disposal of "black liquor wastes" in the deep sea can be accomplished without determinable effects on marine biota. Ultimate disposal is expected to be accomplished by bacteria. Advisable to monitor each separate load of waste to determine toxicity in laboratory.
Ammonium sulfate (mother liquor)	Waste concentration (0-150')		No fish mortality. No floating oil. Bulk of waste sank. Maximum waste concentration at depth.		9×10^2	No undesirable effects were observed. Diffusion great enough to ensure good dispersion to minimize harmful effects to biota.
Waste liquor	Waste concentration (0-150'), temp., salinity pH, oxygen, (0-600').	Plankton, pelagic fish, *sargassum* weed communities.	No evidence of subsurface maximum waste concentration.	1:10,000-1:100,000 in 2 hrs		Disposal should produce no significant mortality in the biota, nor any prolonged effects.
Chlorinated waste liquor						No significant mortality would be expected from disposal of this waste in the open ocean. It is suggested that a full-scale disposal operation be properly monitored and continued to verify preliminary study results.
Sodium sludge (containerized)	pH	Plankton, fish, *sargassum* weed.	No mortality to fish. Flying debris hazardous to disposal personnel. 30% mortality to plankton due to collection methods.			Explosions caused by reaction of sodium with seawater had no significant effects on *sargassum* and zooplankton populations. Absence of fish-kiln was probably due to barrenness of disposal area.
Pesticides				100:1	0.002×10^3	Consideration of available toxicity and diffusion data from literature sources indicate that the zone of water containing toxic concentrations of waste surrounding each disposal drum will be limited in extent and duration and will not endanger motile aquatic life in the disposal area to a significant degree.
Caprolactam wastes	Temperature (0-60'), salinity (surface), wind, weather, sea state.	Plankton, bottom samples, C_{14}, chlorophyll-A,-B,-C.				Limited scale (one day) of survey precluded any significant results. Recommend future surveys be conducted on 3-day basis for better results.

Because the inherent patchy distribution of the microorganisms at sea makes it difficult to obtain statistically significant results, the effects of various toxicity levels on phytoplankton and zooplankton were tested under controlled laboratory conditions. Inhibition to photosynthesis and respiration caused by the wastes was determined by using standard oceanographic procedures for measuring carbon-14 uptake and oxygen evolution. Comparison of the results with those obtained from uncontaminated control samples maintained under identical laboratory conditions provided an index of inhibition. Acute toxicity tests were also conducted on the plankton and various species of fish. Results of the laboratory toxicity studies showed that the inhibition to photosynthesis and respiration is a much more sensitive and meaningful measure of the toxic effects of organic contaminants in seawater than acute toxicity tests on organisms such as fish.

The effects of wastes on organisms endemic to the disposal sites was assessed in the field by examining the flora and fauna associated with the seaweed, Sargassum, present both within the outside of the disposal area. It was found that direct contact with the undiluted hydrocarbon wastes at the point of discharge was generally lethal to the small crabs, shrimp, fish, snails, etc., but in most cases samples taken in the disposal area showed conditions had returned to normal within three to eight hours.

With regard to diffusion and dispersion of the hydrocarbon wastes, it was estimated that about seven square miles of surface area would be affected by a single disposal operation. The work further indicated that within three to nine hours, the wastes in the surface layers were diluted to levels below the limit of detection by the analytical method used.

In contrast to this, slow dispersion of the wastes occurred at depths from 50 to 500 feet. For example, at one station the waste concentration at 500 feet was roughly equivalent to that observed at the surface in the actual disposal wake 42 hours earlier. Because of the possibility of these waste-laden waters working their way shoreward to bottom fishing areas or regions of upwelling, it was recommended that future disposal areas should be located seaward of the 400-fathom line.

It was concluded that chlorinated hydrocarbon wastes could be discharged at sea with only minor adverse effects on marine organisms within the dispersal area. It should be pointed out, however, that the work did not examine the possible effects of the wastes at depth or of bottom contamination on benthic organisms, even though samples of bottom mud within the disposal area contained from 10 to 100% of the original surface concentration of hydrocarbon waste.

Laboratory tests on the toxicity of two chlorinated hydrocarbon wastes produced by a chemical plant were carried out. This study was in preliminary support for an application for marine disposal off the Texas coast. The tests, which used top-water minnows, brine shrimp, and various phytoplankton, measured both the median toxicity level at 24 hours and the critical concentration range, defined as that dilution range between which there is no mortality on the one hand and no survival on the other. In the case of the test plants, 50% survived at concentrations of 0.05 to 0.5% by volume on 24-hour exposure to the hydrocarbon wastes, and the waste, which included dilute sulfuric acid, was somewhat more toxic, i.e., 50% survival at concentrations of 0.02 and 0.05% by volume.

This greater toxicity should be greatly reduced in a marine disposal operation because of the rapid neutralization of sulfuric acid which occurs in seawater. It was estimated that the waste would be diluted below the 0.006% concentration range within an hour by using a pumping rate between 8 and 15 tons per hour while underway at 5 knots, and was concluded that no significant mortality would be expected from the disposal of either waste in the open ocean.

Waste Liquor: Laboratory and field studies on an industrial waste liquor being discharged off the Texas coast were carried out. The waste consisted of various compounds of sodium, hydrogen, and sulfur, along with sodium chloride in an aqueous solution (specific gravity greater than 1.25) that was completely miscible in seawater. On the basis of the laboratory toxicity studies, they concluded marine disposal would produce no significant mortality, nor any prolonged deleterious effects, nor would the concentrations of the waste persist beyond two hours after discharge.

Because the wastes were about 25% more dense than seawater, the bulk of the materials sank rapidly from the surface layer, but to the depth covered by sampling (164 feet) there was no evidence of plunging of the wastes in an undiluted stream. Further dilution and dispersion of the wastes were expected to occur on the bottom but sampling was not carried out to confirm this fact.

The reported maximum value of the waste concentration 1.5 miles astern of the barge was less than one in 10^5 (.001%), which was within a factor of two of the minimum value of TL_M reported earlier for the phytoplankton used in the toxicity tests. In spite of these favorable results, it was recommended that future operations use a significantly reduced pumping discharge rate per mile of vessel track in order to insure adequate dispersion under all oceanographic conditions.

The toxicity of ferrous sulfate and sulfuric acid was investigated in relation to dumping operations carried out in the New York Bight area. Reported conclusions claim no permanent effect on plant, fish, or animal life. The report mentions similar studies and conclusions for both containerized and liquid wastes.

Ammonium Sulfate Mother Liquor: The effects of ammonium sulfate mother liquor on brine shrimp and minnows in the laboratory, as well as the diffusion of these wastes during discharge were studied. This waste material is saturated with ammonium sulfate and contains about 10% organic carbon as alcohols, amides, esters, etc.; it is generated in obtaining ammonium sulfate from sulfuric acid in a fertilizer manufacturing process. The work included chemical analyses, biodegradability and toxicity studies, and preliminary diffusion calculations, as well as sampling at sea to determine actual diffusion.

On the basis of the preliminary diffusion and toxicity studies, it was concluded that the wastes would not produce adverse effects on the environment or organisms within the disposal area. A trial disposal operation verified the predictions that diffusion would be sufficiently rapid to minimize the harmful effects to the biota.

Sodium Sludge: A study of the disposal of sodium sludge off the Texas coast was performed. The sludge consisted of approximately 75% metallic sodium and 24% metallic calcium, and was a waste product from a petrochemical plant. The barreled sodium sludge was punctured and dropped overboard; when seawater made contact with the sodium, the drums exploded. Field observation and sampling indicated that the explosions and elevated pH (resulting from the interaction of the sodium with seawater) had no significant effects on the Sargassum and zooplankton communities as compared to those in the control area. An absence of floating dead fish may have been due to the barrenness of the area.

Corps of Engineers records indicate that on two occasions unpierced barrels loaded with sodium sludge have been retrieved from Gulf waters by fishermen. In these cases, it is almost certain that the disposal operators did not dispose of the barrels in the designated disposal area.

Pesticides: The probable environmental effects of the proposed marine disposal of wastes from a Gulf Coast plant that manufactures herbicides and fungicides were studied. The waste included chemicals of the following types: anilines (primarily chloroaniline and aniline with small amounts of monochlorobenzene); liquid organic solvents (methanol, p-xylene and chlorobenzene); and dry chemicals including Thiram, Thiram-E, Thionex, Zineb, Ferbam, Monuron, and carbon disulfide. Because of their toxic nature, these wastes are disposed of in weighted steel drums. To facilitate discharge and dispersal along the bottom, a small air space is left in each drum to ensure deformation and rupture of the drum on the ocean floor.

Toxicity data on these waste materials on the basis of the available literature were evaluated, and it was concluded that: The herbicides and fungicides are generally more toxic than the liquid organics. These dry chemicals, however, have a very low solubility so that once in solution a relatively low dilution ratio (on the order of 100:1) will reduce the concentration below the median tolerance limits (TL_M). In addition, the susceptibility to biodegradation and chemical instability in an aqueous environment would further reduce localized toxic conditions.

Drill Cuttings and Drilling Muds: The rock chips produced by the bit in drilling an oil well are termed drill cuttings. On offshore drilling platforms along the southern California coast, these cuttings are generally washed and discharged below the rig. The resulting solid waste accumulation is volumetrically of little significance. The cuttings were found to have no effect, either adverse or beneficial, on the environment. If the cuttings were to be disposed of several hundred feet away from the drilling platform and capped with stones or other rubble, they might provide an artificial habitat for sport fishing.

The disposal of drilling muds and cuttings in deep water from a barge has also been investigated. Observations on the disposal operation consisted of aerial photographs and visual inspection at the sea surface to assess the extent of surface contamination. On the basis of the brief study involved, it was concluded that the disposal operation posed no threat to the environment.

Waste Oil: Although environmental studies have not been made on the single industrial waste oil disposal operation noted during the survey, some very significant work has been carried out on the wastes from tank cleaning operations conducted by oil tankers at sea. Based on earlier work regarding the oxidation of hydrocarbons by marine bacteria, the importance of discharging these wastes in a finely divided form in order to optimize bacterial degradation was shown. Calculations showed that the full quantity of waste oil (at an average concentration of 0.13 mg/l) resulting from cleaning the tanks of a 45,000 dwt tanker underway at 16 knots would be oxidized by bacteria in 2 to 4.5 days, depending on whether the wastes were discharged continuously over a 24-hour period, or discharged as fast as the tanks were cleaned.

Sewage Sludge: Sewage sludge discharged at sea is the residual from municipal sewage treatment plants and is generally 3 to 10% solids by weight. It is discharged from tank barges at one disposal site in the New York Bight and another off Cape May, New Jersey. There are no similar operations on the Gulf or Pacific coasts, although large volumes of sludge are discharged in southern California coastal waters through submarine outfalls.

Philadelphia's disposal of digested sludge to the ocean via barges was initiated in 1961. The dumping is confined to an area of one square mile, ten miles off the coast near Cape May, New Jersey. By 1969 the annual volume had reached 115 million gallons.

The sludge dumping grounds in the New York Bight area, located 11 miles off the coast in waters no less than 72 feet, have been described as a vast desert on the ocean bottom. Another study of the area stated that, "the bottom of the area of the mud, rubble-excavation and sewage sludge dump is so badly fouled that changes in the dump location would be of little help to the immediate area."

As reported on a bacterial study made in the New York and Philadelphia sludge dumping ground, while noted that the coliform concentration decreased quite rapidly in the waters receiving the sludge, high levels of fecal coliform contamination were found in surf clams, forcing the closure of affected areas to further harvesting. It was also noted that there was considerable sludge covering the bottom. In a report on an investigation of sludge dumping in the Oslo Fjord with primary emphasis on the spreading of sinking particles, it was reported that heavy sludge particles sank to the bottom rapidly and adhered to meshes of shrimp trawlers in the area, while a cloud of polluted water was visible at the surface for long periods of time.

There have also been numerous studies of ocean outfalls for sewage sludge disposal on the West Coast which are pertinent because of the reported effects. The report on the effects of digested sludge discharged approximately two miles off the San Diego shore in 200 feet of water estimates a sludge accumulation rate of one tenth inch per year near the outfall with 40% of the solids settling at such slow rates that their accumulation within a five mile area could be considered negligible. Grease and floatables were identified as potential problems but were not defined quantitatively. Apparently, little consideration was given the ultimate fate of the solids which are carried away from the discharge site.

In a study of the sedimentation and dilution of digested sludge in Santa Monica Bay, it was concluded that sludge accumulation rates should average 2 to 3 inches per year within a 500-foot radius decreasing to 0.25 inch per year at a two-mile radius from the outfall. This analysis assumed a constant current of 0.2 knot with equal frequencies in all directions. These rates of accumulation were considered unobjectionable based on 1956 standards. Other studies of this area have reported that the disposal of approximately 4,000 tons of solids per day has had no apparent effect on fish abundance. Another study, however, has indicated that California's giant kelp is being adversely affected by increases in sea urchin poulation apparently fostered by waste disposal operations along the coast.

The environmental effects of barge disposal of sludge in the designated sites in the New York Bight and off Cape May, New Jersey were studied at the U.S. Public Health Service's Northeast Technical Services Unit, FDA Laboratory with respect to the possible contamination of the commercial surf clam beds adjacent to the sewage sludge disposal grounds. Of particular concern was the possibility that the surf clams might accumulate and concentrate bacteria, viruses, and other toxic substances present in the sludge; these substances could, in turn, be transmitted to consumers of the shellfish. Results of this study indicate that sludge settles rapidly to the bottom and that, except in the immediate wake of the barge discharge, the highest coliform contents occurred in the samples from the near-bottom water.

Preliminary results of the current Corps of Engineers study of the New York City sludge disposal area indicate that, because of the high bacterial contamination found in surf clams adjacent to the sludge disposal grounds harvesting clams should be prohibited from within a 6 mile radius of the center of the disposal grounds.

It was also concluded that the sludge from the disposal area off Cape May poses a potential threat to existing commercial surf clam beds and shellfish beds located to the south and west of the present disposal site, as they are in a direct line to receive sludge carried by the prevailing tidal currents. In addition, it was pointed out that the location of the New York sludge disposal area within a few miles of the Hudson Submarine Canyon, and the apparent southerly drift of the sludge on the bottom towards the head of the canyon, constitute a potential threat to the lobster and red crab populations which may breed in the shallower portions of the canyon.

The work on several major southern California submarine sewage outfalls by various investigators demonstrated the presence of coliform concentrations in the upper layer of bottom sediment adjacent to the outfalls. In connection with these studies, the various factors, which are important in reducing bacterial populations in seawater have been divided into three categories:

(1) bacteriologic factors endemic to the marine environment—these include competition for food, predation, salinity, sunlight, temperature, pressure, etc. All of these factors may be combined into a single factor—mortality;

(2) settling behavior of suspended solids—this is dependent on the size, shape, flocculating characteristics, and density of the sludge particles, which, in turn, are a function of the type of sewage treatment process. Particle characteristics are also important because the bacteria are concentrated within or on the particles;

(3) dispersion of the effluent field—this takes place as the sludge field moves downstream from the outfall.

A simple mathematical expression relating these three factors was set up. Using this relationship in an investigation of the persistence of coliform bacteria in the primary effluent from the Hyperion Outfall off Los Angeles, sedimentation was shown by far, to be the most important factor in the disappearance of coliform bacteria from the surface layer of the water column.

The results of work done to investigate the environmental effects of sewage disposal from the Whites Point outfall in 65 to 195 feet of water off Los Angeles reveal that, shoreward of the 10-fathom curve, large-scale ecological changes have occurred. For example, many species that normally occur over rocky substrates at these depths (e.g., kelp, lobster, abalone, and many species of fish) are either rare or no longer present. The length of coastline affected by the sludge deposit has spread from two miles in 1954 to six miles in 1969, at least a threefold increase in length in 15 years. In contrast, it was reported elsewhere that disposal from the Hyperion outfall in 320 feet of water in Santa Monica Bay has apparently had no measurable effects on fish abundance.

In summary, studies have indicated that sewage sludge in large concentrations destroys the marine habitat in the immediate vicinity of the sludge field; that the sludge drifts slowly along the bottom because of currents; that coliform and related toxic substances are potential threats to shellfish within a radius of 5 to 10 miles of the site; that the toxic substances and coliform bacteria associated with the sludge are concentrated in bottom sediments; and that a great deal more field and laboratory work is required to predict accurately the detailed behavior of the sludge and the probable environmental response.

Refuse: Although there have been no sizeable municipal refuse disposal operations at sea from the United States in the past 25 years, several U.S. coastal cities (including New York, Oakland, and San Diego) had previously dumped their refuse at sea. These operations were generally unsatisfactory because of the associated fouling of beaches and resultant public disfavor and eventual preventative legal actions.

The results of the present survey show, however, that in 1968 there was small-scale disposal at sea of refuse from military installations in Long Beach and San Diego and from a cannery in the San Francisco Bay area. It was also reported that the city of Charlotte Amalie on St. Thomas Island, Virgin Islands, in 1962 began disposing at sea a daily average of 280 cubic yards of refuse.

Studies associated with the past disposal of refuse at sea have been either investigations to establish the sources of wastes found on public beaches, or examinations of the economic and technical feasibility of conducting such disposal operations. Recently there have been serious attempts to determine the possible environmental effects of disposal of refuse and related solid waste in the ocean. These studies are an outgrowth of the several new methods that have been proposed to facilitate effective marine disposal.

A five-year study conducted jointly by the Harvard University School of Public Health and Rhode Island's Graduate School of Oceanography investigated waste incineration at sea and the subsequent disposal of the nonfloating residue. The results of this study indicated little or no toxicity to a series of marine organisms and concluded that a depth of 200 feet was sufficient to keep the material from reaching the beaches in the test area. However, it was reported that waves alone have been responsible for sediment movements in water of this depth. To date, the burning of garbage and refuse at sea has not proven to be economically justifiable and current and future air pollution regulations may prevent it from becoming technically acceptable.

Besides defining costs and operating characteristics of the incinerator vessels, the study examined what effects the incinerator residues would have on a variety of marine organisms. Both acute and chronic toxicity tests were carried out on species ranging from phytoplankton and lobster larvae to flounders and clams. Oceanographic studies were made in order to predict the movement of incinerated solid wastes along the bottom under various conditions. Meteorological studies were conducted to determine safe burning sites for every weather pattern likely to be experienced in the general offshore region proposed for the disposal operations.

Short- and long-term toxicity tests were performed on a series of organisms including clams, shrimp, scallop, lobster larvae, and fish. The quahaug (*Mercenaria mercenaria*) was the most resistant; 90-day quahaug bioassays in residue concentrations up to 10% by weight failed to produce mortalities or significant changes in growth. The common mummiehog survived with no mortalities in 40-day bioassay tests with residue concentrations up to 30% by weight. First- and second-stage lobster larvae and the common prawn did not experience significant mortality in residue concentrations of 1% or less.

Of all the fish species tested, the juvenile menhaden was the most sensitive with less than 50% surviving 50 days of exposure to the incinerator residue in concentrations exceeding 1% by weight. The sea scallop was the most sensitive bottom organism tested and significant mortality occurred in concentrations greater than 3% by weight.

The toxicity effects reported were not considered a serious drawback to the proposed disposal program because calculations show that about 25 years of daily disposal of 500 tons over a 1-square-mile area would be required to equal the toxic level of one present residue concentration.

Oceanographic studies showed that incinerator residue (including cans) in 50 feet of water would drift under storm conditions with the net movement caused by the combined effects of the wave orbital velocity associated with passing waves and the tidal currents found in the area. Once the cans and other debris begin to rock back and forth as a result of wave orbital motion, weak but steady tidal currents cause a net movement in the direction of the current. Calculations indicate that wave heights greater than 3 feet can move debris down to 50 feet, wave heights above 9 feet can move debris at 100 feet, and wave heights above 12 feet can move debris at 200 feet. On the other hand, burnt cans disintegrate at a fairly rapid rate so that, at depths beyond 100 to 200 feet, disintegration was likely to occur long before the object moved a significant distance.

Radioactive Wastes: Radioactive wastes are a potential hazard to man because of radiation received from the immediate environment and by substances taken into the body by ingestion, inhalation, or absorption through the skin. It is feared wastes may reduce the life span, impair the functioning of parts of the body, or increase the mutation rate altering the inherited characteristics in future generations.

Transuranic elements have been introduced to the environment in dispersed form in three ways: (1) Local and worldwide fallout from nuclear explosive testing; (2) Atmospheric burn-up of plutonium power supplies and nuclear weapons (e.g., Thule, Greenland); and (3) Fluid wastes from chemical reprocessing and reactor operations. The disposal of radioactive wastes is generally accomplished either by containment—allowing for natural radioactive decay, or by dispersal—diluting the radioactivity to permissible levels, or a combination of the two. In the past, ocean disposal required containment and placement in waters exceeding 6,000 feet in depth. The increased use of radioactive materials by universities, hospitals, and research facilities has resulted in a corresponding increase in low-level radioactive wastes.

Recent findings indicate an insufficiency of information regarding the chemical and biological behavior of the transuranics in the marine environment. Without such information meaningful quantitative assessment of either current or future impacts is not possible. The possibility that marine concentrations of one or more transuranics may approach levels of concern through naturally-acting mechanisms which are not understood cannot be dismissed. Present evidence indicates that transuranic elements introduced to marine waters are rapidly transferred to the sediments.

If this holds true, then any input from waste treatment or reprocessing will mainly affect the coastal zone near the point of introduction. These sources would, then, not be expected to contribute significantly to open ocean pollution which would be largely due to fallout from atmospheric releases. Predictions of the effects of transuranic elements on the marine environment cannot be formulated until more extensive knowledge of their modes of transport and their biogeochemical and radiobiological behavior is available.

Military Explosives and Chemical Warfare Agents: Surplus or obsolete explosives and chemical and biological warfare agents (CBW) have been disposed of at sea for many years. Although the environmental effects of explosives disposed in the ocean are probably quite limited, they certainly pose hazards to man and his equipment. The hazards that explosive wastes present to operations conducted with research submarines were studied and indicated that, although most of the explosive disposal sites are situated in at least 6,000 feet of water, large amounts of explosive ordnance have come to rest in the shallow waters of the Continental Shelf as a result of combat operations.

CBW agents, on the other hand, because of their inherent toxicity, constitute potential hazards of considerably greater significance. Unfortunately, the nature and scope of these hazards are not well understood. A study by a panel of the National Academy of Sciences to consider proposed marine disposal of CBW agents by the Army noted that while various chemical warfare agents have been repeatedly disposed of in the oceans by the United States and other nations, there is no information regarding possible deleterious effects of these operations on the ecosphere of the seas.

On three occasions the U.S. Army disposed of chemical weapons in the ocean, but made no effort to determine whether or not there would be harmful effects on the environment. These reported instances do not include a 1958 disposal operation in which the Army disposed of 8,000 tons of mustard and lewisite gas by loading it aboard a surplus vessel, towing her to sea, and scuttling her.

The National Academy of Sciences' panel on disposal methods for CBW agents stated that certain CBW agents already embedded in concrete could be disposed of in the sea without serious environmental effects, but specified that this was accepted as a last resort solution in the event land disposal would pose unnecessary hazards to disposal personnel. The panel also recommended that in the future the Army should assume that all chemical weapons will require eventual disposal and should build disposal facilities that will not require dumping at sea.

Public Law 92-532 of October 23, 1972, however, states under prohibited acts:

> (a) No person shall transport from the United States any radiological, chemical, or biological warfare agent or any high-level radioactive waste, or, except as may be authorized in a permit issued under this title and subject to regulations issued under section 108 hereof by the Secretary of the Department in which the Coast Guard is operating any other material for the purpose of dumping it into ocean waters.
>
> (b) No person shall dump any radiological, chemical or biological warfare agent or any high-level radioactive waste, or, except as may be authorized in a permit issued under this title, any other material, transported from any location outside the United States, (1) into the territorial sea of the United States, or (2) into a zone contiguous to the territorial sea of the United States, extending to a line twelve nautical miles seaward from the base line from which the breadth of the territorial sea is measured, to the extent that it may affect the territorial sea or the territory of the United States.
>
> (c) No officer, employee, agent, department, agency, or instrumentality of the United States shall transport from any location outside the United States any radiological, chemical, or biological warfare agent or any high-level radioactive waste, or, except as may be authorized in a permit issued under this title, any other material for the purpose of dumping it into ocean waters.

Understanding of the short- and long-term responses of the more important marine food chain organisms to various types of waste is extremely limited. For example, wastes may have a detrimental effect through alterations in the natural environmental conditions (i.e., temperature, pH, etc.) or through physiological and other changes resulting from the addition of the wastes, or through both. Thus, in determining the toxic effects of wastes, it is important to consider environmental, physiological, and accumulative effects. Not only are marine organisms directly affected by wastes, but also indirectly through their interaction with other forms of organisms which comprise their food, competition, and predators.

The situation is further complicated by the fact that different species and different developmental or life stages of the same species may vary widely in sensitivity or tolerance to different wastes. Thus, unfavorable conditions which may be tolerable for long periods by adults may be entirely unfavorable for spawning, and thus possibly endanger survival of the species.

From the foregoing, it can be seen that a great deal of work remains to be done in order to establish both the short- and long-term environmental effects of various classes of wastes in the marine ecosystem. Laboratory tests on specific organisms, both on a short- and long-term basis, are required in order to establish safe discharge rates of the various wastes. The results of these tests must then be tested in the field to determine their adequacy for protection of marine life in the disposal area.

Areas for Study

Substantial research efforts related to the environmental effects of waste in the sea are required in three major areas to insure that the ocean environment is not damaged by discharges of waste and that the pollution problems that currently exist in most major U.S. rivers, lakes, and estuaries are avoided.

Baseline Environmental Data: In order to properly evaluate the effects of introducing any given foreign substance into the marine environment, it is essential to have an adequate understanding of the various natural fluctuations of the biota and water mass characteristics that are normal for the area in question. Without such an understanding, it is impossible to distinguish between normal variations and those resulting from the presence of the pollutant.

Baseline studies carried out prior to discharging wastes are the most effective means of providing reference data for use as a standard in measuring the effects of introduction of wastes. For an existing discharge, a control study area is set up in the general vicinity of the disposal operation but far enough removed so that it is not affected by the discharged material. In either case, a broad-spectrum study program is required. Physical and chemical studies of the control area serve to identify the natural processes responsible for the observed distributions of oceanographic properties, such as temperature, salinity, etc. Biological studies concentrate on both the quantity and quality of the biota, as well as the natural diversity of the fauna in the area.

Laboratory Studies of Waste Toxicity: The bulk of the toxicity test programs carried out on the various wastes being discharged at sea are limited both in scope and number. Substantial work is required not only on acute toxicity, but on chronic or sublethal toxicity as well, particularly for industrial wastes.

Nearly all toxicity bioassay experiments performed have been acute toxicities (TLM_{50}) which are only 96 hours in duration. Little has been done to observe the survival, reproduction, and behavior of successive generations to determine chronic toxicity levels of various wastes. Data resulting from some acute tests indicate although test animals survived certain concentrations, there was little or no reproductive capability in the first or second generation.

Because of the diversity of the marine fauna and the wide variety in types of wastes disposed at sea, short-cut methods are needed for determining the toxic effects of wastes both on a short- and long-term basis.

Another important item for research is the development of standard test procedures which can easily be performed without expensive or elaborate instrumentation or highly trained personnel. The Marine Water Quality Laboratory has developed a bioassay technique that uses brine shrimp (Artemia) whose eggs are easily obtainable throughout the country. To further standardize these tests, sea-salts now on the market in convenient packages are recommended for the preparation of artificial seawater.

Research in toxicity effects on the marine environment resulting from barge disposal is nonexistent and urgently needed. Results of investigations around submarine outfalls in 400 feet of water or less over the past 20 years cannot be extrapolated to the much deeper truly oceanic conditions. Nevertheless, the data from the outfall studies provide useful guidance.

Fate of Wastes and Effects on the Biota: Understanding the fate of the waste after discharge requires an understanding of how the waste physically mixes and disperses in the sea, how the waste degrades chemically and biologically, and what components of the waste tend to concentrate either in the bottom sediments or in various plants and animals in the food chain.

Waste Dispersion – At present, the understanding of how wastes mix and disperse after discharge at sea is limited; generalized theoretical models are too imprecise to allow effective prediction for a specific disposal situation. Environmental studies are required to verify predictive models. Present disposal operations afford excellent opportunities for conducting important full-scale, at-sea experiments on mixing.

Improved sampling methods and equipment should be developed in order to obtain the necessary synoptic data to verify mixing models at all depths in the sea. Because of the irregular frequency of many discharge operations, and the high costs of ship time for monitoring programs, research is also required to determine the most desirable spatial and temporal at-sea sampling patterns for various types of discharge operations.

Effects on the Biota – Research studies on biologic response to barge-discharged wastes carried out to date, although extremely valuable, merely scratch the surface of the problem.

Beneficial Uses of Solid Wastes

One beneficial use of solid wastes in the marine environment has been found to be artificial habitats for fish. Studies have shown that properly constructed habitats are a very effective means in congregating the available fish from a given area. It has been postulated that the artificial habitats constructed with solid waste serve to increase the populations of other migratory fish by providing additional spawning sites for adults and protection and food for the juveniles.

Because of the ease in handling, availability, long life, and low cost, automotive tires are the most attractive material for constructing artificial reefs. Research has shown that of several materials tested (wood, glass, concrete, metal, etc.) in the environment, rubber was found to be the most desirable substrate for colonization by the majority of invertebrate organisms in the area.

During 1969, over 30,000 tires were implanted on three different experimental reefs. Each reef is inspected by biologist-divers to assess the effectiveness of the tires in increasing the productivity of the area and to inspect the development of new fish around the habitat. Several types of artificial habitats have been constructed with such wastes as car bodies, tires, and rubble, and the various marine life have been attracted to them.

It has also been suggested that baled refuse would provide both food and habitat as a means of enhancing the production of the environment. Other potential beneficial uses of solid waste materials in the marine environment include its use as a building material in artificial islands, surfing reefs, and floating breakwaters. Another concept noted during the survey was the proposed use of scrap tires for the construction of a moored, floating offshore breakwater to attenuate wave action. This structure would consist of large truck tires joined to form a flexible floating barrier.

Information indicates that the total volume of waste dumped in the ocean will probably increase rapidly, primarily due to increases in sewage sludge and dredge spoil loads. Disposal of sewage sludge on the East Coast is expected to increase rapidly, with New York City increasing at the rate of 6% per year and the Baltimore-Washington area by 7% per year. Polluted dredge spoil will become a major disposal problem.

Although new municipal and industrial waste treatment facilities will lessen the problems for dredge spoil in the future, it is certain that polluted dredge spoil will remain a problem for at least the next decade. As this is the period of increased dredging activities to meet new marine commerce requirements and energy crises, the development of safe methods of disposal of polluted dredge spoil will be a major concern of EPA and the Corps of Engineers.

DREDGE SPOIL

In this discussion of the work done by the Corps in developing and maintaining the nation's navigable waterways and harbors, the quantities cited refer to dredging done by both Corps-owned plants and private plants on Corps projects, but do not include dredging done by other agencies or private interests under the permit program administered by the Corps. The volume of new work dredging is quite variable from year to year, as a result of Congressional authorization and funding decisions.

However, dredging concerned with new projects or new project dimensions has averaged about 80,000,000 cubic yards per year over the last three years. About 85% of this work is done by contract. The volume of maintenance dredging required annually to restore project dimensions by removing natural or man-induced shoal material is a somewhat more stable quantity and is about evenly divided between Corps-owned and private plants. Dredging required to maintain authorized project dimensions now totals about 300,000,000 cubic yards per year.

The work required on individual projects may vary from the removal of a few thousand cubic yards of sediment deposited in a critical area of a harbor to the removal of millions of cubic yards over an entire project. The cost for moving this material may vary from about 20 cents to several dollars per cubic yard depending on a number of factors including type of material, type of dredge, disposal practice, distance to disposal site, etc. However, a rough nationwide average of about 40 cents per cubic yard emerges when the total annual cost of approximately $150,000,000 for CE dredging operations is divided by the total annual quantity of material dredged.

Future trends in either dredging costs or volumes of material cannot be predicted for more than a few years with any reasonable degree of accuracy. Although trends during the last decade or so have proven (in hindsight) to be relatively constant, there are several significant factors or eventualities that could cause future trends to alter dramatically and unpredictably. Whereas some factors such as the need for channel enlargement for deeper draft vessels and the need for new port and harbor developments are predictable and could form a basis for trends, they could be completely negated by such factors as Congressional authorization and funding rates for new projects, national defense considerations, offshore terminal development trends, and the development of new shipping concepts. Perhaps the greatest potential for change in future trends of both cost and volume relates to the magnitude and patterns of environmental impact of dredging and disposal operations and developing national pollution control policy and legislation. These factors also are currently considered to be the most difficult to predict as far as consequences to future trends are concerned.

Different types of dredging equipment have regional importance. A wide range of dredging plants is available, but the vast majority of the work is done by pipeline and hopper dredges. In the past, the Corps has been able to deposit the material removed by dredging activities at selected disposal sites near enough to the dredging site to minimize disposal costs, but in locations which would have a minimum direct effect on other important activities in the area (beach areas, water intakes, commercial fishing areas, etc.). Decisions concerning open water disposal or land disposal have been based primarily on economic considerations.

Maintenance dredging data have also been used to provide an indication of the relative importance of the common disposal methods (open water disposal, confined disposal, and unconfined upland disposal) in various geographical regions. This data indicates that open water disposal is by far the dominant disposal method with some two-thirds of the material dredged during maintenance operations being disposed of in this manner.

As a result of the continued rapid industrialization and population concentration contiguous to some of the navigable waterways, the materials dredged from some harbors and channels have become polluted. As a result of concern over the possibility of adverse effects on water quality, or aquatic organisms by dredging and disposing of this material in open waters the Corps as early as 1966 began investigating the feasibility of alternative spoil disposal methods at selected harbors in the Great Lakes.

A pilot investigation on the Great Lakes which was accomplished under the guidance of an independent board of consultants in cooperation with the Federal Water Pollution Control Administration (now the Environmental Protection Agency) indicated that assessing the impact of open water disposal of polluted dredged materials was more difficult than anticipated and that a continuing study would be desirable. Pending the availability of more definitive information on the environmental impact of various disposal practices and in anticipation of the elimination of many sources of pollution, it was believed that the containment of polluted dredge spoil for a period of years would result in an improved situation in the Great Lakes.

In the River and Harbor Act of 1970 (Public Law 91-611), the Congress authorized the Secretary of the Army, acting through the Chief of Engineers, to construct, operate, and maintain contained disposal facilities of sufficient capacity to handle polluted dredge spoil from the Great Lakes and connecting channels for a period not to exceed ten years.

Another example of the nation's increasing concern over the environmental impact of dredging and spoil disposal is the National Environmental Policy Act of 1969 which requires a detailed statement of the environmental impact of proposed new navigation projects and projects requiring maintenance dredging along with all other proposed activities affecting the environment.

Composition and Characteristics

Dredge spoil characteristics can be identified in terms of physical, engineering, and chemical properties. Basically, dredge spoil is composed of solids and liquids. The complicating factor is that the solids and liquids consist of a wide range of constituents, many of which are classified as pollutants. Normally, dredge spoil solids are composed of fine- and/or coarse-grained soils. The soil particles generally are mineral grains of various sizes and shapes, occurring in every conceivable arrangement. Dredge spoils normally contain considerably more fluids than solids, a factor that complicates alternatives to open water disposal. In addition to soil, dredge spoil often contains other solids such as rock, wood, pieces of metal, broken glass, and other debris.

Based on the assessment of the reported materials being dredged by the Corps, there is no standard method used to identify and classify dredge spoil material. Terms used to identify dredge spoil range from the basic terminology (gravel, sand, silt, clay, or combinations thereof) to less descriptive terms such as mud, topsoil, and muck. The basic terminology is derived from the Unified Soil Classification System which is based on the size of the particles, the amounts of the various sizes, and the characteristics of the fine grains. This system provides a standard method of classification and should be used for reporting types of dredge materials.

Dredge spoil from new work often has appreciably better physical, engineering, and chemical properties than maintenance dredge spoil. Maintenance dredge materials from the coastal zone generally comprise the fine-grained materials such as silt and clay from surface runoff and sludges from municipal and industrial sewage. Consequently, there are probably more environmental problems involved in dredging and disposing of maintenance dredgings. However, land disposal of any fine-grained material with the characteristics of dredge spoil poses problems that can only be dealt with by utilizing the most modern scientific and engineering skills.

Physical Properties: Grain sizes, moisture contents, and plasticity are the important physical properties of dredge spoil. These physical properties affect dredging rates, types of dredging equipment needed, and method of spoil disposal. Grain sizes also determine the amount of turbidity associated with the disposal operation and the rate of solids settleability. For engineering purposes dredge spoil can be divided into three principal types: coarse-grained, fine-grained, and organic materials. The coarse- and fine-grained types can be further divided into descriptive terms on the basis of grain size. The fine-grained and organic dredge spoils normally cause the problems related with dredgings. Coarse-grained dredge spoil usually does not cause dredging problems.

Grain Sizes – Dredge spoil types reported by the Corps Divisions were grouped into five related groups as follows: Group A- Mud, clay, silt, topsoil, shale; Group B- Silt, sand; Group C- Sand, gravel, shell; Group D- Peat, organic muck, sludge, municipal and industrial wastes; and Group E- Mixed types.

The problem dredge spoils are contained in groups A, D, E, and possibly B. These materials are fine grained and are smaller than the No. 200 sieve size. The No. 200 sieve size is about the smallest particle that can be individually distinguished by the unaided eye. These fine-grained materials usually are poor fill materials for a number of reasons. Groups B (with low silt content) and C are reasonably good fill and foundation materials.

Water Content – Hydraulically dredged materials contain considerably more water than solids and exist as a slurry. Because of the high water content, the slurry can be readily pumped through pipelines to hopper bins and other storage areas or directly to disposal areas. At this point the high water content becomes a disadvantage and must be reduced. High water content in hopper bins and storage areas results in smaller payloads. Also, dredge spoils which retain a high water content cause land disposal problems.

The coarse-grained materials drain rather freely while fine-grained and organic materials retain a high water content and drain slowly. The consistency of these materials is markedly affected by changes in water content. Normally the degree of firmness of a fine-grained soil increases as the water content decreases.

Organic Content – The characteristics of dredge spoil are even poorer when high organic contents are present. Organic materials often produce objectionable odors when disposed of on land and retain a high water content resulting in poor undesirable materials. Organic materials can also cause adverse effects such as temporary dissolved oxygen depletion when disposed of in open water.

The assessment of maintenance dredging according to types of material indicates that more than half of the material dredged in the New England Division and Chicago District has a high organic content. All Districts reported the presence of organic materials in maintenance dredging but in most cases, not in very large quantities.

Engineering Properties: Engineering properties of soils include shear strength, compressibility, and permeability. Fine-grained dredge spoils are known to have poor engineering properties in that they are generally very compressible and very weak in their natural state. During the dredging operation the material is saturated to a liquid state, the structure destroyed as particles are rearranged, and sometimes oils, grease, and other pollutants in the water are mixed with the soil. The rate at which dredge spoil engineering properties improve after being dredged depends on a number of factors, most of which are related to drainage.

Coarse-grained materials can be stable without drainage if they are relatively dense and no appreciable seepage forces exist. However, fine-grained dredge spoils are generally unstable in an undrained condition. Coarse-grained materials recover rather rapidly if drainage is provided. However, fine-grained dredge spoils have much lower permeabilities and drainage occurs at a much slower rate. They are also weaker and more compressible than coarse-grained dredge spoils. Therefore, the engineering properties are normally poor and tend to improve at a much slower rate as drying and consolidation occur.

Open water dredging operations are defined as those operations that result in the disposition of dredged (spoil) materials into the open ocean, bays, estuaries, and inland rivers and lakes. From the standpoint of water quality effects, this definition is expanded to include those materials placed on beaches, marshes, along river edges, or any other type of unconfined land disposal in which the placed materials are subject to the influence of tides, river stage fluctuations, or are readily washed back into the water by rainfall.

The average annual amount of maintenance dredging materials readily identified as being disposed of in open water is approximately 182,000,000 cubic yards (1971). Approximately 5,000,000 cubic yards per year of maintenance materials, considered "unconfined disposal," should be added to the open water total in order to conform with the above definition of open water disposal. Because many of these projects are in the planning stages, it is impossible to determine what portion of these materials will be placed in open water. For the purposes of this report, an average annual value of 250,000,000 cubic yards is used to estimate the minimal amount of materials disposed of in open waters.

Of the 250,000,000 cubic yards per year used to represent open water disposal, it is estimated, based on the OCE inventory, that 200,000,000 cubic yards represent maintenance dredging and, of this amount, 110,000,000 cubic yards are disposed of in fresh water and 90,000,000 cubic yards in salt water. Freshwater disposal is almost totally confined to rivers, the major exception being approximately 6,000,000 cubic yards per year disposed of in the Great Lakes.

Saltwater disposal is limited almost entirely to the coastal zone—intracoastal waterways, bays, estuaries, etc. By far the largest category (approximately 153,000,000 cubic yards per year) is that classified as mixed sand and silt. About half this value is associated with the coastal areas of the United States, and the other half the inland rivers. Approximately 30,000,000 cubic yards per year of that category including sand, gravel, and shell is dredged from the nation's inland waterways, while the remaining 22,000,000 cubic yards are dredged from the coastal zone.

The ill-defined materials, mud, clay, silt, topsoil, and shale account for 80,000,000 cubic yards per year, all but 8,400,000 cubic yards of which are dredged from the eastern one-third of the United States. Finally, although the group including organic muck, sludge, peat, and municipal-industrial wastes accounts for only 1,400,000 cubic yards per year, some of the more pressing environmental problems are associated with this group.

Generally speaking, the materials dredged and disposed of in inland waterways are sand and gravel. The moving sand bottoms of many of the nation's navigable rivers have been a supply of sand and gravel for construction purposes for years. Again, speaking quite generally, in lakes, harbors, and many areas of the coastal zones where the carrying capacity of the water is quite low, the dredged materials often consist of small, light particles such as clays and silts.

One of the most extensive studies on the engineering characteristics of hydraulic dredge spoil was reported by the Philadelphia District. Four disposal areas for Delaware River maintenance dredging were selected for investigation. The investigation included field exploration of subsurface conditions, laboratory testing of the materials, and preliminary engineering studies of stability and settlement characteristics, types of foundations adaptable to the deposits, and methods of improving the spoil. These studies were concerned almost entirely with fine-grained spoil materials. A summary of typical spoil properties for selected Delaware River sites is shown in Table 2.8.

TABLE 2.8: TYPICAL SPOIL PROPERTIES AT SELECTED DELAWARE RIVER DISPOSAL SITES

Disposal Sites	Water Content %	Liquid Limit	Plasticity Index	Liquidity Index	Dry Density pcf	Specific Gravity	Void Ratio e_o
Edgemoor 1911-1967	62-133 (93)	56-164 (109)	19-94 (65)	0.4-1.40 (0.75)	34-72 (49)	2.26-2.56 (2.47)	1.54-3.68 (2.62)
Oldmans No. 1 1946-1963	44-155 (107)	63-193 (125)	33-103 (74)	0.42-1.16 (0.75)	31-71 (44)	2.23-2.61 (2.46)	1.28-3.53 (2.62)
Darby Creek 1956-1967	24-139 (80)	24-184 (101)	4-93 (58)	0.37-1.0 (0.60)	34-102 (55)	2.29-2.65 (2.50)	0.63-3.23 (1.86)
Pigeon Point 1948-1966	41-105 (69)	40-136 (91)	8-93 (54)	0.19-1.5 (0.63)	42-75 (56)	2.52-2.57 (2.54)	1.38-2.20 (1.88)

Disposal Sites	Compression Index	Coefficient of Consolidation $in^2/min \times 10^{-5}$	Shear Strength Q tons/sq ft	Shear Strength R c tons/sq ft	Shear Strength R ϕ deg	Shear Strength $\bar{R}$ ϕ deg
Edgemoor 1911-1967	0.51-1.45 (0.98)	67-150 (102)	0.05-0.28 (0.14)	0.05-0.25 (0.14)	12-15 (14)	30-37 (33)
Oldmans No. 1 1946-1963	0.90-1.28 (1.09)	65-100 (83)	0.07-0.48 (0.20)	0.1 (0.1)	12-16 (14)	30-32 (31)
Darby Creek 1956-1967	0.39	120	0.05-0.42 (0.29)	0.07-0.35 (0.21)	11-15 (13)	34-38 (36)
Pigeon Point 1948-1966	0.75-0.96 (0.83)	60-100 (84)	0.09-0.55 (0.25)	0.15-0.25 (0.21)	13-15 (14)	35-38 (36)

Notes: (1) Numbers in parentheses are average values.
(2) Shear strengths are given in cohesion (c), tons per square foot and angle of internal friction (ϕ) in degrees.
Q denotes unconsolidated-undrained triaxial compression tests.
R denotes consolidated-undrained triaxial compression tests.
$\bar{R}$ denotes consolidated-undrained triaxial compression tests with pore pressure measured.
(3) All data based on laboratory results reported in U.S. Army Engineer District, Philadelphia, CE, *Long Range Spoil Disposal Study; Part II, Substudy 1, Short Range Solution,* June 1969, Philadelphia, Pa.
(4) Laboratory tests made in 1967.

Chemical Properties: Data on the chemical properties of dredge spoil soils are quite limited. Most attention in the past has been focused on the water chemistry and the nature of pollutants. Studies of the Great Lakes and the lower Hudson River indicate that dredge spoils are largely silicates with small percentages of organic materials both from the natural environment and from human activities, however this is not necessarily representative of all dredge spoil. Since dredge spoil types vary regionally, it is not unreasonable to assume that chemical properties of dredge spoil soils also vary regionally.

Chemical properties have a direct bearing on the physical and engineering properties of dredge spoil soils. Therefore, research on the chemical properties of dredged soils is necessary to provide a better understanding of their behavioral characteristics. The fine-grained group known as clay minerals can be divided into three subgroups known as kaolinites, illites, and montmorillonites. Each of these subgroups has different behavioral characteristics which affect the behavior of dredge spoil and the uptake and release of pollutants.

Environmental Effects

In order to assess the effects of dredging and disposal of bottom materials in open waters and to classify such materials as either polluted or unpolluted, the EPA issued *Criteria for Determining Acceptability of Dredged Spoil Disposal to the Nation's Waters*. The criteria were circulated to Corps offices by OCE in Engineering Circular (EC) 1165-2-97. The responsibility for accomplishing the needed sediment sampling and analysis was set forth in the EC as follows:

> Where EPA has the capability for taking and analyzing the samples, it can be expected that this agency will accomplish this work in making their evaluation of whether or not the materials at Corps projects are to be considered polluted or unpolluted. In those regions where EPA does not have such a capability, an in-house capability to perform the required testing may be developed, but only to the extent of acquiring the minimum equipment necessary for this purpose for work on Corps projects. Where it is impracticable to establish this capability at District or Division level, arrangements may be made to have the necessary sampling and testing accomplished by qualified testing laboratories.

For additional guidance and quality control of techniques for sample collection, preservation, and preparation, the EPA criteria referred to the Federal Water Quality Administration (FWQA) analytical manual on sediments.

The EPA criteria for determining acceptability of dredged spoil disposal to the nation's waters are presented below. These criteria were developed as guidelines for FWQA evaluation of proposals and applications to dredge sediments from fresh and saline waters. The decision whether to oppose plans for disposal of dredged spoil in United States waters must be made on a case-by-case basis after considering all appropriate factors; including the following:

(a) Volume of dredged material.
(b) Existing and potential quality and use of the water in the disposal area.
(c) Other conditions at the disposal site such as depth and currents.
(d) Time of year of disposal (in relation to fish migration and spawning, etc.).
(e) Method of disposal and alternatives.
(f) Physical, chemical, and biological characteristics of the dredged material.
(g) Likely recurrence and total number of disposal requests in a receiving water area.
(h) Predicted long and short term effects on receiving water quality. When concentrations, in sediments, of one or more of the following pollution parameters exceed the limits expressed below, the sediment will be considered polluted in all cases and, therefore, unacceptable for open water disposal.

Sediments in Fresh and Marine Waters	Concentration % (dry wt basis)
*Volatile solids	6.0
Chemical oxygen demand (COD)	5.0
Total Kjeldahl nitrogen	0.10
Oil-grease	0.15
Mercury	0.001

(continued)

Sediments in Fresh and Marine Waters	Concentration % (dry wt basis)
Lead	0.005
Zinc	0.005

*When analyzing sediments dredged from marine waters, the following correlation between volatile solids and COD should be made:

$$\text{TVS\% (dry)} = 1.32 + 0.98(\text{COD\%})$$

If the results show a significant deviation from this equation, additional samples should be analyzed to insure reliable measurements.

The volatile solids and COD analyses should be made first. If the maximum limits are exceeded the sample can be characterized as polluted and the additional parameters would not have to be investigated. Dredged sediment having concentrations of constituents less than the limits stated above will not be automatically considered acceptable for disposal. A judgment must be made on a case-by-case basis after considering the factors listed in (a) through (h) above. In addition to the analyses required to determine compliance with the stated numerical criteria, the following additional tests are recommended where appropriate and pertinent:

Total phosphorus
Total organic carbon (TOC)
Immediate oxygen demand (IOD)
Settleability
Sulfides
Trace metals (iron, cadmium, copper, chromium, arsenic, and nickel)
Pesticides
Bioassay

The first four analyses would be considered desirable in almost all instances. This may be added to the mandatory list when sufficient experience with their interpretation is gained. For example, as experience is gained, the TOC test may prove to be a valid substitute for the volatile solids and COD analyses. Tests for trace metals and pesticides should be made where significant concentrations of these materials are expected from known waste discharges. All analyses and techniques for sample collection, preservation, and preparation shall be in accord with a current FWQA analytical manual on sediments.

Effects of Dredging: Generally speaking, the concerns associated with dredging all fall into two categories: (1) direct effects on biological communities, and (2) indirect effects on biological communities. Undoubtedly, the greatest concern is over the destruction of benthic communities. The extent and significance of benthic destruction vary considerably, depending on many circumstances. Generally, the potential for benthic destruction is greater in new work projects than in maintenance activities. This is primarily because the areas subjected to maintenance dredging have been previously dredged and the light, shifting substrates often found in such areas are not conducive to extensive benthic growth. Also, whether the project is of the maintenance or new work type, the direct effect of the dredging activity on benthic communities is usually confined to the project area.

The indirect effects restricted to the dredging actions of the project are much more difficult to define than the direct effects. This does not mean such effects are of less significance; they could be much farther reaching, both in extent and significance. It is not possible to say whether such effects are detrimental, beneficial, or, as is more likely, a combination. The needed information is simply not available, although there is no doubt that changes do occur as a result of dredging activities. The potential for the indirect effects on biological communities is usually attributed to physical alterations of the environment due to dredging. These physical alterations include changes in bottom geometry and the creation of deep-water regions, new open water, changes in bottom substrates and habitats, alterations in water velocity and current patterns, changes in future sediment distribution patterns, alteration of the sediment-water interface with subsequent release of biostimulatory or toxic chemicals, and the creation of turbidity clouds.

Effects of Disposal: There can be no doubt that most of the concern over the Corps' dredging operations is directed toward the open water disposal of dredged materials. In order to assess the effects of open water disposal, it is necessary to know, as a bare minimum, how much of what kinds of materials are being disposed of in what kind of aquatic and marine ecosystems.

The short-term effects associated with disposal operations can usually be detected during the disposal operation or shortly thereafter. The most common short-term effects that have been reported are:

(1) Turbidity—aesthetically displeasing, reduce light penetration, flocculate planktonic algae, decrease availability of food;
(2) Sediment buildup—destroy spawning areas, smother benthic organisms, reduce bottom habitat diversity, reduce food supplies, reduce vegetation covering;
(3) Oxygen depletion—suffocate organisms in area, release of noxious materials such as methane, sulfides, and metals.

The magnitude and duration of these effects can vary significantly depending on the amounts and types of materials involved and the manner in which they are returned to the water. Generally speaking, hopper dredges and barges practicing dump disposal minimize the extent of spoil-water contact. Therefore, the intensity of the short-term effects might be high, but the duration and range relatively small. This is particularly true for stationary dumping as opposed to the slowly moving type dumping.

On the other hand, sediment buildup is usually intensified by this type disposal but also, better confined to one location. Of course, all of the above comments must be qualified in terms of the water depth, current velocities and patterns, and bottom configuration in the disposal area.

According to *Ocean Dumping—A National Policy* (1970), dredge spoils account for 80% by weight of all ocean dumping and 34% of this amount is estimated to be polluted. In *Ocean Disposal of Barge-Delivered Liquid and Solid Wastes from U.S. Coastal Cities* (1971), the main environmental effects attributed to ocean dumping of dredge spoils were bottom sediment buildup and turbidity. The effects on fish and shellfish due to sediment buildup were listed as destruction of spawning areas, reduction in food supplies and vegetational cover, trapping of organic matter (with resultant development of anaerobic bottom conditions), and the absorption or adsorption of organic matter (including oil).

Both direct and indirect effects on fish and shellfish due to turbidity were cited. The direct effects listed indicated possible mortalities due to suffocation or the reduced growth and survival of fish and shellfish larval stages. The listed indirect effects included reduced photosynthesis, reduced feeding visibility, destruction of spawning areas, reduced food supplies, reduced vegetation cover, the trapping of organic matter with subsequent anaerobic conditions, flocculation of algae, and the sorption of organic matter or inorganic ions. The report did point out that all these effects are dependent on the characteristics of the spoil site. As an example, it was pointed out that materials discharged in waters that are normally turbid would produce almost negligible effects.

Sedimentation Factors in Open Water Disposal: According to data contained in a 1971 study of ocean waste disposal, there are 161 active disposal sites for dredge spoil in coastal waters, over 90% of which are along the Atlantic and Gulf Coasts. This number does not include disposal sites in coastal bays, sounds, or estuaries. Of the 161 sites, 60% are located within 3 miles of the coast in water depths less than 100 feet.

The principal criterion in the selection of these sites, many of which have been in use for 40 or more years, was disposal costs or economics. Selected locations were usually those closest to the dredging project that offered reasonable assurance that (a) the disposed material would not return directly to the dredged channel, (b) accumulating spoil would not restrict navigation, (c) known fishing grounds would not be destroyed, and (d) spoil would not adversely affect beaches, water intakes, and other coastal facilities. Considerations of environmental impact seldom were employed in site selection procedures. Moreover, few or no investigations were made of the hydrologic regimes to determine if or at what rates the spoil would disperse. In fact, these questions still cannot be answered for most offshore disposal sites.

Now that dredge spoil contains increasing amounts of pollutants and is potentially more toxic to marine or aquatic life, far more consideration has to be given to both the direct environmental impacts of disposal operations and the long-term fate (and effects) of the spoil. To permit selection of new offshore disposal sites and consideration of new disposal concepts, and to enable unpolluted or lightly polluted spoil to be used for certain beneficial or productive uses, appreciable information is needed as to the degree of compatibility between the introduced sediment and the naturally occurring substrate conditions.

It is felt that the nature and distribution of bottom sediments and coastal processes hold important clues as to the location for environmentally compatible disposal operations. Sediment distributional and compositional data also are quite indicative of prevailing energy regimes which will directly govern patterns and rates of spoil dispersal. From the standpoint of biological effects, dispersal or concentration (or containment) of spoil can be selectively employed to minimize adverse effects depending on the nature of spoil constituents.

Substrate Considerations: Inner Shelf Zone – The continental shelves of the United States encompass an area of over 275,000 square miles; however, their distribution and width are highly variable. For example, the continental shelf along the Pacific Coast averages 20 miles in width and includes only 6% of the total shelf area of the nation. Shelf widths average 80 miles along the Atlantic Coast and 65 miles along the Gulf Coast.

The inner shelf zone as discussed herein is an irregular belt of variable width which includes the beaches, barrier islands, and related features of the shore zone and which extends outward to the seaward edge of the nearshore modern sand prism. The seaward limit of the latter is definable in terms of sediment origin, composition, and abundance and may lie anywhere from several miles to several tens of miles offshore. It is generally present in an unbroken belt along the Atlantic and Gulf Coasts except in the area of influence of the Mississippi River (generally coincident with the coast of Louisiana) and is intermittently present along the Pacific Coast.

Included within the nearshore modern sand prism is the greatest relative concentration of sand and coarse-grained sediments of the entire coastal zone. Reworking of geologically older deposits on the outer continental shelf during the latest episode of worldwide (eustatic) sea level rise plus contemporary coastal headland erosion have been the prevailing sources of sediments to this zone. Since estuaries are effective sediment traps, very little fluvial sediment from the mainland enters this zone.

Under present geologic conditions and controls, the sediment budget of the nearshore modern sediment zone is finite and slowly diminishing. Sediment is being lost from the system through onshore and offshore transport by winds and currents (much during storms) at a time when the impetus for renewal from offshore erosion of older formations has largely ceased. Consequently, as far as shoreline stability is concerned, transgressive (eroding) or stable conditions prevail over regressive (building or prograding) conditions.

The inner shelf zone is also an area of high energy in comparison with the estuarine and outer shelf areas. The most dynamic environments are in and along the shore zone where sediments are almost constantly being affected by waves, currents, winds, and severe storms. Energy levels decline seaward but remain sufficient for active sediment transport throughout the nearshore modern sediment zone as evidenced by prevailing bottom ripples, waves, shoals, and related features. Suspended sediment loads in the inner shelf zone are significantly less than in the estuarine zone and reach their peak when land runoff is of such magnitude that sediment-laden water discharges from the estuaries and enters the littoral drift regime.

Outer Shelf Zone — The greater part of the continental shelves on all coasts is included in the outer shelf zone. This zone extends from the contact with the inner zone seaward to the edge of the shelf. For the most part, the outer shelf area is characterized by a moderately uneven erosional topography lightly and discontinuously veneered with a blanket of relict, coarse-grained deposits. These deposits attest to the effective winnowing action of tidal currents and permanent ocean currents that prevail across much of the outer shelf zone.

Sand characterizes much of the relict sediment veneer; however, gravel, shell, and various finer constituents occur. Heaviest gravel accumulations occur in formerly glaciated regions off the New England Coast. Fine-grained sediments are almost all of modern or geologically recent origin and seldom occur outside of enclosed shelf basins. The basins are generally restricted to the New England shelf area where they are of glacial origin and to the Pacific Coast shelf area where they are of structural or tectonic origin. The only other zones of significant modern sediment accumulation are in the submarine canyons which occur sporadically on the outer shelf and shelf slope.

Substrate Compatibility – It would appear that the greatest potential for environmentally compatible disposal of unpolluted or nontoxic fine-grained spoil (clays and silts) would be in estuarine waters. Relatively high turbidities that would result would be more environmentally compatible here than elsewhere and deleterious impact could be minimized by seasonal dredging timed to coincide with periods of naturally high turbidity. Adverse sedimentation effects may not materialize if care is taken to select sites well away from important or commercially valuable benthic communities. Equal care would have to be exercised in selecting sites that would not result in return of the spoil to the dredged channel. Because of the known or suspected adverse effects of highly polluted spoil and the biological importance of estuaries, disposal of this type of material in estuarine waters is inadvisable.

The diversity of hydrologic conditions in estuaries should permit selection of disposal sites for either concentration or dispersal of the spoil depending on which is desired. More so here than elsewhere, the nature of the substrate should be indicative of the energy levels present in the area. Sandy dredge spoil would most certainly be compatible in many estuarine areas; however, because of the value of this material as a limited resource, disposal in the inner shelf zone would be more logical where it could be accomplished economically. Sand may be of additional value when used in either estuarine areas or the outer shelf zone to create artificial habitats.

Neither the inner nor the outer shelf zones offer much potential for finding disposal sites conducive to concentration or containment of spoil, particularly fine-grained (silty and clayey) spoil. Only the enclosed basins such as have been mentioned could be used for this purpose. Particular basins in favorable locations may afford environmentally compatible disposal potential for even moderately polluted spoil; however, outside of these basins, dispersal and winnowing are probably too prevalent to permit this. Where the desired effect is maximum dispersal, the inner shelf zone and the outer shelf zone both offer excellent potential. Areas of sandy substrate and particularly shoals or submerged ridges are most desirable in this regard.

The development of environmentally compatible spoil disposal practices requires an intimate knowledge of the biological communities associated with a specific disposal site. This includes not only a knowledge of ecologically and/or commercially valuable species that may be subject to smothering, but also a knowledge of the types of benthic species that may eventually recolonize the modified substratum.

The concept of sand-on-sand, mud-on-mud spoil disposal may have merit in minimizing environmental impact of spoil disposal. The natural sediment occurrence in a specific site is indicative of the existing physical environment, and the indigenous benthic community is highly adapted to these physical conditions. For example, on a sandy substrate, suspension feeders are usually dominant due to their sessile or semimotile nature and their requirement for a continuous current to import plankton which they strain from the overlying water as food. Benthic detritus feeders are usually dominant on a mud substrate with minimal currents that allow organic detrital food to settle out onto the substrate.

Sand-on-sand, mud-on-mud spoil disposal may impose a short-term burial of some benthic species, but due to the environmental adaptation of the indigenous benthic species and similarities in sediment types of the spoil and disposal site, the spoil substrate should be more conducive to rapid recolonization. For example, in spoil disposal studies conducted in Chesapeake Bay, spoil was disposed along the channel and was distinguishable from natural sediment only by bulk properties. Localized smothering occurred, but the area was recolonized within two years by the same species that had inhabited the area before disposal.

Substrate Enhancement – In many aquatic areas, it might be desirable to modify the substrate of a disposal site to improve productivity of the area. An example of potential beneficial substrate modification with dredge spoil is the creation of shellfish habitat. This is a common practice in commercial fisheries and dates back at least to the Roman Empire. It involves distributing hard objects such as shell or rock on the bottom surface to provide attachment areas for shellfish larvae. The feasibility of using dredge spoil for this purpose is demonstrated by several coastal Districts' mention of inadvertent colonization of spoil mounds by shellfish.

One of the largest oyster reefs in Mobile Bay was once a spoil disposal site. The Norfolk and Galveston Districts have reportedly ceased disposal of spoil in some areas due to colonization of spoil mounds by shellfish. The Seattle District presently utilizes dredge spoil in some areas to provide habitat for commercially valuable clams.

Another potential use of dredge spoil is as a construction material in artificial fishing reefs. Commercial and sport fishermen operate on the premise that certain fish species are found only in certain areas. The concept behind artificial reefs is obtaining knowledge of environmental factors, such as adequate shelter or food species availability, that make certain areas suitable for fish species valued by man. This knowledge can then be applied to other areas of sea floor to manipulate the environment and create productive areas.

Marine biologists have had much success with reefs constructed from different solid wastes such as car bodies, construction debris, and old tires. Some of the biggest problems encountered in reef construction, however, are transportation of these wastes to the reef site and not providing sufficient area to prevent overfishing. Dredge spoil, especially in large quantities associated with new work projects, might be useful here as a supplemental construction material not only in creating larger areas for colonization, but also in providing a suitable substrate for additional benthic food species. Substrate enhancement is also a possibility in rehabilitating abandoned disposal sites, particularly where high pollutant levels prevent or greatly decrease the probability of natural recolonization.

Deepwater Disposal: The apparent attractiveness of the deep ocean basins for solid waste disposal, including dredge spoil, has drawn much attention and comment, but very little definitive research or investigation. Differences in opinion as to the environmental impact of deepwater disposal are widespread and reach the proportions of major controversy when polluted and/or toxic substances are involved. Investigations aimed at the controversial issues have proceeded slowly largely because large-scale deepwater disposal operations have not been economically feasible to date. However, with alternative solutions to currently unsatisfactory disposal operations standing to increase unit costs by several orders of magnitude, deepwater disposal appears to reenter the realm of economic reality.

Certain aspects of deepwater disposal are beyond controversy. Beyond the edges of the continental shelves in water depths of over 12,000 to 15,000 feet lie tremendous areas of ocean bottom in the form of abyssal plains and trenches. These involve approximately 90% of the total area of the world oceans. Since these areas are below the permanent oceanic thermocline, they are composed of relatively still masses of cold water existing under conditions of high pressures and densities.

Nearly all scientists agree that these are regions of relatively uniform and constant physical and chemical environments and areas of slow biochemical decomposition. Many scientists also agree that these areas are well below the zone capable of supporting plant growth (the euphotic zone) and hence have very low productivity. From the standpoint of commercial fisheries, these areas are of virtually no value either now or in the future.

These facts have led some scientists to conclude that the deep ocean areas may be ideal places to dispose of many types of wastes, including dredge spoil. They argue that the deep oceanic areas are natural sinks that are removed from food chains and biological life cycles. Although there is disagreement over dispersal and assimilation capacities, it is contended that degradation rates are actually irrelevant. Proponents of this school of thought argue that unpolluted spoil is a waste compatible with the marine environment and can be disposed of safely even in certain unproductive areas on the continental shelf.

The opposing school of thought is that no type of spoil disposal should be allowed in the deep ocean, at least until more information is available from research. It is contended that the deepwater fauna are finely tuned to the environment and might be vulnerable to sudden changes induced by the introduction of spoil. Fears are expressed that if deleterious effects occur, opportunities to reverse the process are minimal. Inherent in this opinion is the belief or suspicion that the potentially affected fauna are important in food chains or the marine ecosystem.

Since the long transport distances involved in deepwater disposal and the consequent high unit disposal costs are severely limiting factors, submarine canyons manifest themselves as features worthy of careful consideration. The term submarine canyon includes at least 30 features off the coasts of the United States which are generally steep-walled valleys that cut into the continental slope and frequently head on the continental shelf. Canyons off the Pacific Coast usually head close to shore, sometimes within a mile of land, whereas those off the Atlantic Coast head considerably farther seaward, often 50 miles or more offshore.

Although submarine canyons vary widely in size, shape, and configuration and are probably of a variety of origins, most are important for the fact that they serve to funnel large volumes of sediment from nearshore shelf areas downslope into deep oceanic areas. Mechanisms of sediment movement in canyons are the subject of much current research, and appear to involve slumping, turbidity currents, traction transport, and others. Sediment movement rates are being investigated by direct observation by underwater television and through the use of fluorescent sand tracers.

An attractive aspect of submarine canyons is that the indigenous flora and fauna are adapted to a physical environment characterized by abundant modern sediment actively moving downslope. How much more sediment could be transported by the canyons without creating an excessive environmental impact? What would be the direct and indirect ecosystem responses to various pollutants that might enter the canyons? These are questions that require answers forthcoming through research before submarine canyons could be used for spoil disposal.

The importance of navigable waterways and harbors to the nation's economic growth dictates that large volumes of material will continue to be dredged each year. Increasing concern over the environmental impact of man's activities is resulting in significant controls over methods used to dredge and dispose of this spoil material.

Because of the highly variable environmental situations in which dredging and dredge spoil disposal must be accomplished, there is no universally acceptable method for doing these tasks. Consequently, it is extremely important that there be a number of proven systems from which an appropriate dredging and spoil disposal method can be selected, depending upon the environmental situation involved. Alternative methods designed to make the dredging and spoil disposal operations more environmentally compatible in most cases result in increased direct initial costs. Efforts aimed at reducing dredging volumes could materially help the problem of spoil disposal. The prudent use of structures and techniques can effectively reduce shoaling in dredged channels and/or control the location of shoaling to minimize spoil disposal problems.

Constraints on the type of material which can be disposed of in open water are being made. The constraints will probably refer to the pollution status of the spoil material. There are strong reservations within the scientific community over the guidelines which rely upon the chemical composition of spoil material as the sole indicator of pollution status, and thus as the basis for deciding the acceptability of the spoil material for disposal in open water.

Better understanding of the nature and magnitude of the effects on water quality and aquatic organisms due to varying methods of disposal and types of materials deposited must be developed. Constraints should then be based on sound technical information with provisions for periodic review and update and for the regional variability found throughout the nation's marine and fresh waters.

The pollution status of spoil and the spoil composition and characteristics indicate the need for extensive spoil sampling and analysis in conjunction with the Corps' dredging activities. Sampling programs should be well planned and adequate guidelines furnished for sampling, sample preparation, and analysis so that the results of this work are as meaningful as possible. Large volumes of the spoil material dredged each year are not polluted even when evaluated according to very conservative pollution criteria. This is particularly true in the riverine environment.

A ban on open water disposal of unpolluted material would not be in the nation's best interest even though it is recognized that such disposal can, in some cases, result in adverse biological effects. One reason for this conclusion is the fact that beneficial spoil disposal practices such as marsh creation, spoil island development for wildlife habitat, and beach nourishment require open water disposal. A second reason is concern over confined disposal which encroaches upon estuaries through landfill operations or which takes valuable wetlands, farmland, or land needed for commercial development. Also, land disposal of unpolluted material would in most cases greatly increase direct disposal costs. Studies in which attempts have been made to determine the effects of open water disposal have documented certain short-term effects. These include temporary increases in turbidity, temporary decreases in dissolved oxygen, and the smothering of benthic organisms. The possibility for detrimental long-term effects due to open water disposal exists, but attempts to document these effects have, to date, been largely inconclusive. Although such effects may be significant, they appear to be quite subtle and, therefore, hard to assess.

Many lines of evidence indicate that dredge spoil can and should be considered as a manageable resource. Numerous cases of beneficial environmental effects of spoil and disposal operations can be documented. Potential beneficial uses of spoil in open water include marsh creation, spoil island development, beach nourishment, and substrate enhancement.

Marsh creation and enhancement through regulated spoil disposal and plant recolonization offers outstanding potential as a disposal practice alternative. One such project on Nott Island in the Connecticut River is reported as taking hold very successfully.

Planned development and colonization of spoil islands should permit disposal of large volumes of spoil in a manner that will produce desirable habitat for terrestrial wildlife. Existing spoil islands have become valuable habitats for local and migratory waterfowl and other wildlife through natural colonization. Beach nourishment and spoil disposal are related problems that often can and should be approached in a combined and coordinated manner. The exploitation of offshore sand resources for beach fill may present an opportunity for backfilling subaqueous borrow pits with undesirable spoil.

Spoil offers distinct possibilities for enhancing substrate conditions in open water areas. Both sport and commercial fishing have been and can be improved through selective spoil disposal. Spoil can also be used effectively in creating new habitats for shellfish. Dredging also has been shown to create beneficial changes such as cold weather refuge areas, new spawning grounds, and the release of nutrients.

Selection of offshore disposal sites requires detailed evaluations of environmental conditions. Historically, disposal sites were selected largely on the basis of location and economics. New sites must be selected largely on the basis of substrate characteristics, biological communities, hydrologic conditions, and related considerations as compared with and evaluated against spoil composition and characteristics. Deepwater (oceanic) spoil disposal is an attractive potential spoil disposal alternative once all the questions regarding biologic effects of pollutants in a largely unknown environment are known (e.g., the role of submarine canyons as natural transport mechanisms).

GLOBAL MARINE ECOSYSTEMS

There are many sources of marine pollution. Industrial wastes and municipal sewage effluent reach the sea through streams and rivers. These wastes are also transported in barges to offshore disposal sites or discharged directly into coastal waters through outfalls. Rivers carry sediments and polluted materials from agricultural operations and land-development activities to the sea. Pesticides, heavy metals, and other contaminants reach the ocean through surface runoff and the atmosphere. The introduction of these contaminants into the marine environment has caused serious pollution problems in coastal areas and in entire seas (e.g., Mediterranean, North, and Baltic Seas). Coastal and regional pollution in recent years have become the subject of many studies and remedial programs, both at the national and international level. Generally, however, these problems have tended to be relatively localized or regionalized in nature.

Here the emphasis is on the broader aspects of ocean pollution, including the possible long-term changes in ocean ecosystems. There is a growing concern that buildup of certain materials in the oceans may be a problem in that they are suspected of showing up widely distributed in the marine environment. These classes of materials generally recognized to be of global concern are treated in this section.

Conferences and workshops set up to examine the issue of which pollutants might represent a serious global threat to the oceans concluded that petroleum, metals, and synthetic organics are the principal pollutants of global significance. The subject of oil pollution is covered in *Oil Spill Prevention and Removal Handbook,* Noyes Data, Park Ridge, New Jersey, 1974.

Heavy Metals

Metals are found throughout the biosphere in greatly varying concentrations. Table 2.9 gives average background ocean concentrations of metals that could be considered toxic at high concentrations.

TABLE 2.9: INORGANIC CHEMICALS CONSIDERED AS POLLUTANTS OF THE MARINE ENVIRONMENT (1971)

Element	Natural Concentration of Seawater, μg/l	World Production 1968 mt/yr	Routes of Entry into the Sea*	Pollution Categories**
Beryllium (Be)	0.001	250	U	IV c
Titanium (Ti)	2	1,000,000	A	IV b
Vanadium (V)	2	9,000	A	IV a
Chromium (Cr)	0.04	1,500,000	R (U)	IV c
Iron (Fe)	10	480,000,000	D, R	IV c
Copper (Cu)	1	5,000,000	D, R	IV c
Zinc (Zn)	2	5,000,000	D, R	III c
Cadmium (Cd)	0.02	15,000	A, R	IV c
Mercury (Hg)	0.1	9,000	A, R	I b
Selenium (Se)	0.45	1,000	U	III c
Lead (Pb)	0.02	3,000,000	A, R	I a
Phosphorus (P)	-	?	D	IV c
Arsenic (As)	2	60,000	D	II c
Antimony (Sb)	0.45	60,000	U	IV c
Bismuth (Bi)	0.02	3,800	U	IV c

*D, dumping; A, through atmospheric pollution; R, through rivers (runoff) or pipelines; U, unknown.

**I-IV probable order of decreasing menace; a - worldwide, b - regional, c - local (coastal, bays, estuaries, single dumpings).

The concentrations of the elements given in the table above are approximate.

The natural levels of the metals given in the table are being augmented by mining and other extractive and industrial processes. However, direct relationship between industrial production and increased total levels in the sea has not been established. The metals contributed by man's activities reach the sea by the same routes as other pollutants, that is, through industrial waste discharges, municipal sewage systems, runoff, ocean dumping, and atmospheric transport. Some metals in pesticides are organically bound and reach the marine environment in this highly toxic form.

The presence of metals in the marine environment was brought to worldwide attention in the 1960s when 111 people died or suffered serious neurological damage in Japan as a result of Minimata Disease caused by mercury poisoning. World production of mercury is currently about 9,000 metric tons per year; about 5,700 tons per year are released into the environment as a result of man's activities (Table 2.10). This artificial input has caused such problems as: elevated mercury levels in fish and water birds in Scandinavian countries; accumulation of mercury in eels from the Rhine River in Holland; and fish contamination in the Great Lakes.

In the marine environment, metallic mercury and most mercury compounds can be biologically converted into toxic organic compounds such as methyl mercury that is concentrated by some food chains. This is important because in addition to the known toxicity of methyl mercury, there is evidence that it can also cause genetic effects.

TABLE 2.10: ESTIMATES OF ENVIRONMENTAL MERCURY FLUXES, 1971

	Tons/yr
Natural flows	
Continents to atmosphere (by degassing of the earth's crust	
Based on precipitation with rain	84,000
Based on atmospheric content	150,000
Based on content in Greenland Glacier	25,000
River transport to oceans	3,800
Range of totals: 28,800 - 153,800	
Flows involving man	
World production (1968)	8,800
Entry to atmosphere from fossil fuel combustion	1,600
Entry to atmosphere during cement manufacture	100
Losses in industrial and agricultural usage	4,000
Totals: 5,700	

Lead is another important environmental contaminant. The progressive contamination of some remote areas by lead has been recorded by some investigators. However, these data cannot be taken as representative of worldwide circumstances. In their isotopic analysis of lead in polar snows in Greenland, it has been calculated that concentrations of that metal have increased from less than

0.0005 micrograms per kilogram (ppb) of ice in 800 BC to 0.20 micrograms per kilogram (ppb) by 1965. This implies a hemisphere-wide atmospheric transport and dispersion of industrial lead over land and sea. What happens to such lead in the sea is unknown. Since worldwide lead production is increasing each year, there is a need to know more about how this lead is cycled.

An increase in surface concentrations of lead in some waters has been established by analysis of isotope ratios. Investigators have found that surface waters in the Pacific Ocean and the Mediterranean Sea contained 0.3 and 0.2 ppb of lead, respectively, compared to less than 0.1 ppb in deeper waters; however, no difference was noted in the Atlantic Ocean. These puzzling results further point up our ignorance on the natural recycling of lead.

Furthermore, the investigators suggested that the widespread use of leaded gasoline is the principal source of industrial lead found in the atmosphere and in the surface waters of the oceans. Yet, to ascribe this lead specifically to manufactured tetraethyl lead, it would be necessary to know whether the isotopic composition of tetraethyl lead differs from that in all the other products of industry that contain lead and which also enter the marine environment by weathering, burning, and other routes.

It is not known whether lead is being concentrated by marine organisms, nor what the effect of any accumulation might be. However, in contrast with mercury, lead is apparently not biotransformed. Inorganic lead is nonetheless toxic in excessive quantities and can accumulate in mammalian tissues.

The situation in regard to lead can be briefly summed up as follows: (1) industrial lead is entering the oceans via outfalls, dumping, and aerosols at increasing rates; (2) the natural accumulation rate and effects of oceanic lead have not been adequately investigated; (3) lead is toxic at low concentrations; and (4) there is no known adverse effects of lead at its present level in the oceans.

Other metals may be potentially more hazardous to man's health than mercury and lead. For instance, cadmium has been shown to accumulate in tissues and cause larval mortality in crabs and certain fishes (e.g., the tautog). Relationship to human disease is yet to be demonstrated, however. Silver, copper, zinc, and chromium are other metals which require further investigation.

Synthetic Organics

The major groups of synthetic organic chemicals that may have an impact on the marine environment are: chlorinated hydrocarbons such as DDT and polychlorinated biphenyls (PCBs), and volatile organic liquids and gases, such as Freon and dry cleaning solvent. These compounds are toxic; moreover, many resist chemical and biological degradation and thus persist and accumulate in the environment. Another characteristic of synthetic organics is that they are volatile and readily escape into the atmosphere; and large quantities are transported to the oceans by wind currents.

Perhaps the best known synthetic hydrocarbon compound is DDT. Vast quantities of this pesticide were produced and applied in this country until it was banned in 1972. DDT itself has a very low solubility in water and tends to be adsorbed on soil and silt particles. The characteristics of its application and its volatility combine to make atmospheric transport a more significant pathway for DDT and DDT residuals than soil erosion and surface runoff. In soil and sediments, the half life of DDT has been estimated to be around 10 to 15 years.

DDT degrades in the presence of oxygen to DDE. It also can be metabolized in the absence of oxygen to other forms such as DDD. It is DDE that is the most persistent of the degraded forms of DDT. In fact, DDE appears to be the most abundant of the synthetic organic pollutants now in the oceans. DDT compounds have been detected in both snow and marine organisms in the Antarctic, strengthening the hypothesis that air transport and subsequent fallout is an important dispersal mechanism and pathway into the marine environment.

It is estimated that DDE usually comprises at least 80% of the DDT residuals found in marine organisms. Also, there is no known marine organism that metabolizes DDE, thus it tends to accumulate. DDT and its residues are known to accumulate in the fatty tissues of marine organisms, and its concentration increases as it moves through the food chain. The magnification is such that fish-eating birds (e.g., the herring gull) have been found to contain DDT concentrations which exceed the concentration found in the water by a factor of one million.

It appears that there is considerable variation in the ability of different species to tolerate and concentrate chlorinated hydrocarbons. In 1969 a shipment of 28,000 pounds of coho salmon taken from Lake Michigan was embargoed because the fish contained 19 parts per million of DDT. However, other fish in the lake, such as herring and whitefish, have not been shown to carry high residues.

Utilizing 20 years of plankton collections from intensive sea surveys of the California Current region, it has been shown that there has been a continual accumulation of DDT and its metabolites in planktonic fishes near Los Angeles from 1949 to 1970. The source of the pesticide was shown to be one chemical plant dumping DDT wastes into the Los Angeles sewer system; this dumping was stopped in 1970.

PCBs have been in use in the United States and elsewhere for over 40 years. Although its sole U.S. producer is the Monsanto Company, it also enters the United States from other manufacturers from abroad. Because of their superior insulating as well as fire and explosive resistant properties, PCBs are widely used in heat transfer systems and in electrical devices, particularly capacitors and transformers. PCBs have many other industrial applications, such as additives to plastics and paints and as coating compounds. As in the case of DDT and its residues, PCBs are highly persistent and ubiquitous. They are distributed in the marine environment in a manner that suggests that atmospheric transport and subsequent deposition on surface waters is an important pathway into the oceans.

Although little is known of the effects of PCB on marine organisms, there is some evidence that they can have adverse effects on human health. As with organochlorine insecticides, PCBs are fat-soluble and resist metabolic change. Thus, they tend to concentrate at succeedingly higher levels in the food chain. PCBs have been shown to accumulate in fish and aquatic invertebrates to levels 75,000 times those present in water.

A considerable amount of information is now accumulating on the biological effects of various levels of PCBs on man and other animals. However, more research is needed to determine the magnitudes and the cycling mechanisms of PCBs in the marine environment and on the acute and chronic pathologic, mutagenic, physiologic, biochemical, and behavioral effects of PCBs on marine organisms.

The EPA and NOAA's National Marine Fisheries Service entered into arrangements beginning January 1974 to analyze pesticide residues in fish. This interagency agreement was made in response to the Federal Environmental Pesticide Control Act of 1972 which requires EPA to conduct, in cooperation with other agencies, a nationwide pesticide monitoring program. Under the terms of the agreement, NMFS furnishes fish samples from regularly scheduled cruises to EPA for the analysis of pesticide residues.

Other marine pollution studies of the National Marine Fisheries Service include mutagenic effects of pollutants in which larval marine forms are exposed to varying concentrations of a number of contaminants, including pesticides. Genetic tests are providing an especially sensitive means of measuring presence or absence of low dosage and long-term effects. The National Marine Fisheries Service has also long had a program to study the pathological effects and mortalities that result from infectious disease. Organ, tissue, and cell systems that are selective targets for certain contaminants found in the marine environment are used in the studies.

These studies seek to determine the distribution and chemical forms of contaminants in sediments, water, and biota of selected estuaries; determine turnover rates and food chain transport of these contaminants; and develop predictive models of the distribution and cycling of these contaminants in marine ecosystems.

Plastic Debris

The National Marine Fisheries Service, in fish surveys in the Pacific and Atlantic Oceans carried out in 1972 and 1973, reported the widespread presence of plastic debris floating near the surface and washed up on beaches. The field survey teams estimated that over 20,000 plastic items, including 12 tons of trawl web and perhaps 7,000 gillnet floats had washed up along 60 miles of beaches of Amchitka Island in the Aleutian Chain.

Floating synthetic ropes and nets entangle in ships' propellers and derelict nets entrap fish, birds, seals and other marine animals. A similar situation was found in the Atlantic Ocean surveys in which oil clumps and plastic fragments were observed from Cape Cod to the northern coast of South America, an area encompassing over 700,000 square miles.

Plastic debris appear to be extensive in the ocean and research efforts are directed toward ascertaining their biological significance, if any. Plastics have not been found in adult fish, indicating that these materials are not being ingested or they are passed through the digestive tract. Plastic material has been found in larval fish from Long Island Sound, however. Initial laboratory experiments show that the plastic does not appear to harm larvae and juveniles. In summary, based on limited studies to date, there is no evidence of biological harm to fish from plastic debris, but in view of the widespread distribution of these materials, continued study is warranted.

SCIENTIFIC PARAMETERS FOR STUDYING OCEAN WASTE DISPOSAL

The following sources were used for the information provided in this chapter.

PB 198 032
PB 204 868
PB 210 710
PB 213 473
PB 224 793
AD 775 826
COM 73 11292

STUDYING THE MARINE ENVIRONMENT

Physical Oceanography

Physical oceanography is that branch of oceanography concerned with the description and interpretation of the physical state of the sea and its variations in space and time, and of the physical processes occurring in the sea. This science deals with water masses and how they are formed, the major current systems dispersing and mixing these masses, and the motions and interactions of oceanic energies. It also includes the study and prediction of temperature, tides, currents, and waves.

Various types of currents may transport waste materials laterally in the water column over great distances, as well as moving materials which have settled to the bottom. Of special significance with respect to waste liquids and sludges is the process of mixing in which dissimilar liquids diffuse into each other—the mixing rate determines how rapidly a waste will be diluted.

Waters of the oceans are typically characterized by their temperature, salinity, and density so that oceanographers can identify both the sources of the water masses and the degree of mixing. A brief discussion of these properties is necessary before considering the ocean's dynamic processes such as currents and mixing.

Distribution of Temperature, Salinity, and Density: Water temperature and salinity are two of the basic physical properties used to measure and describe the processes and circulation of the sea. Density, the third basic property, is a nonlinear function of temperature, salinity, and pressure, and thus changes in these three properties result in changes in water density.

Sea temperature has been measured for centuries and is relatively easy to determine. It varies with time, depth, latitude, incoming radiation and, as discovered more recently, may be subject to variations at some depths because of internal waves, In a very general way, the ocean can be considered a three-layer system: the thin, warm surface layer which is isothermal in the winter; the main thermocline separating a transition layer about 1,500 to 3,000 feet thick where the temperature drops from about 17° to 5°C, and underlying nearly homogenous cold, deep water extending to the bottom. During summer months, warming tends to develop stratification in the surface layer, but little seasonal effect penetrates below 600 to 900 feet.

Salinity, generally speaking, refers to the total concentration of dissolved solids in the ocean. It increases with depth and varies with time, latitude, and to some extent with local mixing. Variations in salinity (and in temperature) generally occur in the upper layers of the sea and decrease with depth. Temperature and salinity data are usually plotted on a T-S curve (Figure 3.1).

FIGURE 3.1: TEMPERATURE-SALINITY CURVE

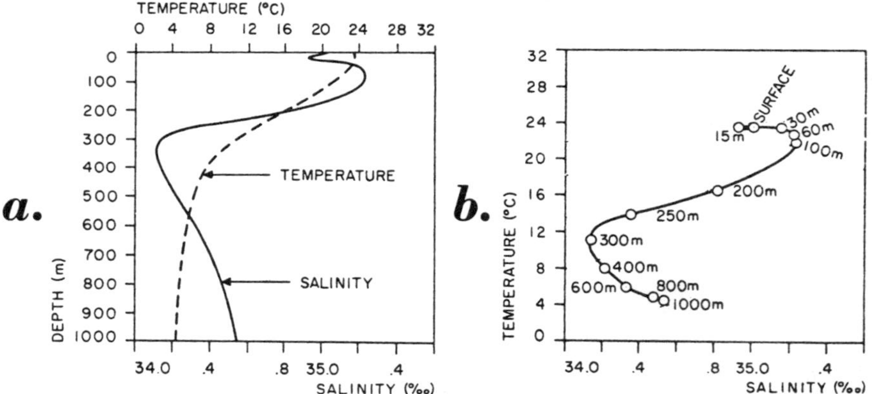

Source: PB-213 473

Figure 3.1a illustrates the variation of temperature and salinity with depths for an area in the vicinity of the Hawaiian Islands. The T-S curve (Figure 3.1b) illustrates temperature vs salinity in profile with depth. This curve identifies a water mass which helps in tracing the origins of distinct layers.

Density of seawater is defined as the mass per unit volume expressed in grams per cubic centimeter. Variations in density are a function of variations in temperature, salinity, and pressure, and values typically range from 1 for freshwater to 1.07 for the greatest depths. The vertical stability of the sea is due to the increase of density with depth. Calculations of density are made indirectly through measurement of temperature, salinity, and pressure (depth). Through the determination of layers of equal potential densities, calculations of water circulation are possible since water moves along these surfaces.

Water Masses: The term water mass refers to a more or less discrete body of ocean water defined on the basis of its physical and chemical characterisitcs, most typically expressed in terms of temperature-salinity relationships. A water mass thus indentified can be traced from its source area as it moves into adjacent oceanic areas. Subsurface water masses form because surface water in a given area becomes dense enough through evaporation or cooling to sink and spread in layers according to their densities.

A water mass is usually made up of several water types, which are represented by single temperature-salinity values. There are many different water masses in the ocean, but in general they may be grouped into five major categories: surface or near surface water (maximum depth approximately 900 feet), central water (900 to 1,500 feet), intermediate water (1,500 to 3,000 feet), deep water (3,000 to 12,000 feet), and bottom water. Within the five broad categories, there are over 23 major water masses known.

Discharges of large volumes of waste materials in a given area may produce a small water mass whose characteristics could be used to describe the ultimate fate of the wastes. For example, off the New York City area the local effect of acid waste disposal on water color, pH and salinity are identifiable and provide a means of tracing the wastes.

Ocean Currents: Both the nearshore local currents and the permanent major ocean currents are likely to play a role in the ultimate distribution of finer or lighter fractions of such wastes as dredge spoil and industrial chemicals once discharged at sea. Because local currents are somewhat stronger, they may be important in carrying materials from the disposal site into shallower or deeper water. Deeper currents offshore at mid-depths can move suspended materials and thus prevent them from reaching the intended sea floor site. Recently collected evidence from great depths using bottom photographs of ripple marks, scouring, and absence of few sediments on high points, has shown that bottom currents have sufficient strength to move sizable particles and thus shift comparable waste solids discharged into the particular site.

The major circulation of the oceans involves large-scale currents of a permanent nature, such as the Kuroshio, Gulf Stream, and Equatorial Currents. The driving force for these currents is caused primarily by differences in density. In this type of water flow, the effect of the earth's rotation, called Coriolis force, is also a significant factor that causes a deflection of the currents. The nonpermanent currents, which are more localized, are related to winds, tides, and waves.

Density Currents – Density currents develop because of differences in water density over horizontal distances where water tends to move down a density gradient. By measuring a vertical profile of temperature, salinity, and pressure at many ocean stations, the calculated densities provide the basis for dynamic computations that show the field of relative motion in a fluid. This technique is known as the geostrophic method and gives a broad picture of the total steady-state ocean circulation. Measurements by current meters, drogues, and neutrally buoyant sonar floats have confirmed some of these calculations. Further, although direct measurements of current at sea are difficult and time consuming, such measurements have resulted in some discoveries such as finding countercurrents flowing in the opposite direction below many of the major currents. The existence of such countercurrents was not indicated by geostrophic calculations, which had satisfactorily defined ocean circulation patterns.

Wind Stress Currents – When winds blow over the sea surface a stress is exerted that causes wind drift of a thin layer of surface water. This transport in turn tends to alter the density distribution and may lead to the formation of density currents as described above. In 1902, theoretical calculations and experiments were made that showed that a current induced by a surface wind would decrease with depth to a point where the frictional forces would become negligible. In developing the concept of the Eckman Spiral effect (Figure 3.2), it was concluded that because of Coriolis effect (the apparent deflecting force resulting from the earth's rotation) in the Northern Hemisphere the wind-blown surface waters move at 45 degrees to the right of the wind direction.

As the current velocity decreased with depth, the current direction deflected progressively to the right, so that at a particular depth the mass transport would be at right angles to the wind direction. Experiments have shown that actual wind drift was 3 to 5 percent of the wind velocity and from 30 to 60 degrees of wind direction. The magnitude of those wind stress currents depends on wind speed and duration. Current velocities range up to 2 knots.

FIGURE 3.2: THE EKMAN SPIRAL

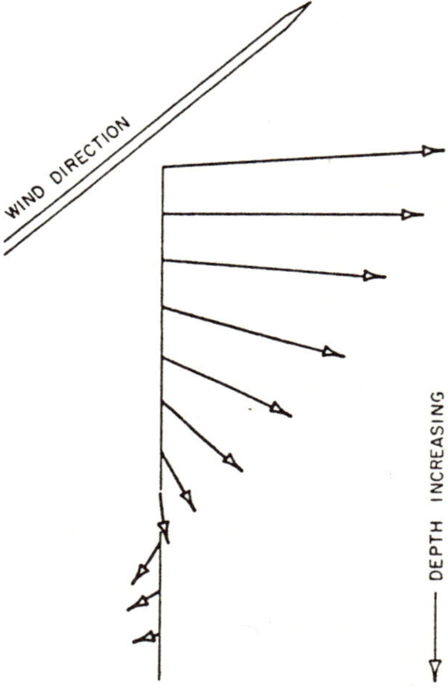

Source: PB 213 473

The Ekman Spiral (see Figure 3.2) shows the effect of surface winds on drift currents. Because of Coriolis force, the net mass transport is 90 degrees to the right of the wind direction in the Northern Hemisphere. The length of the arrows represents current velocities, which decrease with depth as a result of friction.

Near coasts this mass transport of water at right angles to the wind direction becomes extremely important (for example, along the coasts of California and Peru). In both cases, warmer, lighter nearshore water is moved away at right angles from the coast, and colder, deep water is drawn up in the process known as upwelling. Valuable nutrients are brought to the surface during the summer months aiding in the productivity of fisheries. Upwelling is also a possible mechanism which could bring dissolved traces of wastes orginally discharged in deep water back to the near surface waters where man's activities are concentrated.

Tidal Currents – Currents caused by the astronomical forces of the moon and the sun can be extremely strong in certain restricted areas, but for the most part these currents do not bring about the transport of water over large distances. Strong tidal currents move large volumes of sediment along the bottom, although because of alternating directions of the ebb and flood, the net transport is small. Such currents may have important effects on some nearshore disposal grounds by thorough repeated frequent stirring, mixing and winnowing of the wastes.

In the open ocean, tidal currents generally are rotational either on a 12- (semidiurnal) or a 24-hour (diurnal) basis with rotational effects depending on Coriolis force. Theoretically, the tidal current runs in the same direction and with the same velocity from surface to bottom. Actually, this holds true only in shallow water to about 60 to 90 feet off the bottom. In coastal areas, tidal velocities up to several knots are reported. In an area such as San Francisco Bay, tidal currents approaching 5 knots on the surface with 3 knots along the bottom have been observed; while in Cook Inlet, Alaska, where tides are higher, currents may be about 8 knots or more.

Associated with tidal currents is the occurrence of internal waves, which are subsurface waves found between layers of different density. In deep water these waves may be several hundred feet in height and could be responsible for considerable water mixing and transport, although it has been difficult to measure effects from these waves. Velocities resulting from internal waves may be sufficient to prevent sediment deposition on topographic prominences on the sea floor.

Wave-Induced Currents – Surface gravity waves caused by wind stresses have an advancing orbital motion that produces a small net drift or wave drift of water masses in the direction of wave advance. Surface waves in deep water have an orbital motion, which decreases with depth, while in shallow water, the motion becomes elliptical and, at the bottom, oscillatory (Figure 3.3). Oscillatory currents on the bottom become significant at depths less than half the wave lengths. These cause sediment ripples and sorting at depths up to 600 feet.

Currents resulting from wave action in the surf zone occur when waves breaking at an angle to the shore induce a longshore current, or littoral drift. This current forms parallel to the beach and proceeds until a natural break, or passage, occurs allowing return of the water in a rip current. Both of these wave-induced currents are very local, but in times of large storm waves they can be responsible for nearshore movement of large amounts of sediment or waste materials.

FIGURE 3.3: ORBITAL MOTION IN WAVES

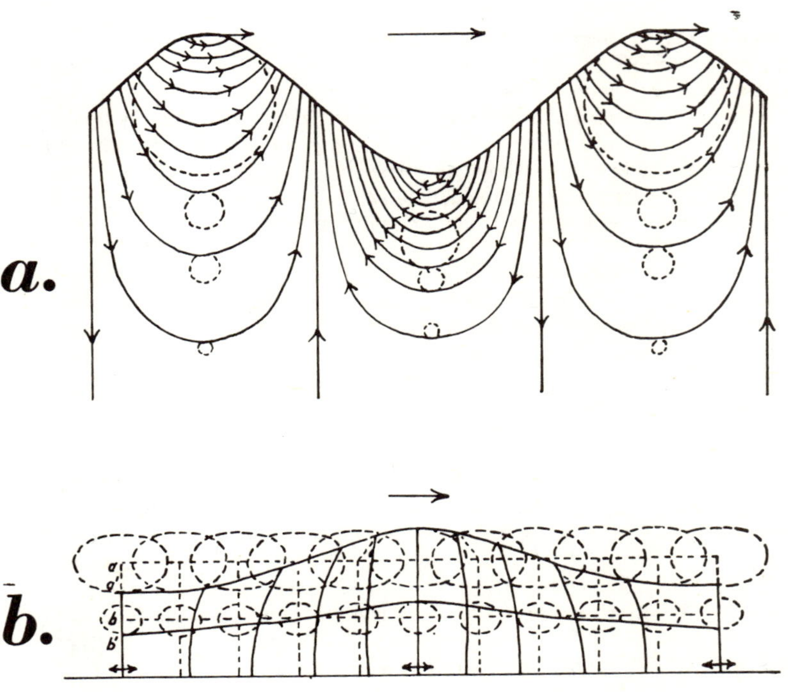

Source: PB 213 473

Figure 3.3a illustrates streamlines and orbits in a wave traveling to the right in deep water. Figure 3.3b illustrates orbital motion and positions of water filaments in a progressive wave traveling to the right in shallow water.

Turbidity Currents – When sediment is suspended in water, the combination is denser than water and this mixture will flow downslope; laboratory experiments in 1950 showed the nature of such turbidity currents. In the past, these currents were believed to be a major cause in the erosive formation of submarine canyons that extend across many of the continental shelves of the world. Recent evidence indicated that although turbidity currents are unlikely explanations for the erosive formation of submarine canyons, they may move large volumes of sediment out through these canyons. Probable velocities in turbidity currents are not greater than those found in lower reaches of rivers. These have been reported to be generally between 2 and 10 feet per second.

From the standpoint of marine waste disposal, it is possible that the introduction of high density sludges and chemicals could initiate a turbidity current and carry a slurry of the sediment and waste mixture away from the disposal area and far out to sea.

Some Observations of Bottom Currents – Although bottom currents are known to operate in both shallow and deep water, relatively few in situ measurements have been made. Photographs of seamount tops disclose absence of sediment, or evidence of current motion in depths greater than 6,000 feet. Off the east coast of the U.S., deep-current measurements have shown maximum velocities of 0.8 knots at 12,000 feet, although most observations indicate velocities far lower. Observations from deep submersibles have given reliable current measurements.

Importance of Currents in Dispersion of Solids: As more direct current observations are made, it is apparent that mid-water and bottom circulation occurs in many more areas and deeper than was once thought. Strong current action along the continental shelf can move sizable particles. Further, the somewhat slower circulation in deeper areas of water over the bottom can be important as a dilution factor. The movement and dispersion of dissolved materials (including any wastes which are present) are controlled by the water circulation in any adjacent to the disposal site and the natural turbulent mixing processes in the sea.

Along the beaches, or littoral areas, and on most of the continental shelf, a mixture of tidal and wind currents with some wave-induced effects will be the most important forces. Off the shelf and down the continental slope, currents become noticeably weaker with occasional turbity current action in submarine canyons. Finally, at great depths currents appear to move fine grained sediment but are less understood as to origin, duration, or measured velocity.

Stirring and Mixing: There are three distinct observable stages that are analogous to the introduction of one fluid of differing density into another (or of a sludge or chemical into the sea). These three stages are: (1) an initial stage in which rather large volumes of the different fluids are distinctly visible with sharp gradients at the interfaces. If no motion is induced, the boundaries persist for some time; (2) an intermediate stage that continues after stirring of the two liquids, is induced. The contrasting masses of fluid are distorted, increasing the extent of the interfacial areas having high concentration gradients; (3) the final stage in which the gradients disappear quite suddenly, and the liquids become homogenous.

This stage is usually referred to as mixing (to distinguish it from stirring). This is presumably caused by diffusion. Thus, it is likely that after the introduction of one fluid into another, the two may show distinct boundries which persist until stirring motions change the average gradients and mixing results.

Geological Oceanography

Geological oceanography, sometimes called submarine geology or marine geology, is that branch of the science that includes the study of coasts and shorelines; the continental shelf (the broad platform surrounding most coasts); the continental slope leading from the shelf edge down to the deep ocean; and the deep ocean floor with its occasional basins and trenches. It also considers the nature and origin of sea floor topography and sediments and includes geophysical studies that provide information about the earth's structure beneath the sea floor. By examining the regions bordering our continents, we can understand the physical description of areas where waste materials are now being discharged, or will be in the future. Further, we find that the same forces that act on natural sediments, either those derived from land or the ocean, act equally on many waste materials, particularly sludges and dredge spoil. Thus, bottom currents that can suspend and transport sand grains, will move equivalent size particles of waste material.

Continental Borders: In the world ocean, the continental land masses are fringed by shallow continental shelves of varying width. At their outer edges the shelves steepen abruptly into the continental slope that extends down to the deep ocean floor. These features and the basins and trenches, which mark the ocean floor and constitute its major bathymetric divisions, have characteristic depths and areas (Table 3.1) (Figure 3.4).

TABLE 3.1: DEPTH AND AREAS OF MAJOR BATHYMETRIC OCEAN DIVISIONS

	Depth (feet)	Total ocean area (%)
Continental shelf	0 – 600	7
Continental slope	600 – 3,000	13
Deep ocean	3,000 – 18,000	80
Ocean basins and trenches	18,000 – 35,000	1

The geologic development of the edges of the continents is complex and involves a combination of repeated sea-level changes and sedimentation processes that have built the continental shelves. The geologic development of the ocean floor with its deep basins and trenches is probably even more complex and includes possible drift of the continental masses associated with sea floor spreading and build-up of major ranges of submarine volcanic mountains.

Continental Shelf: The shallow platform surrounding the continents forms the continental shelf which comprises 18 percent of the land area and 7 percent of the ocean's total area. The shelf generally terminates seaward with a distinct increase in slope called the shelf-break or shelf-edge. As a matter of convenience and because early navigation charts only showed 10, 100, and 1,000 fathoms, the figures of 100 fathoms (600 feet) historically was the definition of the shelf-edge. Most workers agree that shelves can vary in edge depth from 60 to 1,800 feet with a world average of 430 feet.

The average bottom gradient for all shelves is 12 feet per mile (or an angle of 7 minutes), which is a slope less than the human eye can detect. The width of continental shelves varies markedly, ranging from 0 to 750 miles, with an average width of 44 miles. The shelf off the east coast of the United States ranges in width from a mile or two at Miami to over 200 miles off Newfoundland. On the west coast, it varies from 2 to 15 miles. Detailed information is available on the topography of most of the world shelf, but similar detailed data on the nature and extent of the sediment cover is known for only about one quarter of this area. The geologic structure and bedrock relationships are known for less than

FIGURE 3.4: RELATIONSHIP OF MAJOR CONTINENTAL BORDER AND OCEANIC FEATURES

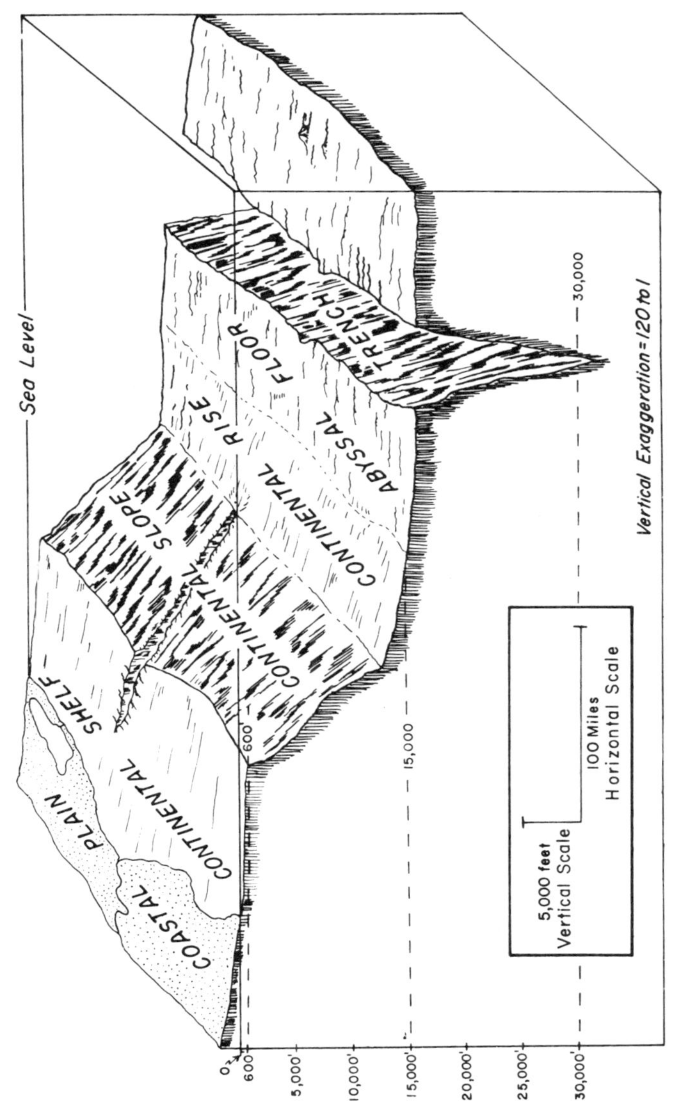

Source: PB 213 473

10 percent of the shelf area. In general, the continental shelves may be viewed as depositional environments where sediments eroded from the continents over millions and millions of years have been built-up in enormous volumes.

The continental shelf has been classified into six types on the basis of origin. These are: (1) tectonic dams, formed by geological uplift or lava overflow as on the west coast of North and South America; (2) reef dams, created by marine organisms, as in the Carribbean; (3) the Diapir dam, formed by salt domes damming sediments, in the West Gulf of Mexico; (4) no dams, simply sediment layers, such as most of U.S. East Coast; (5) ice-eroded, ice erodes any of the above, as in the Arctic and Antarctic; (6) wave-eroded, waves erode any of the above, as on the west coast of North and South America.

Continental Slope: The continental slope is the important transition zone between the two general levels of the earth's surface; namely, the continental surface, which more or less approximates sea level, and the vast ocean floor, which averages about 13,000 feet below sea level. The upper border of the continental slope begins at the shelf break; the lower limit is less clearcut because the slope grades into the surface of the deep ocean floor. The average gradient for the continental slope has been estimated as on the order of 4 to 5 degrees (about 400 to 500 feet per mile), but slope gradients as steep as 45 degrees may occur, and 25 degrees may be common.

Slope gradients vary by region; the Pacific Coast slopes are steeper than the Atlantic Coast. About half the slopes terminate in bordering trenches and depressions, while the rest end in sediment fans, or a rise. This lower region where it approaches the deep ocean floor might be called the continental rise. The general width of all continental slopes is 10 to 20 miles disregarding the rise, which would considerably extend the width.

Continental slopes can be irregular in profile and have such features as hills and basins, plateaus and terraces. With respect to the sediment cover, structure, and bathymetry much less is known about the continental slope than for the shelf. Structurally, the crustal material beneath the continental slope thins from the continental to the oceanic crust. As with the continental shelves, the slopes do not all appear to have the same origin. Possible causes for the formation of slopes include; wave-built terraces, deltaic beds, down-warped continental sediments, and fault zones.

Submarine canyons are certainly the most distinctive topographic features on the shelf and slope. They are not all of one origin, but likely result from several causes. The canyons have been classified into five types: (1) V-shaped valleys with winding courses and dendritic tributaries; (2) straight-walled, trough-shaped valleys, cut with unconsolidated sediments; (3) winding valleys with tributaries cut into sediment fans on slopes; (4) valleys, which run along faults and commonly parallel the coast; (5) continental shelf valleys not extending to slopes. It is interesting that these canyons function as channels that funnel large sedimentary loads from nearshore areas downslope into deep water.

Deep Ocean: The Deep Ocean is usually classified as extending to 18,000 feet and covering 80 percent or the greatest part of the total ocean floor. The provinces within it are the abyssal hills, which stick up through layers of sediment of varying thickness, and the abyssal plains, a generally smooth, featureless floor with low slopes. The plains usually begin at the base of the continental rise and extend seaward into the hills.

Trenches: Although trenches and basins make up only about 1 percent of the total ocean area, the trenches form exceptionally striking features in the floor of the Pacific Ocean and to lesser extent in other oceans. More than 10 major trenches exist in the Pacific, 2 with depths to 35,590 feet. The deepest Pacific trenches are located along the west margin of the Pacific Basin and are associated with the arc-shaped chains of islands such as the Aleutians, Kuriles, and Marianas where earthquakes are frequent. Trenches generally contain some benthic organisms, but have a relatively low modified density.

Sedimentology: The sedimentary materials normally entering the sea from the land, as well as those originating in the sea, are highly varied and include: detrital material from land, sub-aerial or submarine volcanic products, organic and skeletal matter, inorganic precipitates, products from chemical reactions, and extraterrestrial materials. These input materials result in four basic types of bottom sediment: (1) pelagic sediments (organic ooze); (2) terrigenous sediments (silt and clay); (3) marine glacial sediments (sand and gravel); (4) volcanic sediments (ashes and pumice).

The behavior of several types of solid waste materials discharged at sea is nearly identical with that of sediment particles carried to the sea or precipitated in the water column. These particles settle at calculable rates dependent on water characteristics and on the properties of the particle. These computations allow a general determination of the amounts of a given material that should reach the sea floor more or less immediately, or should remain in suspension for a long time. An understanding of the marine conditions governing sea floor erosion, transportation, and deposition is important in order to appreciate what forces may affect solid waste materials.

Deposition, Transportation, and Erosion — The sedimentary debris that has been transported to the sea settles through the water column, at the same time being carried laterally by the effect of currents. The settling velocity of a sedimentary particle depends upon the specific gravity, size and shape of the particle, as well as the viscosity and specific gravity of the water. This relationship is defined in Stokes' law, which states that the settling velocity varies in direct proportion to the square of the particle diameter.

Most of the fine materials (silt, clay and colloid-size particles) are affected by ocean currents during settling; their effective settling velocities probably range from 3 feet per day to many thousand feet per day. In the absence of currents, a particle measuring 4 microns (or coarse clay) settles at a rate of 3 feet per day, and in deep water may be carried about for many years before reaching the bottom. Fine silts and clays may be carried away from shore by large-scale horizontal eddies, such as those occurring off the Southern California coast. A settling velocity of 50 feet per day is estimated for sediments of these sizes. Resulting rates of sediment accumulation may vary from 0.5 mm per year along some a abyssal plains to the relatively slow accumulation of 1 to 10 mm per 1,000 years in red clay deposits of the deep ocean.

After sediment settles to the bottom, it may be picked up and carried along by currents over great distances before coming permanently to rest. Thus, the presence of sediment types in one locality may not correctly indicate that sediment of this type is actually being deposited there; it could easily have been eroded, transported, and redeposited from another bottom area. The factors operative in this connection are (1) mass movement of unconsolidated sediments by mud flows and slides on slopes; (2) the movement of individual particles by rolling, sliding, and jumping (traction) along as a result of bottom current forces; (3) the effect of turbulence, mixing, and diffusion.

The relationship of the three processes of deposition, transportation, and erosion, are described and represented in a classic series of curves (Figure 3.5). Study of these curves shows that a particle size of 0.5 mm (medium sand) is most easily eroded and requires a velocity of less than 0.4 knots. For larger particles, measuring perhaps eight mm (equivalent to a small pebble) the required velocity approaches 2 knots. Erosive pick-up of very small particles (such as fine silt measuring 0.01 mm) requires current velocities of more than 1 knot.

Velocities required for erosion and transport of particles of 2 to 3 mm only occur along the bottom in shallow waters, while in deeper water currents are only strong enough to transport fine-to-silty sand. Currents near the bottom of up to 0.4 knots east of Cape Cod and southwest of Cape Hatteras have been reported. This velocity is sufficient for erosive pickup of medium sand, as well as the transport of all sediment types found on the continental slope and rise. Figure 3.5 illustrates the Hjulström curves showing approximate regimes for erosion, transportation, and deposition and the relation of particle size to stream velocity. Note that medium sized sediments (sands) are more easily eroded and transported than fine clays or coarse cobbles.

FIGURE 3.5: HJULSTRÖM CURVES

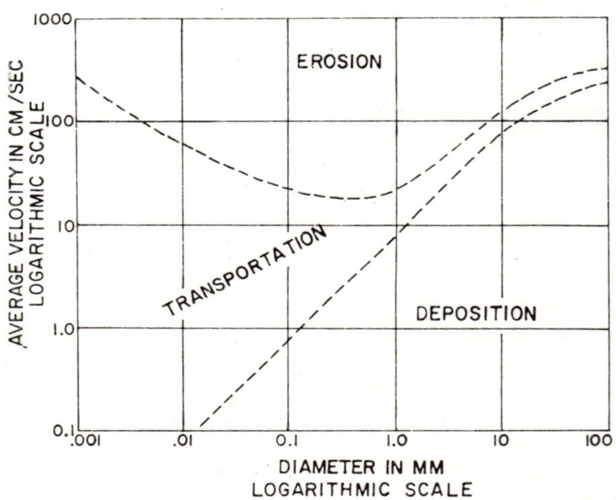

Source: PB 213 473

Erosional and Depositional Features – Such sediment surfaces features as ripple marks and scour marks (similar to ripple marks seen on many beaches) are produced by selective erosion, transport, and deposition of sediment particularly by wave and current forces. In the course of deposition, a series of laminae or layers build up in the sediment; these are termed current bedding structures.

A certain threshold velocity for the initial movement of grains as a fluid flows over a sediment bed has been defined. As this velocity increases, a few particles lift off the bottom into suspension. As the fluid with particles in suspension forms vortices, the characteristic current ripple marks develop on the sediment surface. These are usually asymmetric in cross section with the long, gentle slope in the upcurrent direction, and steep slope facing downstream. Thus, ripple marks indicate current direction. Another type of current feature, longitudinal ripples, also indicates current direction; in this case the current parallels the crest direction of the much larger symmetrical ripple.

A number of other types of ripples have been described that can also be used to determine flow direction. Further, direct observations of large areas using side-scan equipment have revealed large sand ribbons indicating the one-way sand transport by tidal currents.

Flocculation of Fine Grained Sediments and the Nepheloid Layer: Most marine sediments contain a portion of clay minerals. These form the finest particle size ranging from 4.0 to 0.2 microns ($\frac{1}{200}$ to $\frac{1}{4036}$ mm). The principal clay groups are: kaolinite, montmorillonite, and illite. The platelike crystal units of clay tend to stick together, or coagulate in a process called flocculation. Flocculation is enhanced by the presence of salt water; the rate of flocculation may depend on the salinity. Thus, the fine-grained materials that would settle very slowly in a single-grained condition, according to Stokes' law, tend to settle faster when flocculation occurs.

Within recent years, layers of very fine-grained, suspended sediment have been detected near the bottom in waters several thousand feet deep along the continental slope and rise off the northeastern United States. This layer, measuring 500 to 650 feet thick, has been called the nepheloid layer. It has been suggested the layer has its origin in the resuspension of clay-size particles by turbulent flow stirring up the bottom. The occurrence of concentrations of suspended sediment and plankton near the bottom of San Pedro Basin, California has been reported. It is believed to be a residual layer formed by a turbidity current, possibly due in turn to internal waves. Observations of a similar layer of suspended materials have been made by deep submersible pilots.

Interaction with the Benthos – Open ocean or pelagic regions are inhabited by floating (planktonic), or swimming (nektonic) organisms. The bottom or benthic regions are inhabited by sedentary, crawling, or creeping organisms. Among the animals of the benthos are the infauna, which live in the sediments, and the epifauna, which live on the surface or are attached to rocks.

Benthic organisms are involved in a number of sedimentation processes. The larger forms both destroy organic matter as well as ingest sediment causing mechanical abrasion and chemical reactions. In addition, the multitudes of burrowing organisms destroy laminations and related depositional features. Observations from the submersible *Deepstar* confirmed this type of extensive reworking of the upper layers of sediment in the 500- to 2,000-foot-depth range in the Gulf of Mexico. It appears that such reworking may make a significant modification to the surface layers of the lower continental shelf and continental slopes. This reworking by marine animals should also hold true for waste materials overlying or mixed with the sediment.

Areas of Known Marine Geology – The study of marine geology had its formal beginning with the exploratory work of Sir John Murray, resulting from the voyage of the *HMS Challenger* in the last quarter of the 19th century. Since that time, a great volume of sounding, bathymetry, sediment sampling, and, more recently, geophysical exploration has occurred on worldwide basis. Yet, conservative estimates are that less than 5 percent of the marine geology of the oceans is known at this time.

Quite clearly the areas best known are the continental shelves, where there has been greatest activity; in the United States, the entire continental shelf along the east coast, the Gulf of Mexico, and the west coast south of Point Conception are considered in the best known category (Figure 3.6). The only other areas in a similar category are the shelf areas off Alaska, the British Isles, Southeast Asia, Japan, and a small portion off Venezuela.

FIGURE 3.6: CONTINENTAL SHELVES OF UNITED STATES

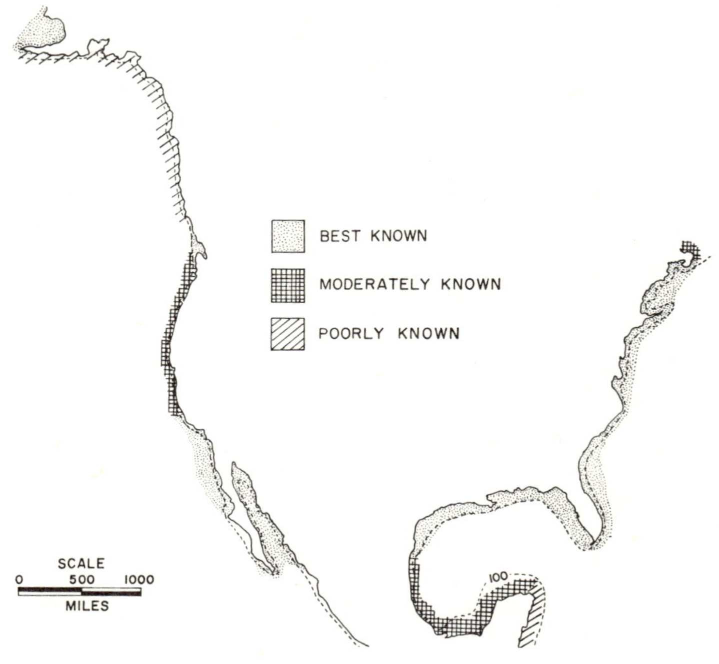

Source: PB 213 473

In the figure, continental shelves underlie 7 percent of the total ocean. The average depth of the shelf edge is 430 feet, and the average width is 44 miles. The most extensive exploration and mapping of sea floor stratigraphic and structural geologic relationships has been carried out on shelf and upper slope areas by private industry in connection with petroleum exploration and by government agencies. Institutional groups conducting deep ocean work generally have been less interested in precise geologic mapping. Most information from exploratory work by geophysical companies is proprietary and, thus, unavailable for other uses.

In summary, though the continental shelves off the United States are relatively well known geologically, additional detailed investigation of the sedimentation processes operating at each proposed disposal site are necessary in order to forecast the ultimate fate of the waste.

Chemical Oceanography

Chemical oceanography is that branch of the science concerned with the description, interpretation, and prediction of the chemical properties and processes of the sea and their interaction with the physical, biological, and geological processes operative in the marine environment. It involves study of the composition of seawater, and its dissolved gases and nutrients, as well as the use of radiochemical/isotope technology.

Chemical oceanography and, in particular, an understanding of the various chemical processes operative in the water column and at the bottom are basic essentials in predicting the fate and effects of waste materials discharged in the sea. These processes are closely interrelated with sedimentation processes and the functions of living organisms in the marine environment. Chemical nutrients and the gases in solution are vital to all life in the sea. In this connection, careful study must be given to the various chemical processes that may affect or be affected by marine waste disposal operations.

The chemical properties and relationships in the sea are such that they are not readily susceptible to direct measurement in situ. This is due in part to certain constituents occurring in high concentrations and others in extremely low concentrations; thus, the former tend to mask the latter. Further, ordinary laboratory techniques for analysis of water samples are generally inadequate for measuring most of the important but low concentration constituents. In this regard, an acceptably precise technique for the direct measurement of salinity using electrical conductivity was developed only recently. Prior to this development, all such measurements were chemically titrated (a precise measuring technique) and related to standard values for the relationship of the total dissolved solids to the chloride ion concentration.

Factors Affecting Seawater Composition: During the 19th century in an attempt to describe the major constituents of the sea, scientists discovered that the total salt concentration was quite uniform. Dittmar in 1887, after a thorough analysis of 77 seawater samples collected from all over the world by the *Challenger* expedition, concluded that the ratio between the more abundant substances was virtually constant. For this reason, the chloride ion is always 55.25 percent of the total dissolved solids. With only minor changes, this relationship, the law of relative proportions, has been proven consistent up to the present. While the relationship of constituents is consistent on a worldwide basis, local factors have some effect. Thus, snow, ice, rain, evaporation, river runoff, biologic activity, and particulate absorption all have effect in peripheral areas, but the total overall change to the ocean is not measurable.

Major and Minor Constituents — Common table salt makes up over 80 percent of the elements occurring in solution in seawater. The first 10 of these are the major elements, and the remaining are considered minor constituents. The major ones are chlorine, sodium, magnesium, sulfur, calcium, potassium, bromine, carbon, strontium, and boron. The minor elements vary from silicon at 4 mg per kg to gold at 0.000006 mg per kg.

Gases in Solution — Atmospheric gases dissolved in the ocean are oxygen, nitrogen, carbon dioxide, argon, helium, and neon; oxygen and carbon dioxide are the most important to the sea and to the life involved. Oxygen is produced by plants in the sea in the photosynthetic process; it also is replenished from the atmosphere in surface exchange. The concentration of oxygen varies from 0 to 9 meq per liter (0 to 14 mg per liter).

Carbon dioxide, on the other hand, is not dependent on surface exchange as the major source, but it is derived from the carbonate system in the form of bicarbonates of sodium, potassium, and calcium. The phytoplankton tend to use carbon dioxide and the zooplankton or animal life produce it through respiration. In addition, the pH (or acidity) of seawater is closely tied to the carbon dioxide system equilibrium.

Nutrients — Among the substances essential to primary production of phytoplankton are the nutrients, such as phosphates and nitrates. Nutrients are also referred to as the nonconservative solutes. These solutes are present in seawater in low concentrations, but a major fraction of this amount enters and leaves the particulate phase each year. Both the phosphates and nitrates have a cycle of replenishment that is directly related to plant production. The biochemical cycle includes a net downward motion of particulate matter that is essentially balanced by a net upward flux of these constituents in solution as a result of water circulation, particularly by vertical upwelling or by the addition from sources outside the oceans.

Radiochemistry and Isotopes — Radioisotopes have been used to study various phenomena and problems associated with marine waste disposal. For example, the addition of radioactive tracers to wastes such as the effluent from submarine sewer outfalls has assisted in the study of mixing processes. Considerable effort has been given to monitoring the uptake of radioisotopes by living organisms some of which tend to concentrate certain isotopes. Isotopes are also used to study the movement of beach sands and to determine the age of sediments.

Oxidizing Environments: An oxidizing environment is characterized by an oxygen-breathing biologic assemblage living in a relatively oxygen-rich water column and sediment cover on the sea floor. Oxidizing environments are typical conditions of open circulation and, thus, predominate over most of the ocean. Although most near-surface marine sediments are generally in the oxygenated state, nearshore areas with high biological producitivity are characterized by anaerobic sediments. In chemical terms, oxidation is a reaction in which electrons are given up. For seawater and sediments, it is generally expressed in terms of electrical potential or Eh. More specifically, Eh is the oxidation-reduction potential (which is also referred to as the redox potential). By measuring the electrical potential of seawater or sediments, the state of oxidation can be determined. For seawater, the Eh is positive or, by definition, oxidizing.

Reducing Environments: A reducing environment is characterized by anaerobic or anoxic conditions, that is, conditions without oxygen. These conditions develop when the accumulation of organic matter is so great as to deplete the available oxygen or when the oxygen supply in the water is diminished by other demands. Such conditions generally occur in relatively restricted areas and are indicative of stagnant water found in oceanic basins with limited circulation. Such anaerobic environments usually have high concentrations of hydrogen sulfide and carbon dioxide, and the sediments are typically black muds containing sulfate reducing bacteria. Chemically, reduction is a reaction in which electrons are taken on. The Eh values for a reducing environment are, by definition, negative.

There are two major types of oceanic basins, the open type, which has outflow over the sill or basin lip; and the closed type with a very shallow sill, which restricts outflow. As would be expected, there are many examples of stagnant basins with shallow sills where all outflow is prevented. The 4,500-foot deep Cariaco Trench off Venezuela is one such basin. Off Southern California, the bottom water flows between basins in a northwesterly direction.

Because variation in sill depths allows sequential flow, most of the basins in this area have sufficient inflow and outflow to maintain oxidizing environments. The Santa Barbara basin appears to have a complete exchange of its water mass in about a two-year period. Even in regions of oxidizing conditions, it has been shown that in many localities samples from a few centimeters below the water-sediment interface show that the sediments contain no free oxygen, and in many cases there is a deficit of oxygen. In the zone of 5 to 10 cm where the bacterial activity is greatest, reducing conditions are typical.

Biological Oceanography

Biological oceanography is that branch of the science concerned with the sea as a biological environment—it is a study of the plants and animals of the sea. The previous discussion has covered the nature of seawater itself, its physical behavior, the sea floor and sediments, and the chemical constituents. All these factors interact vitally with the ocean's vast biologic assemblage. Not only does marine life depend on the particular physical and chemical characteristics of seawater, but also on light, depth, pressure, and water movement.

The animal and plant life in the sea can be broadly pictured as two interrelated series of zones: vertical zones, which range from the intertidal to the deep bottom or abyssal zone, and geographic zones ranging from tropical to polar.

Marine Plants: Inasmuch as light for plant growth penetrates seawater only to depths of 600 to 700 feet, only this thin near-surface band (the euphotic zone) can support plant growth. There are two major groups of marine plants: the attached forms and the floating or planktonic forms. The attached forms are concentrated because they must attach to the bottom. These are the blue-green, green, brown, and red algae.

The floating or planktonic myriad forms are microscopic primary producers, which serve as feed for the lower animals. Prominent among these are: the diatoms, whose siliceous shells form much of the sediment or diatomaceous ooze on the sea floor; the dinoflagellates, which are important to filter and detritus feeders; and the coccolithophores that are important both as a food and as calcareous contributions to bottom sediments.

In the open ocean, the distribution of phytoplankton communities tends to be patchy and irregular. Areas such as the Sargasso Sea are notably sterile. Phytoplankton occur in greater numbers in coastal waters where there is more nourishment. Concentrations of enormous abundance coincide with those coastal areas characterized by upwelling of nutrient-rich cold water.

Marine Animals: The marine animals range from the microscopic zooplankton to the leviathan mammals such as the whales. The smallest animals, the zooplankton, are floating forms unable to propel themselves against currents. There are two distinct types: the holoplankton, which spend their entire life in the plankton community, and the meroplankton, which are larval forms of larger animals that leave the plankton group at maturity.

Probably the two most populous members of the holoplankton group of zooplankton are the copepods and euphausids, which feed on the phytoplankton and in turn are the basic food for larger animals. The zooplankton do not move horizontally any great distance, but some species migrate vertically in a diurnal cycle moving from a depth of roughly 1,000 feet during the day to the surface at night.

At the other end of the scale are the nekton, or free swimmers, which are the most highly specialized group of marine animals and include the vertebrates and some invertebrates. Fish are the largest group among the nekton; they are divided into the sharks and rays (the elasmobranches) and the true (or bony) fishes (the teleosts). Many of the fish are limited to zones determined by pressure, temperature, and salinity. Depth is probably the greatest limitation and keeps the fish with swim bladders from large vertical migrations. Besides vertical zonation, fish concentrations vary with season, distance from shore, relation to upwelling, and other less understood causes.

Of special interest to the problem of disposal of wastes heavier than seawater is the benthic or bottom community, which includes a number of different phyla of which the most important are sponges (porifera), corals (coelenterata), certain colonial animals (bryozoa), starfish (echinodermata), clams and oysters (mollusca), and lobster and shrimp (arthropoda). These forms exist in varying numbers from shallow to great depths with the littoral zone having the largest benthic population.

Food Chain: The relationships of predator and prey in the ocean is referred to as the food chain or web in which each group feeds on the next lower group, thus making up the many links in the chain (Figure 3.7). The basic driving force of the web is the sun's radiation, which penetrates the euphotic zones. The plants use this energy along with various nutrients and carbon dioxide in their growth process, and produce oxygen as a by-product. These plants in the form of phytoplankton are grazed upon by the zooplankton, which in turn are fed on by larger invertebrates, which after many sequential predations become the food of the larger vertebrates. Classically, the efficiency of the web has been given as 10 percent, that is, for every 10,000 grams of phytoplankton, 1,000 grams of zooplankton are fed, and so on, although it has been also suggested that this efficiency may be closer to 15 or 20 percent.

FIGURE 3.7: RELATIONSHIPS OF PREDATOR AND PREY IN THE OCEAN

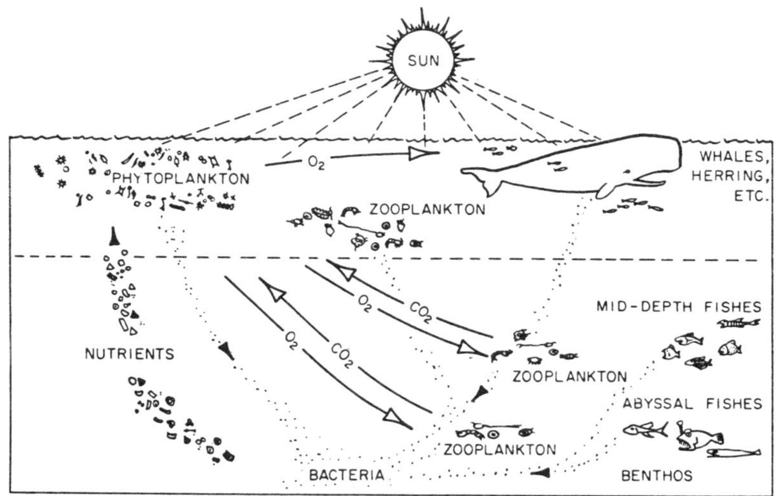

Source: PB 213 473

The above figure is a generalized representation of a food chain or food web showing the primary producers, the phytoplankton, and the animals that feed on these and in turn are food for higher animals. Note the flow of the nutrients, oxygen, and carbon dioxide as elements of this relationship.

CRITERIA FOR DISPOSAL OF DREDGED MATERIAL

Pollution Status

Useful definitive information is still unavailable on the effects of disposal of dredged material known to contain various pollutants on water quality or aquatic organisms. Nevertheless, regulatory agencies, faced with the legislative requirement of establishing dredged material criteria, must strive to establish meaningful criteria based on the best possible knowledge, and avoid the tendency to set forth criteria that precede the current technical state-of-the-art. Furthermore, regulatory criteria should be based on laboratory procedures that can be performed satisfactorily in routine testing laboratories as opposed to complicated procedures that can be conducted only in sophisticated research-level laboratories. Finally, in order to be equitable, the criteria that are required should not be prohibitively expensive. Most of the difficulty usually found surrounding the establishment of criteria centers on the definition and the determination of polluted dredged material; that is, to conform to the definition of polluted, it is necessary to determine prior to the dredging operations whether or not a particular sediment, if dredged, will cause an adverse environmental effect subsequent to disposal. The confusion arises over the fact that the mere presence of a constituent (toxin, biostimulant, etc.) in the sediment does not indicate or predict the nature and significance of adverse effects following disposal. This is because many chemical constituents found in sediments are unavailable and do not react as pollutants.

The question toward which criteria should be addressed thus becomes: do the dredging and disposal of sediment known to contain various potential pollutants cause these constituents to be released to the water column (dissolved), or in any other way become more available to the biological food web?

In response to the requirements set forth in the Marine Protection, Research, and Sanctuaries Act of 1972 (Public Law 92-432) the following laboratory procedure was developed by the EPA in conjunction with the Corps of Engineers (CE) to determine the pollution status of dredged material prior to ocean disposal. This procedure was published in the *Federal Register* dated October 15, 1973.

> Dredged material will be considered unpolluted if it produces a standard elutriate in which the concentration of no major constituent is more than 1.5 times the concentration of the same constituent in the water from the proposed disposal site used for the testing. The "standard elutriate" is the supernatant resulting from the vigorous 30-minute shaking of 1 part bottom sediment with 4 parts water from the proposed disposal site followed by 1 hour of letting the mixture settle and appropriate filtration or centrifugation. "Major constituents" are those water quality parameters deemed critical for the proposed dredging and disposal sites taking into account known point or aerial source discharges in the area and the possible presence in their waste of the materials in Subsections 227.22 and 227.31.

Aquatic Ecosystem

Before further discussion of the criteria, attention is directed to the schematic of the hydraulic dredging process shown in Figure 3.8.

FIGURE 3.8: HYDRAULIC DREDGING

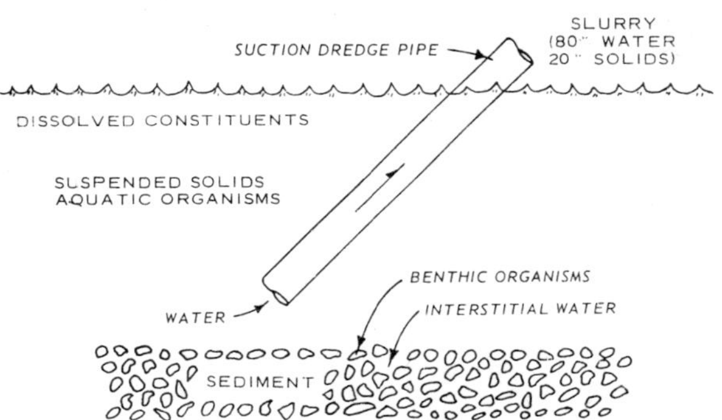

Source: AD 775 826

This schematic is intended to represent any type of hydraulic dredging whether it be pipeline, hopper, or sidecaster. As shown in Figure 3.1 hydraulic dredging causes the rapid mixing of two components of the aquatic ecosystem: water and sediment. Sediment is a complex composite of physical, chemical, and biological substances.

Physically, sediments are comprised primarily of soil materials such as clays, silts, sands, etc., and interstitial water (IW). IW is defined as that water occupying the void between individual sediment particles. Chemically, sediment is known to contain a large and complex variety of constituents that, in one form or another, cover most of the known chemical elements. Bottom sediment is also known to serve as a habitat for a large and complex community of biological organisms collectively known as the benthic community.

Bottom Sediment Chemistry: In order to determine the effect of disposing bottom sediment known to contain various chemical contaminants on water quality and aquatic organisms, at a minimum the chemical form and location of the contaminants within the sediments must be known. A contaminant can exist in many chemical forms in the natural environment and these forms vary from unstable to extremely stable compounds or complexes. The unstable forms are subject to active migration into the biological food web and may range from highly ionized species to soluble, readily available organic complexes. The stable chemical forms may range from highly insoluble inorganic precipitates, complexes, compounds, and minerals to very nonreactive organic complexes. These stable and usually nonreactive forms constitute the major fraction of most sediments and rarely enter into biological cycles.

The chemical form of the contaminant can affect both its relative toxicity and its availability to influence biological communities. An example would be to determine whether or not mercury is in the elemental or the methylated form. The location of chemical constituents within the sediments, which is closely related to the chemical form, also determines the availability of these constituents to influence biological communities and thus their ability to react as pollutants. The chemical constituents and contaminants can be located within the sediment in a variety of positions.

Water Column: The water column above the sediment is also a complex system of physical, chemical, and biological processes. As within the sediments, chemical constituents either may be dissolved or affiliated with suspended sediment particles in any of the manners listed above.

It is generally assumed that the surface layer of sediment is in dynamic equilibrium with the overlying water. There may be movement of dissolved chemical constituents from the bottom sediments to the overlying water or vice versa. This depends on the dynamic physicochemical state of the sediment and the overlying water at the sediment-water interface.

However, deeper sediment may not be in equilibrium with the overlying water; therefore, in dredging, these deeper sediment particles are exposed for a short period to the overlying water during hydraulic transport. It is thus reasonable to assume that when these sediments are exposed to the overlying water during dredging, there could be changes in aqueous chemical concentrations. It is generally assumed that chemical constituents in the dissolved state are in a form most available to influence biological communities. One form of dredged material criteria can thus be based on the changes in dissolved chemical concentrations.

Application of Criteria

The criteria for the ocean dumping of dredged materials described in the *Federal Register* are based on the amount of change of dissolved chemical concentrations that might be expected to occur due to hydraulic dredging. The standard elutriate test is designed to measure that amount of any chemical constituent that is dissolved in the IW and also measures those constituents that, due to dredging, migrate from the solid phase (sorbed, organic, etc.) to the dissolved phase. A schematic of the criteria is shown in Figure 3.9.

FIGURE 3.9: ELUTRIATE TEST

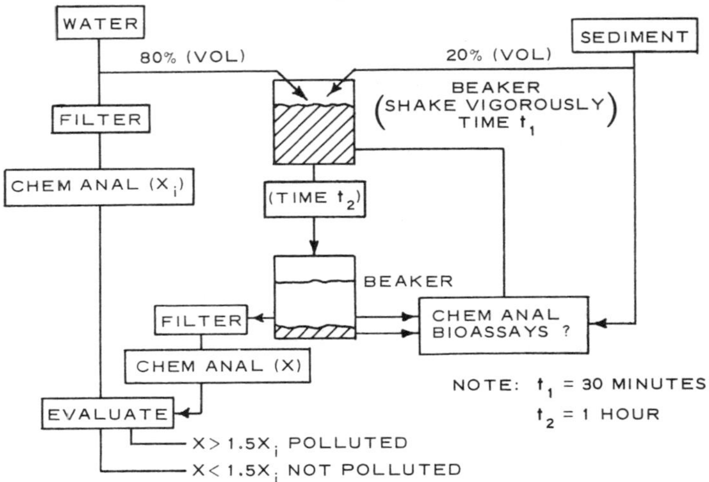

Source: AD 775 826

It is stressed that the criteria are considered interim and through intensive research should be continuously improved and updated. Improvement of the criteria not only should include the mechanics of the operation but address interpretation of the analytical results and any additional analyses that may be run on the elutriate. Bioassay should be considered when further research has developed an implementable and reliable procedure for sediments.

Currently, research is under way by WES personnel and with contract efforts to characterize thoroughly the elutriate test for identification of problems, to develop guidelines for implementation, and for the further development of criteria to determine the pollution capability of dredged material. It should be borne in mind that ultimately there are at least two reasons for sampling a dredge site; namely, to characterize the dredge site, and to help assess the environmental impact associated with the release of dredged material into a receiving system.

With this in mind, it is obvious that the number of samples to be taken at the dredge site needs to be considered rather carefully. Some of the factors which determine the optimum number of samples in a study of this type include:

(a) Cost and availability of adequate sampling and analysis facilities.
(b) Time available for sampling program.
(c) Geographical location of the dredge site.
(d) History of the dredge site.
(e) Area of the dredge site.
(f) Depth of the dredge site.
(g) Homogeneity of the dredged material.
(h) Precision of sampling methods.
(i) Precision of analytical methods.

The optimum number of replicate analyses of a given sample for each constituent should take into account the expected precision of the analytical method. In no case should less than a duplicate analysis be made for a particular constituent in a given sample.

Sampling and analysis programs are generally time-consuming and expensive. The results of such programs are major decsion factors in dredging operations; therefore, all available information regarding a dredging site and receiving site should be gathered, kept up to date, and used to the fullest possible extent.

Insofar as possible, this should be done prior to, or at least in conjunction with, the designing of any sampling program. Wherever possible, when individuals are familiar with the region, dredge site, and/or disposal site location, their advice and information should be considered when designing a sampling program.

CRITERIA FOR SELECTING A DISPOSAL SITE

In order to evaluate an ocean disposal site, certain meteorological, oceanographic, and biological data are necessary to allow a clear assessment of the dump area. Only then would the potential exist for locating the best disposal site and the ability to exercise judgment as to the potential damage to the marine environment.

Meteorology

The meteorological parameters of interest in evaluating ocean disposal sites include average wind speed and direction, average daily solar radiation, and precipitation, over a time period which is significant for the particular site under consideration.

Knowledge of wind velocity and direction in the lower levels of the atmosphere is required when computations of wind-driven circulation are to be made. Wind acts in several ways: wind driven circulation in the uppermost layers of the sea produces transport mechanisms for any floating or suspended material. The wind also generates waves which produce orbital motion in the upper layers of the water which leads to additional mixing and dispersion of the pollutants. In selecting a site for ocean disposal, it is preferable to have wind blowing offshore or such that the transport of the surface layer is away from the coast or critical adjacent areas, rather than onshore. Average daily solar radiation is an important parameter in that light (available light) is a factor for potential biological growth.

Precipitation is an important parameter in two ways: first, if in large enough volumes, it would produce a thin layer of low density water at the surface which would cause overall changes in water density when mixed within the mixed layer depth. Secondly, precipitation is of major significance in that associated land runoff via major rivers, will affect certain nearshore dump sites. The volume of the runoff can completely change an area within a very short period of time (such as caused by the 1972 storms on the East Coast, notably Chesapeake Bay), totally altering oceanographic conditions and completely masking the effects of any disposal activities.

Oceanography

Geological: When selecting an ocean disposal site, there are several geological parameters that should be considered. These parameters are not related to the materials that are being disposed of, but are of significant importance to the benthic marine life in the area.

The depth to the sea floor must be considered as this is important in determining whether dumped materials will actually reach the bottom at the disposal site. Depth is also important because at the greater depths of the ocean there is little or no sunlight and, therefore, less marine life. Bottom relief (flat, undulating, sloping, or steep), is of significance in that the flatter the bottom, the more likely the build-up by material that sinks to the sea floor.

Rock and sediment type as well as sediment thickness are of importance in understanding the benthic populations. Sediment type is important in predicting potentials for increased turbidity when currents or the orbital velocity of waves cause scouring and resuspend the finer particles.

Physical: The physical parameters of importance in assessing an ocean disposal site are temperature, salinity, density, current speed, current direction, depth of the mixed layer, turbidity, and sea state.

Temperature – Temperature is important in the determination of the stability of the water column. Seawater temperature varies with time, depth and latitudes, and solar radiation. It is important to look at the temperature of thin surface layers, the main portion of the thermocline separating the transition layer, and the underlying deep water, which is typically cold and extends to the bottom. A knowledge of stratification is required to the sea bottom when the water depths are shallow or through the mixed layer to the base of the thermocline when water depths are great. The vertical gradient within the thermocline provides an indication of the state of stratification, or stability. Surface and subsurface, through the thermocline, and bottom, temperature data are considered a minimum.

Temperature can be a significant factor in evaluating the disposal site for the effects on marine life. Higher temperatures usually reduce the solubility of dissolved oxygen and thus decrease the oxygen availability for marine life. At the same time, higher temperatures typically increase the oxygen demand of fish and other marine life. Maximum temperatures can be reached at which marine life cannot function. Rapid changes in temperatures are very deleterious to marine life. Temperature shock factors have been developed for various species of fish. Temperature is an important consideration for the disposal of certain acid waste and other chemicals due to accelerated reactions at high temperatures.

Certain biological species from northern waters usually require low temperatures for spawning. These species usually spawn in the winter months, and temperatures are critical. Warming of the water during this period could be damaging and might cause a decrease in some of the more northerly species. In some instances, a rise in temperature of only 3°F might be sufficient to reduce spawning.

Salinity – Salinity is expressed in grams per kilogram of seawater, that is, in parts per thousand (‰). Salinity increases with depth and varies with time and latitude. Variations in salinity occur in the upper portions of the ocean and decrease with depth. Salinity is an important factor in evaluation of the halocline and seawater density layering characteristics. Salinity is an important factor in the flocculation of waste particles, typically of clay material in dredge spoil.

Density – The density of seawater is the mass per unit volume, commonly expressed in grams per cubic centimeter. Density is a function of temperature, salinity, and pressure (depth) and is calculated rather than measured. Typically, density increases only slightly in the upper layer, increases very rapidly through the thermocline, and then has a slow, but steady increase from the base of the thermocline to the sea floor. Density is of primary importance in the determination of waste particle buoyancy characteristics.

Mixed Layer – The mixed layer is a layer of surface water and usually exhibits much less stability than the underlying waters. The mixed layer is from the surface to the depth of the base of the pycnocline, the zone wherein water density changes appreciably with increasing depth. The mixed layer acts as the base for most water processes; below the mixed layer depth, seasonal variations are quite small or nonexistent.

The mixed layer depth is a function of the thermocline, stability factor, mechanical mixing, density structure, convective mixing, and surface currents. It is an important factor in evaluating the initial mixing of a waste product with depth. The variation of the depth of the mixed layer, both in time and space, is an important factor in understanding pollutant dilution processes.

Turbidity – Turbidity is the measure of the extent of light attenuation caused by suspended and colloidal materials in water. It is typically measured by the passage of a light beam through a known-length water path and is usually expressed as Jackson Turbidity Units. The nature and concentration of suspended material in the sea can be highly variable. The suspended particle size is an important property and is critical in a number of phenomena ranging from light scattering to surface chemical activity and sedimentation rates.

Typically, there are both inorganic and organic suspended materials in the water which vary widely from area to area and also with time at the same location, especially the shallower waters on the continental shelves.

The amount of organic detritus is greatly reduced with depth below the euphotic zone. Excessive turbidity inhibits feeding and restricts the growth of certain species of fish and shellfish and is directly related to low planktonic productivity.

Currents – Ocean currents are large scale water movements which occur almost everywhere in the ocean. The forces causing major ocean currents are a function of the mass or density distribution, wind, and from unequal heating or cooling of the ocean waters. The currents may be conveniently divided into three groups: (1) currents that are caused directly by the stress that the wind exerts on the sea surface; (2) currents that are related to distribution of density in the sea, and (3) tidal currents. Wind effects penetrate to depths in the ocean as a function of wind speed and in part by the stability of the water column. The speed of surface currents set up by wind are about two to three percent of the speed of the wind that generated them.

Slight differences in water density set up forces strong enough to cause major water mass movements called geostrophic currents. Some of the major currents off the shores of the United States are related to the distribution of density. Examples of these would be the Gulf Stream, California Current, and the Davidson Current. Typically, these are too far offshore to be of much significance to the ocean disposal sites; however, in a few cases, these currents do directly affect the sites.

Very near the coast tidal currents typically reverse in direction, whereas slightly farther offshore, in the open ocean, where they are not restricted by the coastline, tidal currents exhibit rotary patterns quite different from the reversing currents observed in coastal areas. Open ocean tidal currents are continuous and continually change direction, and create a "current ellipse."

Tidal currents close to the coast may be substantially altered by changes in wind or river runoff. Several classic examples of this would be in the area of the mouth of the Columbia River, mouth of the Mississippi River, and mouth of the Hudson River. In areas where there is a mixed tide, and where there is a substantial amplitude inequality, the current ellipse is more complicated.

As an example the San Francisco Light Ship shows two different ellipses changing direction and speed. Typically, drift bottles or other floating objects as well as drogues, current meters, swallow floats, or bottom drifters are used to measure currents. Interest in currents is primarily for the sake of establishing the direction and rate of movement for materials that are disposed of at the dump sites. For most of the ocean disposal sites, wind drift currents would be the prominent factors in determining the movement of dumped materials, the exceptions are sites located near strong tidal flow, or major ocean currents.

Waves – Basically, there are two types of ocean waves to be considered in the study of ocean waste disposal sites. Wind waves (sea) are those waves that are generated by winds acting on the ocean surface. They are typically sharp crested and frequently break or have their crests blown off by the wind. Given the wind speed, direction, and fetch, it is possible to calculate the size of wind waves generated by a given storm. Waves of many different sizes and periods are present in a fully developed sea. As the waves travel out of the generating area they are gradually transformed and are called swell. These are, in general, smoother, long crested, longer period waves.

Statistical data is available for wave height, period, and direction, for what are known as significant waves (which are the average of the highest ⅓ of the waves present). Waves are important because of the orbital motion and displacement of water particles caused during the passage of the wave. This leads to some dispersion and turbulence within the upper layers of the ocean.

Chemical: Chemical parameters are important in assessing an ocean disposal site due to their impact on the biological life process. The most significant are: dissolved oxygen, biochemical oxygen demand, carbonate, nitrites, nitrates, pH, phosphates, and silicates.

Dissolved Oxygen (DO) – The content of dissolved oxygen in the water at equilibrium with one atmosphere is a function of the depth, temperature, and salinity of the water. The ability of water to hold oxygen (solubility) decreases with increasing temperature and dissolved solids. Ocean waters are seldom at equilibrium and/or seldom saturated with dissolved oxygen because all of the above factors are continually changing.

Inadequate dissolved oxygen may contribute to an unfavorable environment for fish and other aquatic life. The absence of dissolved oxygen may give rise to anaerobic decomposition. Excessive dissolved oxygen, arising from algal growth, may adversely affect beneficial uses of the water.

One major concern in disposal of wastes at sea is to ascertain that the minimum dissolved oxygen concentration necessary to sustain healthy aquatic life is maintained throughout the disposal operation. A general statement cannot be made to give the minimum dissolved oxygen concentration required to support various fish life because the requirements vary with species. It is important that the thresholds of dissolved oxygen for the important species of fish and other aquatic organisms at the disposal site be evaluated, and established.

Biochemical Oxygen Demand (BOD) – BOD is the measure that a combination of substances and conditions have on the dissolved oxygen content of water. BOD is not a pollutant and exercises no direct harm. BOD is important only insofar as it produces a decrease in dissolved oxygen. BOD exerts an indirect effect by depressing the dissolved oxygen content to levels that are adverse to biological life and other beneficial uses of the water.

Carbonate – The reserves of carbonate in the sea and sediments are so great, that the sea acts as a buffer and the disruption of the carbonate content of seawater by biological activity appears to be negligible.

In the ocean the nitrogen is released from the decomposition of the biomaterial, mostly as ammonia (NH_3), which is then oxidized first to nitrite (NO_2) and finally to nitrate (NO_3).

Nitrites (NO_2) – Nitrites are generally formed by the action of bacteria upon ammonia and organic nitrogen. Nitrites are quickly oxidized and are seldom present in surface waters in significant concentrations. Nitrites stimulate the growth of plankton.

Nitrates (NO_3) – Nitrates are the end product of aerobic stabilization of organic nitrogen. Photosynthetic action constantly utilizes nitrates and converts them to organic nitrogen in plant cells. In spite of their many sources, nitrates are seldom abundant in surface waters. High nitrate concentrations stimulate the growth of plankton and aquatic plants.

pH – The pH encountered in the sea is typically between about 7.5 and 8.4. The higher pH values are generally encountered at or near the surface. Where the water is in equilibrium with the carbon dioxide in the atmosphere, the pH is between about 8.1 and 8.3, but higher values may occur when the photosynthetic activity of plants has reduced the content of carbon dioxide. Below the euphotic zone, the pH shows a certain relationship to the amount of dissolved oxygen in the water.

In regions where virtually all the oxygen has been consumed and, consequently, where the total carbon dioxide is high, such as at depths of about 800 meters in the eastern portions of the Equatorial and North Pacific, the pH approaches a minimum value of 7.5. This is a limiting value because no more carbon dioxide can be formed. Below the minimum oxygen layer there is generally a gradual increase in pH with depth. The permissible range of pH for fish depends upon many factors, such as temperature, dissolved oxygen, prior acclimatization, and the content of various anions and cations. The presence of carbonates, phosphates, borates, and similar ions gives water a buffering power so that the addition of an acid or base is less likely to be deleterious to marine life.

Phosphates – In addition to the phosphates formed from ortho-, meta-, and pyrophosphoric acids, anhydrous salts may be produced in which the hydrogen of the acid is replaced only partially by univalent or bivalent metals. The sodium, potassium, and ammonium phosphates are soluble, but most of the others are only slightly soluble.

Usually, dissolved orthophosphate is the predominating form, but the partitioning of the phosphorus content is seasonally dependent. Phosphorus can be considered a primary nutrient of phytoplankton. Phosphates are seldom found in significant concentrations because they are utilized by plants and converted into cell structures by photosynthetic action. Excessive amounts of phosphates may result in an overabundant growth of algae and be detrimental to fish.

In themselves, however, phosphates seldom exhibit toxic effects upon fish and other aquatic life, and may be beneficial to fish culture by increasing algae and zooplankton. Organic phosphates used extensively in pesticides exhibit selective toxicity to many forms of aquatic life.

Silicates – The element silicon is not found free in nature, but it occurs as silica in sand or quartz and as silicates in feldspar, kaolinite, and other minerals. Silicon dioxide, or silica, is relatively insoluble in water or acids, except hydrofluorates. The concentration of silicon in seawater is affected by geological processes. Silicon is a biologically significant element and, like phosphate and nitrate, exhibits a strong seasonal dependence reflecting the waxing and waning of the life processes.

It may occur in natural waters, as finely divided or colloidal suspended matter, in concentrations of 1 to 40 mg/l. An abundance of silica in the water, along with other necessary nutrients, favors the growth of diatoms. Blooms of diatoms synthesize silica into the rest of the organisms and thereby lower the content in the water. The cycle typically begins in the spring with the uptake of silicon by the growing phytoplankton population, resulting in a depletion of the silicon content of the seawater. In the summer the rate of this growth slackens and the silicate is replenished somewhat. But, in the fall, there may be a second spurt in phytoplankton growth. With the approach of winter, the organisms die, and as their remains slowly sink, the silicon is restored to the seawater by processes of redissolution.

Biological – A biological study of an ocean disposal site should establish the nature and value of the resource, provide a data base which can be used for future comparison, provide a basis for projection of the consequential effects of wastes, and provide information for establishing controls to minimize detrimental effects.

The primary biological aim to be considered in evaluating new or existing ocean disposal sites is to establish the nature and value of the existing biological resources in these areas, and to predict the impact of the disposal operation on the resources. Significant economic losses have resulted from ocean pollution. Losses of commercially valuable fish and shellfish, whether killed directly, indirectly, or rendered inedible, may present serious social and financial problems. The economic or commercial importance of the marine life should be a decisive factor in the determination of locations of disposal sites.

Marine ecosystems are affected by wastes depending on how the wastes are dispersed. Elements of the wastes frequently enter living organisms. Phytoplankton organisms absorb nutrients, trace metals, and other materials. Organisms that feed upon the phytoplankton successively pass the pollutants on to higher organisms.

As this process moves through the food web, concentrations reach their highest levels in predators such as large fish, birds, marine mammals, and can eventually affect man. The marine organisms which make up the food web can concentrate toxic chemicals, heavy metals, and other hazardous materials. The ability of marine life to concentrate materials varies from a few hundred to several hundred thousand times the concentrations in the surrounding environment. The concentration of pollutants by the biota can have sublethal and lethal effects.

The marine environment can be divided into two major realms: the benthic zone, which refers to the ocean bottom; and the pelagic zone, which refers to the overlying water. The three major categories of aquatic plants and animals that live in these zones are the plankton, nekton, and benthos.

Plankton – Plankton constitute the bulk of life in the sea and may be animals (zooplankton) or plants (phytoplankton). Plankton are organisms which may be microscopic or macroscopic in size and are free drifters that travel at the mercy of the currents.

Nekton – Nekton are marine animals with the power of locomotion which enables them to swim freely, independent of currents. Nekton have the ability to search actively for food, avoid predators, and migrate extensively. Many species are commercially harvested.

Benthos – Benthos are organisms, including plants, that live on or in the ocean bottom. These organisms may or may not be attached to substrates, and some may have planktonic larval stages. Benthos are important because they modify the physical and chemical properties of the sediment or waste material and some are of commercial significance.

Photosynthesis – In the sea, most of the primary energy production takes place near the surface as a result of photosynthetic activities of chlorophyll-containing, microscopic, planktonic algae. Very little light penetrates below 80 meters. Photosynthesis is the production of organic matter by plants using water and carbon dioxide in the presence of chlorophyll and light; oxygen, a critical by-product, is released in the reaction.

Chlorophyll is the group of green pigments, found in plants, that are essential for the photosynthetic process. Marine algae are responsible for nine-tenths of the world's energy conversion. Secondary sources of conversion in shallow water are attached macroscopic algae (kelp and others) and rooted higher plants (eel grass and others), and to a lesser extent, benthic bacteria.

Interpretation of data on the biota in relation to waste disposal sites requires a knowledge of the normal biota and its variations. Natural factors affecting the marine life must be understood before conclusions can be reached as to the effect of any single variable such as disposed wastes. Marine biota is directly affected by toxicity, oxygen depletion, biostimulation, and habitat changes.

Toxicity – The ecological effects of toxic wastes are complex and not well understood. Additional research is required to determine their effects on deferred and long-range ecological change in the marine environment. Toxic wastes include materials such as pesticides, oil and refinery wastes, heavy metals, and paper mill wastes.

Although concentrated toxic wastes kill marine life, organisms may be adversely affected by concentrations far below the lethal level. Such effects include reduced vitality or growth, reproductive failure, and interference with sensory functions. Pesticides and other toxic materials have been a major cause of fish kills in freshwater systems, and probably have had similar effects in marine waters.

Oxygen Depletion – Oxygen supports marine life and is necessary for the biological degradation of organic materials. The principal sources of dissolved oxygen in water are: (1) directly from the atmosphere through the water surface and (2) from the photosynthesis of chlorophyll-bearing plants. When plants and animals die, oxygen is used in their decomposition.

Because organic wastes require oxygen in order to decompose, if waste loads are too large, oxygen levels required to support marine life will become depleted. If the oxygen content is depleted, the diversity and life functions of marine organisms will be altered, and anaerobic bacteria begin to flourish. Oxygen deficiency in a waste disposal area can be self-perpetuating because the accumulation of organic matter, sulfides, and some metals acts as a reservoir of future oxygen demand.

Biostimulation – The accelerated fertilization of plant life, or biostimulation, can be caused by excessive nutrients, such as nitrates and phosphates. (Sewage sludge is particularly rich in these nutrients). Excessive blooms of algae, caused by biostimulation, indirectly change the nature of bottom sediments and thus the communities of organisms, such as fish, bottom fauna, and aquatic plants. Sediments adjacent to some disposal areas have shown many such greatly increased concentrations of organic matter.

In the past, biostimulation has been recognized as a serious problem in fresh waters, but not in the oceans except for an occasional red tide; however, biostimulation is increasingly affecting estuaries and bays and, occasionally, portions of the continental shelf.

Habitat Changes – Habitat changes are the most common changes that affect entire ecosystems. Accumulations of various kinds and quantities of wastes or sediments on the bottom can drastically alter the marine communities. The most pronounced ecological changes have been caused by dumping sewage sludge, dredge spoil, and toxic wastes which buried or rendered the substrate unlivable.

The estuarine zone has felt a major part of the impact caused by man's development activities. Changes in the ecological system in our estuarine areas has serious environmental effects upon aquatic organisms found in coastal and oceanic waters. Estuaries are used by many marine species as habitats and nursery grounds for breeding, rearing young, and as a food source. Most of the commercially important marine organisms depend upon the estuaries at some stage of their life cycle.

Biological samples of the major groups should be collected periodically and identification, abundance, diversity, distribution, and variations of species should be determined approximately seasonally. Inasmuch as many marine organisms have relatively short life cycles, the sampling program must be carried on throughout the year to detect seasonal variations consistent with the particular species.

The following are biological parameters that should be measured at ocean disposal sites, the first at all possible sites, and the second at sites where possible pathogenic material accumulations are expected.

Biological Parameters –

(a) Biological samples of the plankton (zooplankton and phytoplankton), nekton, and benthos. This will include fish eggs, fish larvae, fish fry concentrations, and the aquatic flora. The plankton samples should include the microscopic as well as the macroscopic organisms.

(b) Identification and density of coliform organisms and other bacteria, which includes total and fecal bacteria counts (MPN coliform and total bacteria). Also, the identification and abundance of micropathogenic organisms such as viruses, fungi, parasites, and others may be important.

Additional Parameters – The following parameters should be calculated for the foregoing biological groups: species identification or composition, species range or diversity, species distribution, species or seasonal variations, and relative abundance of species.

(a) Bioassay or toxicity tests should be conducted to measure the toxicity level of the dumped materials on certain biological groups, especially the commercially important species. These studies should include short and long-term effects. The basic accepted bioassay test is a 96-hour exposure of an appropriate organism, in numbers adequate to assure statistical validity, to an array of concentrations of the substance that will reveal the level of pollution that will cause:

    (1) Irreversible damage to 50% of the test organisms, and

    (2) The maximum concentration causing no apparent effect on the test organisms in 96 hours. (This data will be reported as 96 hours.)

(b) The biomass or standing crop of zooplankton, phytoplankton, nekton, and benthos should be ascertained.

(c) Primary productivity/chlorophyll pigments measurements should be determined.

(d) Statistical data on commercial fish and shellfish catches at sites should be accumulated.

(e) BOD of the wastes and sediments should be determined.

(f) The nature, type, and rate of accumulation of the bottom detrius (organic) material should be ascertained.

COASTAL WASTE MANAGEMENT: PHYSICAL PROCESSES

Transport Mechanisms

There is no single analytical method currently available that will completely and accurately predict the dispersion and dilution of a waste material discharged into the marine environment.

The waste characteristics are infinitely variable; however, in analyzing the initial distribution, only the physical characteristics need be considered. The bulk specific gravity of a waste slurry is one characteristic that, within physical limitations, can be controlled. Changes in the specific gravity of a waste slurry or sludge, whether through the use of a thickening process or through the use of additives, can affect both the resulting dilution and penetration.

Such changes may result in changes in either the concentration, density or size of the solid constituents or any combination of the three. Studies lead to the conclusion that suspensions with grain sizes less than 60 microns can be assumed to act as pure liquids. Interrelating effects are shown between such parameters as settling velocity and particle concentration with an increase in particle concentration increasing the diffusion while causing a decrease in the settling velocities of the particles.

Several other controls are available and are of a physical nature. Included among these are the depth, size, and orientation of the discharge outlet, as well as the barge speed and direction, which can be used to influence and control the final distribution to some degree. However, the optimum discharge method must be separately determined for each waste and should include consideration of possible biological and chemical effects. These relationships are not generally independent and a change in one of the parameters without a compensating change in the discharge rate will necessitate the use of a continuity relationship.

The transport of waste materials dumped into the sea depends, in general, upon:

(1) What is introduced – its physical, biological, and chemical properties.
(2) Where it is introduced – its position with respect to local ambient-density and velocity distributions.
(3) How it is introduced – its residual buoyancy and momentum.

Both the immediate mixing and dispersion of wastes over periods of time that are relatively short when compared to the circulation times of the oceans as a whole are emphasized here. The physical oceanographic processes, whereby wastes can be diluted and dispersed from one part of the ocean to another is not considered. These are known to continually vary with both time and space as well as with changes in boundary condition. This aspect will be discussed only qualitatively to aid in visualizing the applicability and limitations of the analyses subsequently presented.

It is often assumed or theorized that although the ocean is in continuous motion the rates of motion and exchange cover such wide ranges that they can be separated into nearshore horizontal and vertical exchange, intermediate and deep circulation exchange and the exchange associated with coastal and enclosed basin circulation.

In most coastal and open waters agitation and turbulence generated by wind stresses on the surface result in a surface or mixed layer characterized by a near uniform density gradient. This layer varies between 60 and 1,200 feet and is separated from the colder deeper waters by a stable layer exhibiting a sharp density gradient–the thermocline or pycnocline. The magnitude of this gradient can vary in both time and space and characterizes the relative stability or strength of the layer.

Wastes introduced into the mixed layer generally will be rapidly distributed vertically throughout this layer due to convection, wind-stirring or mixing, density differences, and internal currents. If they fail to penetrate the pycnocline they will be transported from the area of introduction primarily by wind-driven surface currents which, in general, extend throughout this layer. The analysis of wastes which do penetrate the pycnocline will be influenced, if not controlled, by large-scale global currents such as the Gulf Stream and the Kuoroshio. The average location, magnitude, and direction of these currents has been documented and in lieu of on-site determinations their use would produce approximate but reasonable results. Estuarine and nearshore currents have also been studied, although to a lesser degree.

The presence of eddies resulting from turbulence can act to vertically disperse waste materials in addition to mean current dispersion. The rapid increase of density with depth in the thermocline inhibits vertical transfer, and eddy diffusion is small compared to that of the mixed surface layer with its near uniform density gradient.

There are other localized phenomena that can influence the exchange of material between the surface and subsurface layers. This occurs in areas where:

(1) The pycnocline is shallow and subject to disturbances, usually wind generated.
(2) Offshore transport of surface waters results in an upwelling of colder subsurface waters.
(3) Downwelling exists caused by an increase in the density of surface waters due to evaporation or cooling.

The first step in a complete analysis of the effects of a waste discharged to the ocean is to predict its physical fate. The objective of the analysis usually dictates a time scale that varies as a function of the waste material itself. For example, the time required to reduce a toxic waste through dilution to a nontoxic concentration may be on the order of hours, if it is susceptible to chemical and biological destruction but on the orders of weeks if it is refractory. The subsequent analyses utilize a variety of simplifying assumptions and are limited to environmental conditions variable only with depth and totally exclude biological and chemical effects.

The total transport of waste materials can be divided into four basic transport phases.

(1) Convective descent
(2) Collapse
(3) Long-term dispersion
(4) Bottom transport and resuspension

These phases are graphically presented in Figure 3.10. It can be noted that the first two phases, convective transport and collapse are of short duration, when compared to long-term diffusion and are important in determining the initial conditions for the long term diffusion stage.

FIGURE 3.10: BASIC TRANSPORT PHASES

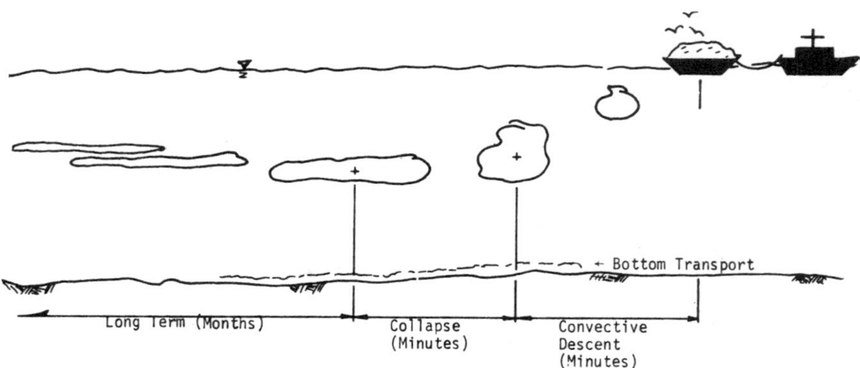

Source: PB 204 868

Convective Descent: A waste material discharged to the ocean from the surface generally possesses an initial downward momentum and a density greater than that of the receiving fluid. These result in forces that cause the waste to settle in the form of a cloud. As the cloud settles shear stresses are developed at the interface between the moving cloud and the receiving fluid. These stresses result in a dispersion of momentum and in the creation of turbulent eddies that entrain ambient fluid.

Entrainment of the less dense ambient fluid reduces the density differential and tends to slow the descent of the cloud. The descent speed is, at the same time, being reduced as solids with settling velocities greater than the descent speed of the cloud settle out, further reducing the cloud's density. The waste cloud may, in a stably stratified fluid, eventually reach an equilibrium level where the descent velocity is zero and the density of the cloud is in approximate equilibrium with the ambient fluid. This zero velocity state is considered the end of the convective descent stage.

To solve this problem accurately the size and density distribution of the constituent waste elements would have to be known. If solids are considered to settle continuously, the end of the convective descent stage would theoretically never be reached for more than an instant because, due to continual particle settling, the cloud would become positively buoyant and begin to rise. Many studies of this phenomenon have shown that when both the concentration and size of the solids are small the waste slurry will tend to act as a pure liquid. For the wastes commonly dumped into the ocean this assumption is currently made for most classes excepting dredge spoils and industrial wastes having high solid concentrations.

Bottom Transport and Resuspension: This final transport phase assumes that the solid constituents of the waste slurry do indeed settle and reach the bottom. This could be accomplished entirely during the convective descent stage if the water depth were less than the predicted or theoretical total penetration depth. The resulting distribution of the solid constituents for this case would become a function of the residual momentum of the cloud as well as the existing density disparity.

When the residual momentum is high, a dynamic turbulent rebound effect might be expected, but for near zero momentum the density disparity would cause a spread similar in nature to that of the collapse phase. Two questions are raised regarding the solids once they reach the bottom:

(1) Will movement occur?, and
(2) If movement occurs, what form will occur—bed load or resuspension?

In general, it is accepted that motion will be initiated in a flow field when the shear stress on the particle creates a lift force in excess of the submerged weight of the particle. The direction of the initial particle movement will be nearly perpendicular to the plane of the applied shear stress and for a horizontal bed will be near vertical.

This force will lift the particle off the boundary and it will resettle, subject to currents and eddies, either to a position in the flow where the resultant lift on the particle just equals its settling velocity (suspended transport), or when the settling velocity exceeds the lift force, to another position on the boundary (bed load transport). Suspended transport is most common when the current shear is nearly constant with bed load motion resulting when an additional shear resulting from turbulent eddies is superimposed.

Long Term Dispersion: Standardized differential equations are available that describe dispersion and convection in turbulent flows. To date, only simplified solutions have been used. These are usually applicable for open ocean conditions only and presuppose a waste that acts solely as a liquid.

Collapse: The second phase of transport, the collapse phase, is the transition between convective descent and long term dispersion. The analysis of this phenomenon assumes that the cloud has come to rest at some equilibrium position and that a dynamic vertical collapse characterized by horizontal spreading occurs. This collapse is driven primarily by a pressure force and resisted by inertial and frictional forces.

Many complex actions may be occurring simultaneously here and an analogy to a density underflow is useful in attempting to describe them in qualitative terms. A three-dimensional form of surge head accompanied by the corresponding reverse flows should exist, with the circulation increasing the diffusion in the area of the spread.

The internal density structure of the cloud relative to the ambient density structure will exercise control over the magnitude of this spreading rate and instability criteria should exist, similar to that of the two-dimensional case, that will predict breaking interfacial waves for some velocity level. If this occurs, mixing and entrainment will further increase. The near zero vertical velocity of the cloud should allow for an increase in the number of solid elements that can settle out interjecting another action that may foster not only a decrease in the driving force but also an upward motion in the cloud itself.

Motion Studies

The coastal area is herein defined as the region between the coastline and the 100 fathom curve; it does not include estuaries. The ultimate sinks for all of the wastes due to man and his activities that are not burned, buried, or spread on the land are the oceans. The ultimate sink for waste heat is, of course, the atmosphere.

The path is by river and/or estuary to the coastal areas and thence to the oceans. At steady state, all that is added to the system is ultimately discharged to the oceans except that which is removed by geological, biological, or chemical processes enroute. Each part of the system has a capacity to receive wastes without undesirable effects which depends upon the distribution of the sources, the physical processes of advection and diffusion and the geological, biological, and chemical processes of removal. The determination of this capacity is the science of waste management.

There are approximately 860,000 miles (statute) of coastal area surrounding the United States. By region, 16% are off the Atlantic Coast, 15% off the Gulf Coast, 3% off the Pacific Coast, and 65% off the Alaskan Coast. Because of the narrowness of the coastal area off the Pacific Coast, i.e., the proximity of deep water close to land, waste management is largely a problem of proper location and introduction of the waste. Because of the higher population and industry density in the coastal area off the Atlantic Coast, and the more immediate requirement for proper waste management, these studies for the most part will be limited to that portion of the coastal area off the Atlantic Coast between Boston and Norfolk.

Mean Motion: The motion in the ocean can be considered as a continuous spectrum encompassing scales ranging from the molecular free path up to the ocean wide circulation. The division between that part of the motion which is assigned to the advective processes, and hence for which the spacial and temporal distribution of the velocity field must be known, and that part of the motion which is assigned to the nonadvective, or diffusive processes, and hence for which only certain statistical properties are required, depends on how much detail, in time and space, are required of the distribution of the introduced waste. A part of the motion which is assigned to advection is also dispersive.

Since there is developing an almost continuous urban concentration along the Atlantic Coast, the wastes are or will be introduced, to all intents and purposes, as a continuous line or plane source extending from Boston to Norfolk. As a result, one could probably consider all scales of motion in the coastal area as dispersive. However appealing such a division might be because of the lack of detailed information on the circulation, circulation patterns shall be considered which are shelf-wide in extent and at least seasonal in persistence as the mean motion; all other motions including tidal motion will be considered to be dispersive.

Certain general rules of coastal circulation, reviewed as they applied to the area in question, suggested that the two primary variables in coastal circulation were the river inflow and the weather. The situation in coastal oceanography was then summarized by posing several unsolved problems. That little progress has been made toward solution of these problems since that time, seems to reside in the following.

With respect to surface circulation, the general rules are: In the northern hemisphere the average flow is parallel to the shoreline with the land on the right-hand side. It follows that the average motion of the surface water is anticlockwise in a bay or gulf and clockwise around a bank. The reverse is of course true in the southern hemisphere. There is a tendency for the most pronounced surface current to be located near the 100 fathom curve, but it is also characteristic to find a second–somewhat shallower and fresher–band of current near the beach. The driving mechanism for the above described surface circulation is, of course, the river inflow.

That is, because the relatively fresher water being discharged by the estuaries is lighter than the offshore waters, a cross-shelf density gradient is set up which gives rise to a circulation pattern with outflow at the surface and inflow at depth. As a result of Coriolis force, the offshore flowing surface current is deflected to the right in the northern hemisphere producing the observed surface currents parallel to the shoreline with the land on the right-hand side. Near the bottom, about all that can be said with certainty is that, owing to bottom friction, the flow near the bottom must have a component in the direction of the pressure gradient which is onshore at that level. Unfortunately, very little can be said concerning the circulation patterns at depths intermediate between the surface and bottom.

Insight into the magnitude of the volume of coastal water that is exchanged with oceanic water as a result of river inflow may be obtained by consideration of continuity principles. Continuity of mass at steady state requires only that the volumetric flux of water seaward through a vertical cross section extending from, say, Nantucket to Cape Hatteras be equal to the river inflow over the same period of time.

However, since salinity typically increases with depth everywhere in these coastal waters, continuity of salt requires the upper layers to discharge many times the volume of fresh water drainage. If one assigns a salinity of 33% to the water leaving the shelf and a salinity of 35% to the slope water inflow along the bottom, then continuity requires the offshore component to exceed by approximately 17 times the volume of fresh water drainage.

Since 50% of the total annual discharge of river water is concentrated in March, April, and May, the currents due to river runoff follow a seasonal behavior with the density gradients reaching a maximum in the summer months when the fresh water accumulation on the shelf is at a maximum.

Although coastal currents, unlike oceanic currents, operate without the direct help of the wind, strong and prolonged winds do cause important variations in the foregoing described system of currents. Accordingly, the winds are considered to be a modifying factor rather than a direct cause of circulation in coastal areas. Along the east coast of the U.S., winds with a southerly component will cause upwelling, i.e., offshore motion in the surface layers and onshore motion in the deeper layers, and winds with a northerly component will result in downswelling with reverse horizontal motion.

They will be most effective in modifying the normal current patterns during the wintertime when the onshore-offshore density gradient is at a minimum. (It has been reported that the normal summertime southwest surface drift on the continental shelf of the east coast of the U.S. may be reversed by the prevailing southerly winds if the spring runoff has been substantially lower than normal.)

One of the most critical phenomena, from the standpoint of transport of waste material onto a beach from offshore, is associated with an abrupt reversal of a wind which has been blowing for a prolonged period of time, i.e., approximately 4 to 5 days, from a direction which has produced upwelling. The upwelling results in steadily increasing the nearshore density at all levels, presumably as a result of the offshore transport of light surface water and a subsurface flow of cold dense water toward the coast.

Eventually, the interface between the light surface water and the heavier inshore water slopes offshore and downward. When the stress of the wind is withdrawn as a result of a windshift, this density distribution is unstable and the cold inshore waters sink and run offshore under the light surface layers which in turn move toward the coast. Obviously any waste material contained in the surface layers which participate in this sloshing motion will be carried directly to the beach.

Dispersion: All scales of motion smaller than those described above have been regarded as producers of internal shear and mixing and are lumped together under the term dispersion. As a result, it is sufficient to quantitate the dispersive characteristics of a given environment by statistical parameters. One such statistical parameter that is used is the rate of spread of a patch of waste or other tracer material as it incorporates surrounding water into itself, i.e., an exchange of water parcels with no net flux of water across the patch boundary. Another useful parameter is the rate of decrease of peak concentration in a tracer patch. The usual procedure is to release a known quantity of tracer material such as rhodamine dye and to measure its three-dimensional distribution as a function of time.

The measurements are then fitted to a particular mixing model and the dispersive characteristics quantitated according to the model selected. That is, if the data are fitted to a radially symetrical two-dimensional mixing model, a diffusion velocity or rate of energy dissipation is sufficient to define the dispersion; if a shear model is used, shears and diffusivities are the required parameters. The point is, of course, that it is possible to measure the dispersive characteristics of a given environment and to make a prediction regarding the effect on the environment of the introduction of a given waste either as an instantaneous or continuous source, at least for scales up to a month in time and 100 km in space.

However, it is difficult if not impossible to quantitate the large scale mixing processes that lead to renewal of the shelf waters by oceanic water. That is, the amount of tracer material released is determined by the sensitivity of the measuring instruments which at present levels would require a prohibitive amount of tracer for this large an experiment. These large scale mixing processes are of great concern, however, since they determine the capacity of the region to assimilate wastes from the east coast megalopolis.

Thus far, what has been said about the processes of advection and dispersion applies to wastes which are either colloidal or in solution. Some wastes, however, consist of particulate matter which are denser than seawater and hence have a finite settling velocity, such as sludge. Other wastes such as garbage, trash, oil and grease, etc. float. In light of this, it is clear that models which are developed for materials in solution require modification if they are to apply generally.

Objects which are constrained to move in a horizontal plane because of their density are subject to various organizing agencies when placed in a three-dimensional field of motion. For example, in the case of an onshore surface current, the shoreward component of the velocity must vanish at some point before reaching shore and the water parcels must either sink or change direction, or both. In any case, the result is a line running roughly parallel to the shoreline where floating debris collects.

Theoretically, if you place a group of floatables in a circle around a singularity they will form an ellipse the area of which either grows or shrinks with time depending upon the sign of the divergence. Singularities in a velocity field are points or lines where the magnitude of the vector is zero. The most important ones are points of convergence (or divergence), neutral points, and lines of convergence (or divergence). If a random motion upon the velocity field is not superimposed, the floatables will also be subject to advection-diffusion. As a result, the group size tends to grow against a collecting power of convergence toward the singularity. The practical consequences of the foregoing are obvious. Intentional ocean disposal of immiscible, floating materials is therefore not acceptable.

Particulate matter which is denser than seawater will, of course, sink and eventually most will be deposited on the ocean floor, i.e., except for the fraction removed by biological and/or chemical processes. The foregoing presumes that the material is not so dense that it all falls out but that an equilibrium distribution exists where the sinking is in balance with vertical diffusion.

Of primary concern is the health significance of the material once deposited. Many areas of the continental shelf off the eastern seaboard support commercial populations of edible shellfish, primarily sea clams, that are capable of concentrating and holding bacteria, viruses, and toxic substances which they ingest from their environment. In addition, a sludge blanket is detrimental to their growth as they require a sand or gravel bottom. At this time there are several areas off the east coast in use as disposal sites for sludge, acid waste, alkali waste, and inert industrial waste.

Dispersion Studies

The disposal of liquid wastes by direct discharge into nearshore coastal waters has been practiced in the United States for many years. Not too long ago it was considered adequate practice to discharge essentially untreated wastewater through relatively short single outlet pipes terminating in shallow water. These outfalls were designed principally on the basis of hydraulic efficiency and often little or no attention was given to the effects of the environment on the dispersion of waste material. Certainly little attention was paid, except for health considerations, to the effects of wastewater on the marine environment per se.

As coastal populations increased, the necessity of insuring not only safe but clean and pleasing recreational waters became more apparent. Consequently, in the design of new or expanded disposal facilities it became necessary to provide higher degrees of waste treatment and more efficient outfall systems.

Submarine Diffusers: In response to the above need, longer and deeper outfalls with multiport diffusers were developed which provided a means of maximizing the so-called initial dilution of wastewater with receiving water in the vicinity of the outlet. Initially diffuser design was based, through necessity, on empirical methods. Within recent years, however, a wide variety of technical studies led to a fuller understanding of the theoretical concepts which describe the performance of multiport submarine diffusers.

It is now possible to include in a given design situation the influence of environmental factors such as density stratification on initial dilution. Today it is rather common practice to design diffuser systems which will insure the formation of submerged waste fields throughout most of the year. In general submerged fields are preferred over surface fields because they are not subjected to wind-induced currents which in many coastal areas move predominately on shore.

Oceanic Dispersion: A number of technical publications have dealt with various theoretical concepts of oceanic diffusion, which have led to a better understanding of diffusion phenomena in general as well as the development of relevant mathematical models which describe dispersion in the ocean. (Dispersion here is defined as the combined effect of horizontal and vertical diffusion plus convection due to currents in the direction of mean motion.)

The process of oceanic diffusion is so complicated, however, that theoretical considerations alone cannot be used to solve practical problems, hence, the engineer still must rely quite heavily on data derived from field experiments. Consequently, considerable attention has been paid to gathering empirical information on the mixing processes affecting wastes discharged into the sea. This effort has been greatly aided by the development of sensitive electronic gear for the rapid detection of fluorescent dyes and radioactive isotopes often used as external tracers.

Recent research has contributed definitive information on dispersion in the ocean and other large bodies of water. It appears from these studies that sufficient empirical data now exists from which to estimate, within the limits of engineering accuracy, pertinent horizontal and vertical diffusion parameters. Such parameters when used with appropriate theoretical models enable the design specialist to predict the so-called rates of subsequent dilution of surface or submerged waste plumes.

Dispersion of Nonconservative Material: However, information relative to the fate of nonconservative constituents of wastewater discharged into the ocean is sadly lacking. Consider the case of one particular nonconservative element of domestic wastewaters, namely, enteric microorganisms such as pathogens and indicator bacteria of the coliform or other groups.

The usual engineering situation calls for the design of a treatment plant-outfall system such that the coliform concentration in a waste field which happens to reach a nearby bathing beach will be below some designated maximum level. The rate of physical dilution of the waste plume can be predicted with reasonable accuracy. However, the manner in which the coliform population grows and/or decays in ocean water must be known before an accurate prediction of the space-time distribution of bacterial concentration can be made.

The effect of coliform decay or die-away on the ultimate bacterial concentration at a point in a waste plume can be much more significant than that of subsequent dilution. In this case it is important that environmental factors which influence the life cycle of coliform or other enteric microorganisms discharged into the ocean be known. In addition it is important that relevant empirical relationships between, say, the rate(s) of population change and key environmental factors be established if at all possible.

The problem is compounded by the fact that for years it was assumed that the results of laboratory experiments could be used to predict the fate of enteric bacteria in the ocean. That this is not the case has been well documented. It is now known that rates of decay of enteric bacteria in seawater as observed in the laboratory differ considerably from those observed in the field.

This discrepancy has led engineers to rely increasingly on empirical data obtained from field studies on existing outfalls. The problem here is that such studies have been somewhat sporadic and unsystematic; the results of experiments conducted apply only to local situations and under rather restricted environmental conditions. Improvements in the methods for bacterial assay coupled with the development of sophisticated methods for tracing waste fields provide the means for conducting needed systematic studies of bacterial decay in the sea. Although some work has been accomplished on this subject, much more needs to be done.

It is also important to point out that although relatively little is known about the growth-decay characteristics of indicator organisms in ocean water, even less is known about enteric pathogens. Little epidemiological evidence has been found to substantiate the fact that bacterial pathogens survive as long as coliforms in recreational waters located near marine outfalls. This fact led many to believe that the die-away rate of pathogenic bacteria is much faster than for coliforms.

However, improved methodology in the detection of Salmonella has resulted in studies which showed that this organism as derived from unchlorinated wastewater survives in fresh and tidal waters for approximately the same period of time as total or fecal coliforms. These studies concluded that the isolation of Salmonella from waters where the total coliform density is as low as 2,200 per 100 ml and fecal coliform density as low as 220 per 100 ml strengthens the contention that the coliform test is an excellent indicator of the possible presence of pathogens. These findings support the conclusion that further definitive information on the growth-decay characteristics of coliforms in the ocean should be obtained.

This conclusion may not hold, however, for the case of viral pathogens. Improved methodology, especially with regard to the detection of the infectious hepatitis virus, will enable researchers to conduct pertinent epidemiological and field studies from which to assess the efficacy of present coliform standards as applied to viral pollution of recreational and other waters.

Various recommendations for areas in which additional research is needed, on the basis of a suggested order of priorities, are as follows:

(1) Much more knowledge is needed about the fate of enteric pathogens and relevant indicator organisms indigenous to wastes discharged into the ocean. Greater knowledge in this area will enable regulatory agencies to establish more meaningful biological standards for recreational and other marine waters. In addition, such information will provide the basis for the rational design of more efficient and economical outfall systems.

(2) Little is known concerning the behavior of submerged waste fields in the ocean. Information is needed regarding the parameters describing dilution and decay of various constituents of wastewater dispersing near the thermocline. If sunlight contributes significantly to bacterial decline in surface fields, then submerged fields may pose a greater threat to public health than at first thought.

(3) Although of lower priority, more information on both horizontal and vertical diffusion of waste plumes in the ocean is needed. Additional empirical data could lead to more relevant diffusion diagrams and answer the theoretical questions of how these coefficients actually vary with the size or scale of the dispersion process.

(4) More information is needed with regard to the so-called density spread of waste plumes. Although it is known that differences in density influence the initial size of surface plumes, the significance of density spread relative to other diffusion processes in the ocean is not known.

The technology now exists to conduct definitive experiments on dispersion and decay of specific constituents of wastewater discharged into the ocean. The estimated costs of such studies would be in the neighborhood of $45,000 to $100,000 per year depending on the oceanographic capability of the institution conducting the research. It is estimated that two to three years total time would suffice for this effort.

COASTAL WASTE MANAGEMENT: BIOLOGICAL PROCESSES

Distribution of Waste

Most of the wastewaters that are discharged into estuaries and coastal waters from large municipal and industrial outfalls are weak suspensions and solutions of both inorganic and organic materials. Natural surface waters are extremely dilute solutions of inorganic salts, atmospheric gases and extracted organic materials. Under most conditions these two types of solutions mix and react slowly over relatively long periods. Significantly, the biological organisms that live in surface waters are adapted to these low concentrations and they commonly react adversely to both sudden, as well as long term changes produced by the introduction of waste substances.

Emphasis on the relatively dilute nature of most waste streams is made primarily because of the need to recognize the inherent difficulty of treating both domestic as well as industrial waste effluents so as to produce completely clarified and purified solutions. Under even the best conditions of design and operation procedures, it will be required to disperse treated wastewaters from the most heavily populated areas into surface waters and most of them will ultimately mix with coastal waters.

Because of the volumes involved and the relatively dilute nature of both the waste streams and coastal waters, the physical, chemical and biological processes that occur when such solutions ultimately mix should be considered. The major biological processes which are involved with waste stabilization are bacterial fermentations which occur on wetted surfaces. Exploitation of such processes is the major mechanism of waste stabilization with conventional biological waste treatment systems, and on a somewhat reduced scale in all nonsterile solutions of water.

During conventional biological waste treatment, the wastewaters are distributed over large surfaces. This occurs when soil, sand and gravel beds are irrigated, and when beds of stones are intermittently flooded. Intermittent flooding and controlled spray distribution are used to secure ventilation and adequate oxygen supply for the large biological demand resulting from bacterial fermentations at the surfaces.

In doing so, additional air-water surfaces are provided, and it has been repeatedly noted that these air-water interfaces concentrate organic materials from solution. Large organic capturing surfaces are also produced by entrained air bubbles in the great variety of biological aeration treatments that have been developed. Surfaces do capture materials that tend to depress the interfacial tension of solutions and colloidal suspensions, including nearly all types of soluble organic wastes and very fine organic solids.

Of course, such simple physical surfaces become saturated after the irrigant comes to equilibrium with the wetted interface with relatively little net change with time. It is at this point that biologically active surfaces become important in the natural biodegradation of highly dilute surface waters.

Marine bacteria and other microorganisms are concentrated in direct proportion to the surface area available for such adsorptive phenomena. Bacterial cells approach colloidal dimensions, and the physical systems that favor concentrations of organic molecules at interfaces also favor concentration of organic molecules in the micron and submicron size range, especially when the solution-suspensions are weak, as is the situation with well-mixed waste discharges in coastal waters.

Observation of emergent vegetation along the shores of all surface waters reveals a microhabitat for such microorganisms on nearly any submerged or partially submerged object. Examination of the relatively thick films of microorganisms that actually appear on such submerged surfaces reveals that such growths are commonly a millimeter thick, especially in mixing zones from discharged wastes and the mass of living material in them is much greater than might be anticipated at surfaces wetted with weak organic colloidal suspensions.

Very clearly, surfaces actually favor the growth of microorganisms, and this growth sticks to the surfaces. These films are essential for the biodegradation of organics introduced into coastal waters and they serve to continuously oxidize organics while renewing themselves. These biological surfaces adsorb and temporarily store organic solutes which are available for oxidation by the cells of the film for the yield of energy necessary for the formation of new cells. From a bioenergetics point of view, such energy exchanges are extremely inefficient and it has been observed that less than one percent of original organic matter is converted to cellular material.

Biological films perform with high efficiency the job of energy degradation and commonly can be shown to contain many types of bacteria. Mixed populations oxidize organic wastes more efficiently than single species, and an effective population will contain groups that oxidize the fermentation products of primary species. Protozoa, rotifera, worms, snails and many higher taxa are able to feed on the bacterial film with high efficiency especially in the shore areas of estuaries which contain submerged vegetation and substrates. The transfer of material and energy from one predatory species to another loses energy and carbon dioxide from the aquatic ecosystem, and the total energy removal by such a complex society of aquatic organisms is always higher than that of a less diverse group.

It is important to realize that interfacial phenomena involving biological films will also be expected to occur on all surfaces of suspended solids entrained in wastewater and coastal water mixing areas prior to their settling out of solution. Also, any tidal or man-influenced turbulence will tend to resuspend flocculated organic and inorganic materials and thus aid in the maintenance of a relatively large reactive surface area for the biodegradation of fermentable wastes. Planktonic organisms, such as diatoms, have also been observed to provide excellent surface substrates for bacterial concentration and thus appear to be involved in the process of biological stabilization within the water column.

It should be noted that marine bacteria, fungi, actinomyces and other microorganisms can utilize a wide variety of organic and inorganic materials as food, provided, of course, that it has available energy. Waste discharges from both municipal as well as industrial discharges contain some organic substances that do not occur in nature. Experience with biological waste treatment has shown that many of these materials can be biodegraded by microorganisms.

Bacteria are highly adaptable with time, to even the most complicated organic structures, provided that other more easily biodegradable and energy providing substrates are not available. Problems occur largely because of this inherent preference of bacterial populations for the more energy rich and fermentable domestic organics over more complex and chemically exotic materials often discharged in low concentrations into coastal waters.

Aside from the concentration of microbiological populations adsorbed onto suspended particles of the water columns, as well as the surfaces of the bottom and shore zones, the surface film of water has been shown in some instances to concentrate larger bacterial populations per unit volume than the underlying water. The surface film is considered to be at the uppermost one-millionth to one-tenth of an inch.

Such bacterial accumulations result from the following:

(1) Surface tension of the film itself.
(2) Low specific gravity of bacteria.
(3) The buoyancy of lipids and gasses associated directly or indirectly with the bacteria.

The surfaces of coastal waters are involved in the biological stabilization of wastes (however dilute they may be) through the influences of four major natural phenomena. These are:

(1) Bacteria tend to concentrate at the surface film.
(2) The film moves with wind currents at speeds in excess of measured ocean currents.
(3) Internal waves, wind, and currents cause horizontal compaction and concentration of materials in the surface film.
(4) Dispersion or diffusion of these surface film materials in underlying water appears to be surprisingly small.

Settling of flocculated and adsorbed organics does occur in time and benthic accumulation of such organic debris ultimately occurs. Both aerobic and anaerobic biological stabilization of introduced waste organics, as well as the cellular debris of organisms residing in the water column above, is known to occur in the bottom layers of coastal waters. Provided that preliminary treatment of discharged wastes has occurred (skimming, screening, primary settling) benthic loads of fermentable waste should not exceed the assimilative capacity of these layers.

Experience with inorganic waste substances suggests that benthic filter feeding organisms, especially annelids and molluses, are likely to concentrate nonbiodegradable substances to many times the levels found in the water. Such benthic invertebrates are evolutionarily adjusted to filtering relatively large volumes of highly dilute waters per day in their straining and filtering processes. In doing so, the flocculated organics and organic-inorganic semisolids that settle to the bottom provide benthic populations with essentially all of their energy and nutrient needs. Excessive deposition of settleable solids has proven to be responsible for the disappearance of this most desirable tropic level in the aquatic ecosystem. Significantly, toxic wastes are especially damaging to benthic populations of filter feeders and the relative influence of such waste products is generally poorly understood in natural surface waters.

The role of benthic silts in the biological stabilization of wastes discharged into marine waters appears to be much less than that of waters from which they settled. Marine muds were found to strongly adsorb soluble nitrogenous organic waste materials of planktonic origin. Muds were found to exert little, if any direct effect upon the rates of efficiency of anaerobic bacterial decomposition of the same planktonic materials. The fact that marine muds adsorb nitrogenous organic materials may be significant, however, in determining the normal course of their breakdown in benthic layers.

Moreover, soluble organic matter released by autolytic or digesting processes of benthic invertebrates may also be localized in the mud where anaerobic changes and low temperatures prevail. The adsorption of organic and inorganic materials onto the surfaces of suspended silts from land sources appears then to be a major mechanism in the transfer of waste materials from the open water to benthic layers in coastal waters. Such materials are likely to be available to benthic animals and thus require review and study of possible toxic influences upon benthic organisms as well as their incorporation into biological food chains that involve man either directly or indirectly.

Knowledge of the extracellular products of planktonic organisms is exceedingly limited. Although a variety of materials are liberated from rapidly growing algal cells (carbohydrates, fatty acids, amino acids, polypeptides, growth substances, vitamins, enzymes and antibiotics), the influences upon the survival of enteric pathogens that are associated with coastal waste discharges are largely unknown.

Effects on Receiving Waters

An ecosystem is the result of a long-term natural bioassay which has determined the composition and abundance of the surviving biological community. Any large change will modify that community, and be toxic to some species or groups. The problem is that of learning to predict the nature and quantity of change which will result from the proposed pollution, and to interpret these changes in relation to human uses and values. Rational management of waste disposal and of other coastal problems requires that the objectives of such management be identified, and that values be placed on each potential cost and benefit.

Field Studies: Field methods for toxicity assessment in the coastal environment have the inherent disadvantages of high expense, long time schedules, and of dynamic complexity in the area under study. They have the unique advantage of involving the real site of effects, where all efforts must be meaningful if they are to have value. Field toxicity studies have two principal uses: (1) they can disclose gross effects, and (2) they can provide highly valuable suggestions for laboratory studies and indicate the ranges of environmental changes, species, and life history stages which are appropriate for further research. Several patterns have been used.

(1) *Before and After Surveys* – Ideally, these include well-designed prepollution sampling of sufficient intensity, extent, and duration to establish the composition of the pertinent biological community, the density and distribution of its components, normal variations (or at least their magnitude) and the associated environment. Rarely, if ever, has this been fully achieved. Compromises which include major components of the biota can be of high value.

Follow-up surveys, designed to detect and measure significant changes in the species present, and the diversity, abundance and distribution of the biota, should be designed to provide comparison with preoperational information.

(2) *Comparison of a Disposal Site and Control Sites* – Since continuation or potential enlargement of the use of a disposal site frequently arouses concern about toxicity, studies are sometimes made to compare the biota of the used site with nonreceiving locations which appear to be similar in other characteristics. The sampling designs for the biological community useful in other approaches should be included, and the same problems of duration must be taken into account.

(3) *Preplanned Site and Control Comparisons* – Such a comparison includes all of the design and duration characteristics of the before and after study, and parallels it with comparable observation of a site carefully chosen for demonstrated close ecological and biological similarity but which receives no wastes.

(4) *Wide-Scale Biological Monitoring* – Sampling for toxicants on a continuous program can have exceptional value in some types of toxicity assessment. These have special potentials as field methods which will assist toxicity assessment and establishment of standards.

Laboratory Studies: Laboratory methods are limited to tiny portions of the real ecosystem, and involve serious problems in reasonable imitation of the natural environment (pressure, light, chemical composition of water, substrates, gas concentrations, bacterial populations, turbulence or other water movement, etc). Perhaps these are the reasons why so little useful data exists on marine toxicity and tolerances.

On the other hand, such experiments can partially reduce the complexity of the natural system, provide for adequate replication, and place the experimental organisms under the continuous observation of the investigator. They sometimes permit multivariate experiments, interaction between computerization and experimentation, and other useful tools in ways which are not applicable in the ocean. If properly designed, laboratory studies have potentials for universal application which are rare in field studies.

Acute vs Chronic: The biological effects which are amenable to laboratory investigation are usually related to the time of exposure and to the concentration of the pollutant. It has become customary to divide the time-related effects into short-term or acute toxicity and longer term or chronic toxicity. While these are essentially different magnifications of the time-scale, they are useful terms here because the experimental possibilities and limitations are different. Acute or short-term toxicity assay methods are attractive because the conditions of experiments are (or seem) easier to define, the time span is short, and answers are often more vivid than in longer tests.

Chronic-effects testing is of unusual importance in achieving full comprehension of the biological and ecological effects of introduced wastes. It is, however, especially difficult to carry out under experimental conditions. In addition to considerations already suggested, the following problems arise:

(1) In static systems, metabolic products, bacterial activity and many other changes constantly alter the experimental circumstances.

(2) In continuous-flow assays, maintenance of constant exposure is sometimes difficult.

(3) Ideally, chronic-effects testing should be continued through at least several generations of the test species. While this is now possible for only a small number of marine species, every effort should be made to extend the use of this technique.

Estimation of Allowable Concentrations: Many methods have been utilized for the estimation of allowable concentrations of waste materials, ranging from prohibition of unsightly messes to rather elaborate mathematical factors derived from carefully obtained data on the tolerance of one or more species. Despite all of the limitations of data and difficulties of application, acceptable levels must be set. They should contribute to achievement of the best possible relationship between man and his environment.

For the coastal areas, several elements to be considered in the strategy of waste concentration control would be:

(1) The capacity of the system for receiving the waste without unacceptable destruction of other values should be clearly established. This will determine the upper limit of permitted total toxicity unless the definition of acceptable loss is modified.

(2) Individual introductions must be evaluated with relation to other toxicants of the same or different type since biological tolerances can be lowered by stress and synergistic effects are possible.

(3) The only enduring policy for waste introduction is likely to be one which places the waste back into natural cycling in nondestructive concentrations. Alternatives are usually short-term patches on the problem.

(4) Despite constructive efforts by many competent scientists to establish specific and mathematically determinable safety factor for individual toxicants, there appears to be little evidence to support any type of universal factor related to median tolerance level, acute toxicity concentrations or other measurable response.

(5) In view of the high values of the coastal waters, the crowding of people and industry toward those waters, and the poor com-

prehension of the long-term effects of toxicants, a highly conservative approach to allowable levels is to be recommended.

(a) Maximum concentrations at any site, or at any time should be a small fraction of the minimum level known to be damaging to any important member of the biota.

(b) "Average" concentrations over time should be avoided as a basis for allowance since death from a short incident is quite permanent.

Biostimulation by Wastes: Conversion of problems into opportunities is an objective obviously worthy of pursuit. Wastes from human or industrial origin obviously contain physical and chemical constituents which, in the proper location and concentration, can be of value. The possibilities of biostimulation, especially of useful stimulation, merit consideration.

Addition of a chemical or of a physical change will stimulate biological production whenever that chemical or physical characteristic has been limiting production. When that factor is no longer limiting, further addition will not enhance production and may be detrimental. Within limits which can now be partially defined, there are possibilities for modest biostimulation in coastal waters.

Inorganic Enrichment – Among the very long list of inorganic chemicals and compounds present in coastal waters, nitrogen and phosphorous are the elements which most commonly limit basic production, and nitrogen is more frequently limiting near shore. Other elements such as cobalt and iron may sometimes be important. Addition of the appropriate element, in the useful form, would therefore increase production.

There is evidence that fertilization can increase phytoplankton, benthic populations and fish production but there is also evidence that efforts sometimes produce ephemeral change or unexpected new biological systems. There are several important limitations to the potentials.

(1) Wastes are often chemically complex, so that addition of an appropriate quantity of phosphate may be accompanied by an excessive or detrimental quantity of another component.

(2) Rich ocean water contains about 0.1 ppm of inorganic phosphate and 0.6 ppm of nitrate-nitrogen. New Jersey coastal water in winter has 0.03 ppm or less of phosphate and 0.08 ppm of nitrate-nitrogen. Elevation by enrichment has rather narrow limits, and it has been suggested that phosphate addition above 0.05 or nitrate-nitrogen elevation above 0.5 would be flirting with unknown dangers.

(3) When nutrients are sufficient, available oxygen can establish limits; and any overload may cause anaerobic conditions which may be highly detrimental.

(4) Elements are used in a general ratio which has been estimated at the following atomic ratio: $\Delta O:\Delta C:\Delta N:\Delta P = -276:106:16:1$. This information can be useful in suggesting additions which might be stimulating.

(5) The chemical form of limiting elements can be highly significant,

e.g., nitrogen in amino acid form is more effective than as ammonium or nitrate salts.

(6) Problems of optimal distribution of stimulating wastes are difficult to solve in the coastal environment.

Organic Enhancement – Many organic materials (amino acids, polysaccharides, amino sugars, fatty acids, organic phosphorus compounds, porphyrins, antibiotics, vitamins and unidentified organics) are present in seawater. These may have profound effects by serving as energy sources (glucose, glycine, aspartic acid for invertebrates; many organics for algal nutrition); growth factors like Vitamin B_{12}; bacterial inhibitors; trace metal solubilizers; or physical mask for harmful trace metals.

These are sufficiently varied and important in the marine ecosystem to merit full understanding and continued attention. It is apparent, however, that some of the same difficulties attend constructive utilization of wastes containing organics as have been noted for inorganics; enhancement is only possible when a limiting factor is contributed, wastes are frequently complex, and the chemical form must be appropriate.

The constructive utilization of energy released as waste heat by steam electric stations and, in much smaller quantities, by other industries has intrigued many scientists and engineers. As new power plants are constructed which produce several million kilowatts, vast quantities of heat are available. One of the most frequently suggested uses of such heat in coastal areas is for aquacultural application. When low temperatures place limits on the season of growth or upon the acceptability of water to a desirable species, controlled release could be helpful.

Suggested uses have included extension of the growth season into spring and fall, for instance. This might in New England convert a two-month growing season to eight or ten months. In the Middle Atlantic, oysters do not put on much growth in summer, and additional heat would probably be detrimental then, but the optimal growth observed in spring and fall might be extended considerably. In warmer waters, this appears to be of little potential for oyster enhancement, but shrimp and fish culture may benefit.

Assay – Assay of biostimulatory effects and potentials of wastes and receiving water can be approached in several ways. The effects of a suggested waste can be examined in a limited way by bringing samples of potential receiving waters into the laboratory, adding the waste, and measuring the effects against controls. This is obviously similar to preliminary toxicity testing and has some of the same advantages and disadvantages. More general information can be obtained by proper testing of the stimulating or depressing effects against cultures in chemostats or other controlled systems which are somewhat better understood than the "wild" receiving water.

Limited releases, following the patterns described for studying toxicity, might reveal stimulating effects and permit estimation of increased effects with larger introduction.

Research – There are several avenues of research which seem to be important.

(1) Far greater understanding of the baseline populations in coastal

waters seems to be essential before biostimulation can even be measured accurately.

(2) The physiological requirements of species and communities do not seem to be known well enough to permit reasonable prediction of the effects of additions.

(3) Adequate methods and techniques for sampling populations and measuring changes may be prerequisite to progress.

(4) Better methods of feeding information between field observations and laboratory investigations will be necessary.

Eutrophication: Nitrogen as a Controlling Factor

The role of southern California coastal sewage outfalls in the eutrophication of local seawater was investigated. The outfall effluents have a measureable influence on standing stocks of phytoplankton, and on primary production. Two cruises were undertaken, in July, 1970 and June, 1971. Kinetic parameters for the assimilation of ammonium, nitrate and urea were determined at the outfall sites using ^{15}N-labelled substrates. These parameters will be useful for simulation models of phytoplankton growth as influenced by local sewage effluents.

The utilization of various forms of nitrogen by phytoplankton, mechanisms and rates of nitrogen assimilation, and enzymes of nitrogen assimilation were investigated in laboratory cultures. Ammonium and nitrate assimilation were found to vary from day to night as does the capacity for photosynthesis when cultures were grown on light-dark cycles simulating natural illumination.

In fitting data on rates of nitrogen assimilation vs concentration of nitrogen to the Michaelis-Menten equation, modified to describe nutrient uptake, it was found that the maximum growth rate was a variable while the saturation constant was uniform over a range of dilution rates of N-limited chemostat cultures. The chemical composition of phytoplankton, particularly ratios of carbon/chlorophyll and carbon/nitrogen, varied with dilution rate in reproducible ways. By varying the dilution rate of such cultures one seems to regulate the degree of nitrogen deficiency of the phytoplankton.

As a result of this study, it was concluded that Southern California coastal sewage outfalls have measureable effects on the concentration of nutrients (ammonium) available for phytoplankton growth in the immediate vicinity of the outfalls. Phytoplankton standing stocks are also elevated in these areas.

Nitrogen appears to be limiting for the growth of phytoplankton stocks in Southern California coastal waters, and ammonium and urea exceed nitrate in importance as nitrogen sources for phytoplankton growth except during upwelling periods. Increased ambient nitrate concentrations during upwelling permit increases in the phytoplankton standing stocks along the coast to levels observed about the outfalls.

Laboratory studies of phytoplankton growth and physiology complement the work at sea by providing data and concepts necessary to an understanding of mechanisms and rates of processes to be expected in nature.

The Marine Food Chain Research Group of the Institute of Marine Resources,

University of California, San Diego, is the site of an interdisciplinary program of research into factors which control the food web in the plankton and is concerned not only with the kinetics of phytoplankton growth, but with the more subtle effects that a change in the nature (as well as the biomass) of the flora may have on the whole food chain leading to the production of harvestable marine life. These aspects must be considered, as the undesirable feature of eutrophication is often not so much the high productivity but the production at the wrong time for a food web to be established which is favorable to man.

It is impossible to predict the course of eutrophication in the marine environment with any degree of sophistication without an understanding of how the phytoplankton respond to an added load of nitrogen compounds, in particular ammonia and nitrate. These compounds are nearly always limiting nutrients in the sea and affect the standing stock and nature of the plant crop. A shortage of silicate and organic trace factors may influence the nature of the flora but probably has little effect on the total biomass.

The nearshore coastal area of Southern California has been surprisingly little studied with respect to phytoplankton ecology, despite the fact that the California Current is one of the most extensively investigated regions with respect to physical oceanography and certain aspects of the zooplankton, in particular fish larvae.

Much of the year surface waters off Southern California are depleted of plant nutrients, especially nitrogen, with nitrate undetectable at the surface and ammonium concentrations less than 1 μM. Urea concentrations are likewise less than 1 μM. Enrichment takes place periodically, especially in spring and summer, when the upwelling of nutrient-rich water markedly increases the concentration of nitrate, phosphate, and silicate.

A detailed study carried out April through September, 1967, off La Jolla, California, provided comparative data on phytoplankton crop size and nutrient concentrations in both quiescent and upwelling periods. The physical oceanography of upwelling has been reviewed and the importance of the processes for local phytoplankton production has been recognized for many years. Nutrient enrichment during upwelling tends to increase the size of the phytoplankton crop in local waters.

In some cases, especially offshore, diatoms are the principal components of the resulting blooms, whereas dinoflagellates often form blooms (red tides) within a few miles of shore. At present it cannot be predicted whether dinoflagellates or diatoms will increase in response to nutrient enrichment near shore and further research on the character and mechanisms of species succession is needed.

Enrichment of surface waters also results from sewage disposal off Southern California in outfalls serving Ventura, Los Angeles, Orange and San Diego counties. At present sufficient data are lacking to compare the nutrient contributions from upwelling and sewage disposal to local surface waters. Very preliminary and approximate estimates suggest that natural upwelling may exceed sewage by an order of magnitude as a source of nitrogen for phytoplankton growth over a year. Fairly accurate estimates of the nitrogen input from sewage can be made, but determining the contribution from upwelling would be very costly of ship time and no doubt variable from year to year.

Upwelling provides nitrogen as nitrate while sewage would be expected to supply ammonium as the principle form of nitrogen. Phytoplankton appear to utilize both forms equally well although their chemical composition, especially C/N and C/chlorophyll-A ratios, may vary somewhat with the nitrogen source used for growth. Since upwelling is seasonal and intermittent a survey of the region during a quiescent period (no upwelling) would be expected to show the outfall areas as points of nutrient-rich water, with high phytoplankton crops, against a low nutrient, low crop background.

This was the case, in part, in July 1–15, 1970, for coastal waters between Los Angeles and San Diego; phytoplankton crops were high only at the outfalls but surface nutrient concentrations were low everywhere. In June, 1971, crops were high at all stations and the effects of sewage effluents were less obvious.

Laboratory studies were carried out to aid in the interpretation of the results from cruise work. Intensive studies of diel periodicity in growth (cell division), photosynthetic rate, and nitrogen assimilation were carried out with single phytoplankton species in the laboratory and with cultures of natural phytoplankton aboard ship. Continuous cultures, operated as nitrogen-limited chemostats, were also studied in the laboratory. Such cultures are particularly amenable to studies of nitrogen assimilation rates and the kinetics of assimilation.

Results of the culture studies fall into four categories: (1) periodicity in nitrogen assimilation, and in cell division, chlorophyll synthesis, and photosynthetic carbon assimilation rate; (2) kinetic studies of nitrogen assimilation; (3) the chemical composition of phytoplankton in N-limited chemostat cultures and the influence of rate of nitrogen input on carbon/nitrogen, carbon/chlorophyll, and nitrogen /chlorophyll ratios in the cells; (4) comparison of results of measuring nitrogen assimilation by direct chemical, isotopic ^{15}N, and enzymatic methods.

COASTAL WASTE MANAGEMENT: CHEMICAL AND GEOCHEMICAL PROCESSES

Effects on Receiving Waters

The Exchange Coefficients for the Sediment-Water Interface: Because interstitial water varies so widely in composition from the overlying water, information on the exchange of dissolved species across the sediment-water interface is important. The importance of this exchange increases as the water depth decreases and the flushing time or renewal time of the water body increases.

Use of the word exchange can lead to confusion in certain instances. For example, the flux of oxygen is almost certainly from the overlying water into the sediment because of the high bacterial demand. The sediments act as a sink for oxygen and oxygen is absent from all but the top few millimeters of most sediments. In the ocean, sulfate is reduced in the sediments and behaves similarly.

Manganese is present in sediments as the +2 form, but it is oxidized and precipitates when it contacts oxic conditions near the sediment water interface. Some species, such as phosphorous, diffuse freely into anoxic lake bottom waters from the sediment, but are sorbed and precipitated under oxic conditions.

Many compounds are produced in the sediments under anoxic conditions by bacterial reactions. Ammonia and methane accumulate to high concentrations and develop strong concentration gradients across the sediment-water interface. Methane has been observed to escape as bubbles if its partial pressure exceeds the in situ pressure.

Processes affecting the sediment-water interface have been investigated by several workers from rather specific viewpoints. Studies on physical processes like diffusion have received most attention, but biological mixing by burrowing organisms has been shown to be important in certain cases.

A good theoretical and practical start has been made toward explaining the effects of the sediment-water interface on exchange and on the depth distributions of dissolved species in interstitial water as a result of ongoing studies. The sediment-water interface is not a purely inorganic system and is very difficult to evaluate because of lack of data. The effects of chemical reactions, mixing by convection, and mixing by organisms are examples.

Further work in this area should include studies of the kinetics of reactions taking place in sediments. Sorption effects should be studied in more detail so that corrections can be applied to diffusion coefficients. Depth distributions of many species are necessary. The effects of mixing by organisms should be investigated and the effects of the size distribution of the sediments on compaction and diffusion should be studied.

Biochemical Considerations: The ocean is a large reservoir whose chemistry reflects the reaction velocity of its constituents. Some concentrations are low because the species are labile and reactions proceed with a relatively small concentration driving force. These include most of the nutrients. Other concentrations are low because input concentrations are low. These include heavy metals and remaining trace constituents. Abundant soluble relatively nonreactive constitutents such as sodium, magnesium, sulfate and chloride ions accumulate in the marine system.

Any change in marine chemistry over a period of time reflects the difference between input and output rates. As major as man's terrestrial activities are in quest of food and manufactured products, changes in the input rates of soluble constituents are inevitable. Increased input results in increased precipitation rates of elements near saturation like calcium and barium, accumulation of soluble ions such as magnesium and chloride and changes in biomass due to metabolites such as nitrogen and iron and antimetabolites such as lead and chlorinated insecticides.

Those constituents affecting the marine biomass are probably of most interest because of the economic value of fisheries and the aesthetic values placed on clean water and seashore life. These constituents are metabolites which control the level of biomass and antimetabolites which, together with metabolites control species distribution or ecology. Marine chemistry as it affects the organism involved is perhaps an order of magnitude more complex than a quantitative chemical analysis would predict because organisms are initially responsive to the molecular shapes they contact rather than the elements comprising those molecules.

The next order of complexity is the different effective shapes of the potentially trivalent phosphate ion. *Staphylococcus aureus* and Sacchromyces apparently only transport $H_2PO_4^-$ while the marine yeast *Rhodotorula glutinus* only uses $HPO_4^=$. Copper is a required cation. It is also a heavy metal capable of precipitating protein and modifying enzymes. Relatively highly populated yeast cultures as compared with the marine system are inhibited by the addition of only 10^{-7} M copper in a phosphate limited system. Yet in a phosphate buffered system 10^{-3} M copper has not effect on the growth rate.

The importance of molecular shapes is well known in the utilization of organic molecules. The rotation of one carbon atom in glucose to form galactose, for example, can effect changes in utilization velocities of many orders of magnitude. Here we see the parallel importance of configuration of inorganic molecules as they exist in the dilute but chemically complex marine environment.

A further complexity is introduced to marine chemistry by the relatively dilute concentrations of most biologically important species and the resulting biomass. For example, the vitamin thiamine exists at a concentration of about 10^{-10} M to 10^{-9} M in seawater. Yet this concentration ($K_s = 3 \times 10^{-13}$ M) is of such great sufficiency that many marine organisms have developed absolute requirements for the co-enzyme. Even the K_s of thiamine for growth in phosphate of 10^{-8} M is below usual means of analytical detection.

Thus the nutrient:antimetabolite:biomass relationship at functional marine concentrations is one which requires some sophistication to understand. It is also the relationship that will be perturbed by a large change in input commonly called pollution. The ecological balance is not however beyond the current capabilities of science to understand. Techniques such as genetic modification of transport systems, relaxation kinetics, and steady and transient state reaction analyses are available to understand the flux and reaction mechanisms of chemical species. These are the types of data, currently largely unavailable for dilute systems, that are useful in assigning numbers to pollution control requirements.

The Role of Chemical Scavengers of Waste Components: The river inflow and chemical scavengers in the sea as related to waste component interaction, involves a study of the nature of riverine supply of chemical species and a guess as to what happens to them as the river encounters the ocean. This consideration must include the detrital (absorbed and dissolved loads of rivers), and in addition requires some idea of the sources of all these materials and the relative controls on the distribution by atmospheric precipitation, weathering and human modification as well as processes acting in the river and at their mouths.

Atmospheric Precipitation and Weathering – Rain and snow are the forms by which water is transferred from the oceans to the land. Atmospheric precipitation is nucleated in the presence of aerosols–particles of sufficiently small size to remain aloft for a long period of time. The transport of organic and inorganic material from the oceans to the continents and across the continents is affected through the aerosols, whether as precipitation, dry fall-out or by impacting.

There are two primary sources of aerosols. These are the very fine particles in the atmosphere derived from sea salts and continental dust. The former is composed of highly soluble chloride and sulfate salts whereas the latter may include addition of some soluble salts, dominant silicate and refractory oxide components. The composition of the aerosol is modified by chemical reactions by which the aerosols or water drop may be subjected. Hence, although marine aerosols and marine rains may, to a first approximation, replicate the proportions of ions found in seawater, on closer examination a great deal of variation is found. Most important differences are in the occurrence of nitrate and ammonium ions in high concentrations in the precipitation relative to seawater. Modification of composition of precipitation by pollution is pronounced.

Aside from soot, modern industry releases directly or indirectly a number of chemicals that modify the composition of atmospheric precipitation. Sulfur compounds and carbon monoxide resulting from fuel burning and hydrogen chloride gas may be mentioned. There is a strong increase locally of chloride and bromide ion of rains in areas of extensive industrial activity. The nitrate to chloride ratio also has been shown to be higher than ambient concentrations in aerosols from urban industrial areas near the oceans.

Pollution of rain by trace metals certainly exists as is indicated by the wide distribution of lead in surface waters and snow fields but except possibly for this element, very little of a qualitative nature is available on this subject. Data concerning the composition of atmospheric precipitation and river water in Japan show that some of the trace elements of the surface run-off were actually depleted relative to rain water falling in the drainage basin, indicating an adsorption on particles or biological removal.

Molybdenum alone of the trace elements studied shows a factor of two increase over the rain water supply. Although it is evident that from the Japanese islands as well as the maritime provinces of the continents the trace element as well as the major element concentration is controlled by the supply of atmospheric precipitation, it seems likely however that continental weathering should have a major part in determining trace element composition of large rivers.

It has long been known that not all elements are efficiently mobilized in solutions by the weathering process. Aside from minerals that have a strong resistance to chemical attack, metals that quickly precipitate as insoluble oxides or other compounds after being released from a mineral also form residual deposits. This is the origin of lateritic (iron-oxide-rich) and bauxite (aluminum-oxide-rich) soils. Various attempts have been made to determine the flux of metals carried away from the region of weathering by streams and ground water. It appears that we are still basically in a quandary as to the rate of supply of metals to streams from the weathering of rocks compared to aerosol contributions.

Transportation — Transportation from the continents to the oceans is effected by the four major agents; streams, wind, coastal erosion by waves and glaciers. At the present time the role of streams seems paramount, but the role of other forms of transport may not be trivial in any one locale. The total run-off of the world and the mean concentrations of dissolve and particular species must be assessed in order to understand the chemical burden brought to the sea and its future path.

Effect of Chemical Environmental Factors on the Mineral-Organic and Mineral-Inorganic Reactions at Fresh Water-Seawater Interface: Although the ocean system may be regarded in the general case as a uniform sink of constant chemical properties, it is impossible to generalize regarding local areas of fresh seawater interaction. The variation in composition of the fresh water inflow is infinite, and estuarine and other nearshore marine waters are unlikely to represent average ocean water due to the nature of the various mixing processes. The following conditions are likely to prevail at the fresh-marine interface however:

(1) A decrease in the particulate sediment load which may be related to the following factors among others. Dilution of the inflow; deposition under the changed chemical environment; deposition due to decreased velocity.

(2) A change in the eH/pH environment. Thus river inflow is generally more acid than the buffered marine system.

(3) An increase in the major cation content or a decrease in trace heavy metal concentrations.

(4) Flocculation of inorganic colloids following injection into the high ionic strength seawater medium.

As regards the impact of organic waste products on the marine environment, many of the problems are the same as for heavy metal contamination. However, it may be assumed that, unlike the latter, many organics of immediate origin are not part of a natural cycle with an approach to equilibrium conditions operating between fresh and marine water and sediment. Thus, in these cases, the oceans may, for some time, operate as a diluting sump.

Solid organic waste material sediments may, depending upon the quantity involved and the nature of bottom water-sediment transport, result in an anoxic environment. Many documented examples of this process are well known. It should be noted that within such areas the heavy metal spectra are greatly modified. Thus iron and manganese may become solubilized and those metals which form sulfides of small K_{sp} are effectively removed from solution.

The soluble organic component of seawater is still being cataloged and little is known regarding specific reactions at the fresh water marine interface. Two phenomena would appear to be of particular importance and interest; removal of organics by various sorption processes on both organic and inorganic solids and the solubilization of inorganic components which would otherwise be removed from the marine environment. It is thus necessary to consider heavy metal behavior also.

Colloids are unstable in seawater. The sedimentation of hydrous ferric oxides at the interface is an example. Of importance here is the fact that this process functions as a classic coprecipitation and many components—and particularly trace metals—are scavenged from solution. Increased concentrations of interstitial trace transition metal concentrations in the presence of annual hydrous oxide layers within Alaskan fiordal sediments have been recorded.

Sorption of both trace organic and inorganic soluble species onto inorganic mineral particles is well known in general but minimal data are available for specific reactions in marine water. Any mineral species will adsorb ions; the process is of importance where large amounts of sediment of small size and hence large surface areas are present (for example, colloid precipitation). Where the sediments are permanently removed from the water column, the sorbed inorganic ions and organics are also removed.

Sorption may be treated as a physical process which is a function of the contacting surface area. However, in the case of many minerals–clay minerals are well known in this context–the ions are held on characterized sites on and within the mineral structure. The term "ion exchange"–a concept first recognized by soil chemists conventionally applied in these cases.

It is convenient to consider ion exchange as a mass action phenomenon, but it should be noted that the charge, size, and nature of the exchange ion are important and that all mineral exchange sites are not equivalent. The behavior of particular trace transition metals is difficult to predict since hydrated and other complex ions (chloride species in seawater for instance) are involved. However, it is known that such heavy metals are easily taken up by clay minerals and correspondingly difficult to remove. Organic solids also have been shown to have high exchange capacities, sometimes even higher than clays, but the specific bonding sites involved are generally not known.

The adsorption of organics onto clay minerals in a similiar fashion to the inorganic ions has been much studied. These reactions have mostly been studied under ideal laboratory conditions but there is little doubt that the particulate sediment content of fresh and marine waters is capable of removing large amounts of a variety of organic species from solution. Unfortunately, little to no work has been done on specific compounds in the natural environment.

When the particulate sediment is deposited, the sorbed organic and inorganic component is, for all practical purposes removed from the water column and is more or less concentrated on the sea floor depending upon the areal extent of sedimentation and subsequent transport. It is unfortunately true that little definitive work has been done on chemical and physical changes inducing or accompanying sedimentation. Of prime importance is the fact that the size fraction of the material which can be held in suspension is a function of the water velocity.

However, it has not satisfactorily established whether specific chemical reactions induce deposition at the fresh marine interface. It is often assumed that the changes in exchange ion content cause flocculation of the inorganic material. However, this process is not simple.

The role of dissolved organic species in seawater in complexing metal ions is being increasingly studied. It has long been evident that purely inorganic reactions are incapable of yielding the concentrations of many minor and trace metals. However, it is only in recent years that analytical techniques have been developed sufficiently for confident soluble species determination in the marine environment, although the concentration of heavy metals by the biomass has long been known and studied.

It is presumed that such solubilization phenomena are chelate complexes analogous to the geologically important metal-porphyrine complexes. It is thus likely that serious heavy metal solubilization and possible pollution is possible in marine environments unusually rich in dissolved organic material.

Interaction with and Influence upon Distribution

The total volume of the world's oceans is 1.37×10^{21} liters. The total surface area is 3.61×10^{14} m^2. In contrast, the surface area of the fresh waters (exclusive of ice) is but 2.4×10^{11} m^2. The oceans cover 71% of the earth's surface. It is useful to subdivide the area and volume of the oceans into at least the following broad classes: the deep seas, the continental shelf zone, and the littoral zone. The areas are as follows:

Deep seas (greater than 200 m)	3.34×10^{14} m^2
Continental shelf (0 to 200 m)	0.27×10^{14} m^2
Littoral zone	1.4×10^{10} m^2

In terms of volume, the waters of the shelf are 8% of the total, and the deep waters are 92% of the total. A great deal is now known about the chemical composition of the waters of the deep seas—particularly with regard to the major constituents. These are found to be practically uniform in space. They are believed to have remained fairly constant for the past few hundred million years. The major constituents of seawater are: H_2O, Na^+, Mg^{+2}, Ca^{+2}, K^+, Sr^{+2}, Cl^-, $SO_4^=$, HCO_3^-, Br^-, F^-, and H_3BO_3.

Changes Brought About by Man's Activities: A major change commonly recognized is the increased concentrations of lead (Pb) in the surface waters of the oceans. The cause is mainly the combustion of tetraethyl lead. In preindustrial times 1.1×10^{10} grams of dissolved Pb per year were added to the northern hemisphere oceans. It has been estimated that 2.5×10^{11} grams/year of lead are being washed from the atmosphere to the oceans. The surface waters of the Pacific off California range from 0.08 to 0.4 μg/l, whereas the deep (older) waters contain uniformly 0.03 μg/l.

On the basis of available data on composition of river flows a phosphorus flux to the ocean of 1.8×10^{11} g/yr (0.005 mg/l median concentration and a world river flow of 3.7×10^{16} l/yr) can be computed.

The value of 1.8×10^{11} g/yr for river flow phosphorus on a world-wide basis can be compared with the total estimated used in the U.S. for detergents and water conditioners in 1958: 1.8×10^{11} g/yr. Another figure is the total estimated phosphorus in domestic sewage in 1959: 1×10^{11} g/yr. A 1966 study of human activities and the annual cycle of water indicates a world figure of 1.4×10^{12} g/yr for phosphorus mined. There is a reasonable concordance among the figures for mining, U.S. detergents, and for phosphorus in domestic sewage.

It seems reasonable to infer that man's activities have by now significantly increased the flow of P to the ocean—for North America and Europe, at least. The magnitude will be reduced, perhaps, by phosphorus removal procedures at treatment plants. Also, not all of the increased P in sewage reaches the oceans. There is significant retention in lakes, e.g., Lake Erie, and in estuaries, which tend to accumulate phosphates in their sediments by photosynthetic activity and precipitation processes. Another factor—already reflected in increased rates of phosphorus mining—is increased agricultural use of phosphorus and passage of some of the applied phosphorus to the agricultural drainage waters.

As to the potential influences of human activities on the flux of elements to the oceans, elements of apparent high pollution potential (meaning high rates of introduction to fresh waters or the oceans) include: Ag, Au, Cd, Cr, Cu, Hg, Pb, Sb, Sn, Tl, Zn, Ba, Bi, Fe, Mn, Mo, P, Ti, and U.

If the world supply of copper were dissolved and mixed in the entire ocean, it would have negligible effects on the plankton, but if it were dissolved and mixed in the fresh waters of the world its concentration would exceed the toxic limit for some algae. Both cases are clearly unrealistic. More pertinent considerations pertain to dispersion of significant quantities of any of the potential pollutants in limited regions of fresh waters, estuaries, or coastal waters. If natural removal processes for the particular pollutant are slow, or if biological concentration factors are unfavorable, undesirable effects may result.

For those elements which are now being added (or might at some time be) to the oceans at rates in excess of the natural rate, one ought to ask, "How is the concentration of the element regulated in the ocean under natural (historical) conditions?" It should be possible to extrapolate from the natural state of affairs when it is reasonably well understood. For the trace elements it is pertinent to examine biologic concentration factors and to compare expected local ocean water concentrations, and expected concentrations in phytoplankton, larger plants, and marine animals with toxicity data.

In general, anoxic conditions do not favor the effective precipitation of most elements and their removal in the sediments. Rapid deposition of organic sludge, either from an effluent directly or as a result of productivity, is likely to produce adverse effects insofar as trace metal sedimentation is concerned. Reducing conditions can dissolve iron oxides, manganic oxides sulfate precipitates, etc.

What unnatural substances are being added to the oceans by way of rivers, wastewater outfalls, shipping, mineral extraction, and atmospheric fallout? It is clear that most of the materials that are added to the sea by man's activities are the same as the materials entering the sea naturally, the principal difference being in the rate of addition.

New Combinations of the Natural Inorganic Elements: The elements in most inorganic compounds, even if in combinations not found in nature, are soon converted by ionization, hydrolysis, and other reactions into forms that occur in nature. The additions of these compounds to the ocean is considered under the heading of natural constituents.

Special consideration may be given to compounds like the active oxy-acids, ozone, chlorine, other strong oxidizing agents, strong acids, and strong bases. Such substances cannot have more than rather localized effects. It is possible for them to be discharged in amounts that may be serious in bays and estuaries. The chemistry and biochemistry of such discharges is relatively uncomplicated; studies of the problems arising are likely to be case specific.

Organo-Metallic Compounds: Some organo-metallic compounds are relatively stable, and if the metal is toxic, it may be more dangerous than in natural combinations. The one known example of serious pollution of marine water by a toxic organo-metallic compound is the Minimata Bay incident, where methyl mercuric chloride was accumulated by marine organisms that were then eaten by humans.

Tetraethyl lead is blamed for an increasing background concentration of lead in the world, but it does not seem probable that any unique problem is created by the properties of the compound.

It is conceivable that metals that are not appreciably toxic could become harmful in organo-metallic combinations. It would not be feasible to begin searching for such compounds in the ocean, but a review of all organo-metallic compounds that are industrially important, including both toxic and relatively nontoxic metals, would show if any of them present possible hazard.

Organic Materials: All naturally occurring forms of organic matter are oxidized in the presence of oxygen by natural processes. The earth reveals no accumulation of naturally formed organic matter except in places where it has been stored anoxically. It is not known whether there are any synthetic organics that are permanently resistant to decay. Certainly some of the plastics will last a very long time, and the fluorocarbons (which are not organics) will probably be extremely resistant. Dissolved or finely divided organic matter in the presence of oxygen and microbial life will probably not last more than a few years or a few decades.

It is not to be expected that any biologically significant organic contributions to the ocean will accumulate indefinitely. Serious effects may occur locally, and some of the synthetic organics, particularly the chlorinated hydrocarbon insecticides, may have far reaching effects because of biological concentration.

Synthetic Surfactants: Synthetic surfactants are a class of materials that have sometimes alarmed the public when their presence has been revealed by foam in stream. A search of the literature made a few years ago for reports of harmful results failed to disclose anything more serious than nuisance effects. Even the biologically resistant forms of the sulfonate detergents decompose under natural conditions at rates that prevent any continuing build-up in the ocean. They are not concentrated biologically, and they do not have any biological effects at concentrations less than a few tenths of a milligram per liter.

How Sewage Can Be Traced in the Ocean: Attempts to explore the effects of a discharged waste on the marine environment frequently require information on the concentration of a discharged waste at a given point, and perhaps also its age, that is, the time interval from its discharge to its arrival at the specified point.

The earliest method of measurement made use of salinities. In view of the variability of the seawater itself, the method cannot be applied beyond a dilution of 200:1 even under the best conditions, and where the seawater is quite variable, as near large rivers, it is practically useless. Next ammonia was used. Under favorable conditions, this method can detect as little as 0.1% of sewage or primary effluent, but it is less applicable with well oxidized effluents. Furthermore, the conversion of ammonia to nitrate makes the method undependable for detecting sewage that has been in the sea for several hours, or sometimes as little as one hour.

Coliform bacteria in an effluent that has not been disinfected have been used for rough estimates, but that method is severely limited by the relatively rapid and variable rate of die-off.

Attempts have been made to estimate the amount of sewage by the presence of anionic detergents. Interference by normal seawater constituents prevents this method from being very useful. Ultraviolet absorption at selected wavelengths is characteristic of unidentified components of sewage. Here, too, one cannot reach to very high dilutions before interference by normal seawater constituents. Turning from attempts to detect normal constituents of sewage, investigators have turned to additions of easily detected unusual substances. Two types of materials are obvious choices: dyes and radioactive materials. The dye that has proved most useful is rhodamine-B. It can be detected by its fluorescence at concentrations down to one microgram per liter.

For radioactive tracing, use has been made of scandium-47. Bromine-82, with a half-life of 36 hours, seems quite suitable if the purpose does not require a longer half-life. Both dyes and radioactive tracers have been used mostly in dispersion experiments, where they are added at a point in the ocean, either as instantaneous or continuous releases. In some situations it would be desirable to add a tracer to the total wastewater discharge. If the flow is large, a continuous addition is costly, whether of dye or radioactive material.

Other materials may be found that can be detected subsequently in the seawater by suitable concentration techniques, such as distillation or solvent extraction. Perhaps a substance like a chlorinated benzene could be distilled out of a fairly large sample, extracted from the distillate, and determined by the gas chromatograph. If three or four such compounds were used successively, the age of the waste detected in the ocean could be determined. Perhaps a large volume of the water could be passed through an absorbing medium such as a carbon column, or a solid medium supporting a film of a fatty material, thus imitating the ability of organisms to concentrate similar materials.

The absorbed material would be treated by suitable techniques for separation and then analyzed by chromatography. Such methods would be very slow compared to the practically instantaneous sensing of dye or radioactive material. Nevertheless, there are types of problems for which the slower method might be acceptable if it could detect 0.01 or 0.001% of sewage at selected points.

COASTAL WASTE MANAGEMENT: MONITORING

Waste Components and Their Effects

The Nature of Water Pollution and its Measurement: Water pollution is a multivariate condition which defies definition on a strictly scientific basis. There are, however, certain characteristics which insufficiently treated wastewater may impart to a natural water body. Methods of pollution measurement are generally based either on the determination of the degree to which these characteristics are present or on analyses for substances known to be causative agents of these characteristics. While future pollution criteria and methods of development must not be limited by present capabilities the results of many years of studies on fresh water systems and the methods employed in defining these systems cannot be ignored.

Oxygen Demanding Wastes – One of the most common and widespread effects observed in polluted water is the reduction of dissolved oxygen to subcritical levels. This is generally attributed to the growth and respiration of aquatic microbiota stimulated by the presence of organic components in the wastewater. Domestic sewage and wastes from industries processing natural as well as many synthetic organic products are typically oxygen demanding wastes. The concentration of dissolved oxygen in the receiving water is a measure of its degree of pollution or degree of stress. Dissolved oxygen may be determined either by chemical tests, which unfortunately are subject to a variety of interferences, or by galvanic or polarographic probes, which are probably subject to fewer interferences but may require frequent maintenance and calibration.

The generally accepted measure of waste strength is the five-day biochemical oxygen demand test (BOD_5). This test is intended to measure the potential for wastewater to cause reduction of dissolved oxygen in a receiving water. Even though the BOD_5 is the basis for many legislative standards, it is ill adapted as a control procedure in a monitoring system requiring immediate data. The dilute nature of waste components in receiving streams led to the development of the BOD_5 test as a qualitative measure of pollution. Thus in any situation where waste components collectively are less than 10 mg/l, no other standard test will provide the information given by the BOD_5 test.

Plant Nutrients – Domestic sewage, agricultural land runoff, and wastes from natural product industries may contain ionic or molecular forms of nitrogen and phosphorus. These forms of nitrogen and phosphorus serve as essential nutrients for aquatic photosynthetic organisms. An excessive amount of these substances results in an overproduction of the organisms (algal bloom), which through their death and decay may produce oxygen demanding substances in the immediate environment. The principal difficulty concerned with the measurement of the nitrogen and phosphorus is the fact that they may be present and active in several forms.

Nitrogen may be present as nitrate, nitrite or ammonium ions or within a number of organic compounds. Phosphorus may be present as orthophosphate or inorganic polyphosphate ions or it may be organically bound. Several sensitive colorimetric procedures exist for the inorganic forms of nitrogen and phosphorus but the organically bound forms require conversion under rather precise conditions prior to analysis. An instrumental technique is available for total nitrogen in discrete samples involving high temperature combustion and reduction with measurement of the resulting ammonia.

Organic Chemicals – Many organic chemicals contribute to oxygen demand and their presence will be detected by the BOD_5 test. There is, however, a seemingly infinite number of organic chemicals which may appear in aqueous wastes, and a great number of these are resistant to microbial degradation. As a result, they will not be detected in the BOD_5 test. Such chemicals would also be expected to pass unaffected through biological treatment systems and persist in the aquatic environment. There may be many persistent chemicals which when chronically exposed to an aquatic community render effects as yet undetermined.

On the other hand, compounds such as the organo-chlorine pesticides possess known toxicity and certain marine life possesses the property of concentrating these compounds within their system. Many organic compounds can be measured collectively by the dichromate oxidation (chemical oxygen demand test) or instrumentally by high temperature oxidation of the organic carbon to carbon dioxide with subsequent measurement of the carbon dioxide in a non-dispersive infrared analyzer. At total organic concentrations below 5 mg/l, however, the reliability of these methods becomes suspect and some type of extraction or concentration procedure may become necessary. Improvements in the combustion tube of the total organic carbon analyzer by a satisfactory enlargement will enhance the capability of this instrument for organic concentrations less than 5 mg/l. Certainly extreme care is required in sample handling as contamination becomes a matter of great concern. Due to the activity of pesticides at very low concentration, their measurement almost certainly requires extraction. Identification may be accomplished by the application of chromatographic techniques on the extract.

Inorganic Chemicals – Generally, inorganic chemicals will be of very little concern in estuarine pollution inasmuch as its natural salt content may be relatively high and controlled by water movement from the open ocean. Certain heavy metal ions from metal plating wastes may be of concern, however, due to their low level toxicity. Cyanide ion may likewise be of concern in some instances. Sensitive and specific techniques, wet chemical and instrumental, exist for the detection of most inorganics although total cyanide may require pretreatment to destroy its metal complexes.

The metals themselves, if present at very low levels, may require pre-extraction procedures. Strong acids and alkalis are other types of inorganic pollutants which may cause local areas of low or high pH. Through the use of pH electrodes, this parameter is very easily and reliably monitored.

Solids in Suspension – Sediment from land erosion or significant quantities of suspended solids from industrial wastes may contribute to physical impairment of water quality. In this regard it may act by rendering the water sufficiently turbid that light penetration is decreased to the point that photosynthetic primary productivity is significantly reduced.

Also solids held in suspension by the relatively high velocity of rivers and streams may settle in the low velocity estuaries thus covering the habitats of bottom dwelling organisms. The relative level of suspended solids may be determined by light scattering measurements, adaptable to routine and continuous monitoring, whereas absolute levels generally require filtration and mass measurement. Further physical, chemical, and biological analyses of sediments may be performed on filtered samples where desired.

Radioactive Substances — Waste streams from certain mining operations or nuclear facilities may contain significant levels of radioactive isotopes from a number of elements. In such instances routine and continuous monitoring for radioactivity should be an absolute requirement at the source and throughout the affected environment. Fortunately instrumentation for monitoring of radioactivity is available and adaptable to routine application.

Thermal Pollution — An increasing amount of cooling water is being returned to the environment at elevated temperatures, making thermal pollution and its effects a matter of concern. While much study is necessary to determine the actual effects of thermal pollution, the measurement and monitoring of temperature itself, through the use of thermistors, are within technical capabilities. The wholesale effects of thermal pollution, however, are not as well understood. With large nuclear power stations, thermal pollution of coastal waters may be one of the most significant parameters to be delineated.

Infectious Agents – This category includes those microorganisms (including certain bacteria, fungi, protozoa, or viruses) from domestic sewage or animal wastes which are capable of producing disease conditions in humans. Viruses may be particularly serious due to their greater resistance to disinfection and difficulty of detection. Routine methods for detection of pollution conditions generally require microbiological culturing techniques for the more numerous, yet nonpathogenic, coliform bacteria present in untreated domestic waste. Routine culturing for the pathogens might not be done unless severe pollution or an outbreak of disease were encountered. These methods are unsuited to a monitoring system requiring rapid data handling and response but some form of reliable monitoring will undoubtedly continue to be a requirement for pollution characterization from the sanitary standpoint.

The Use of Biological Indicators of Pollution: In water pollution, as in many other areas today, engineering finds a complicated problem, the answer to which is partially hidden in the enigma of the ecosystem. Those individuals who specialize in the study of biological organisms find that the conversion of their knowledge into terms the engineer can use is often difficult.

Within the science of biology even biologists find no ready answer, but yet continually publish observations and suggested methods of which those presented in this survey are but a representative sample. The biological literature appears to support sufficient material and methods for utilization today. The real problem lies in the separation of significant methodology from those which are limited in scope, redundant, or outdated.

In general the development of meaningful biological indices of pollution will be predicated on the following:

(1) Because of the relatively vast quantities of water in estuaries and the ocean as compared to rivers and lakes, estuaries and nearshore waters possess introduced materials in much lower concentrations. Greater mixing forces serve to increase dilution and dispersion.

(2) Many physical properties are also different, especially those which serve to concentrate suspended and dissolved material (noticeably so in estuaries).

(3) Perhaps the two greatest natural concentrating mechanisms are precipitation and marine organisms. The results of these two processes are bottom deposits and accumulation in animal tissues, most forms being benthic.

(4) Both factors lead to basic design parameters for biological monitoring of the marine environment.

Design of Marine Biological Monitoring:

(1) Bottom deposits and associated organisms

- (a) Organisms associated with bottom deposits are divided into three categories:
  - (1) Epifauna (to include reefs)
  - (2) Infauna
  - (3) Suprafauna (motile forms)

The use of sedentary forms as monitors of pollution is well documented. For obvious reasons the use of motile forms create a problem of greater magnitude, especially the sampling aspects that category three can be reduced to a minor consideration.

(b) Depending upon the geological facies one of the two remaining groups will assume dominance, both in terms of number of species and number of individuals. The epifauna will be in greatest number on rocky surfaces and other surfaces which afford attachment, and the infauna will be greatest on soft, sedimentary and sandy bottoms.

(c) Two more categories can be created, depending upon major modes of feeding:

(1) Suspension
(2) Detrital

Suspension feeders are buried in sediments or upon the sediment surface, but feed upon the water layer immediately above. Detection of foreign materials involves an analysis of tissues.

Detrital feeders take the sediments within themselves and extract necessary nutrients. Detection in this case involves two kinds of analysis, one of the tissues, and one of the material remaining in the gut after advanced digestion. Important groups of animals to consider are molluscs (pelecypods and protobranchs) and polychaeta annelids in soft bottoms. Important epifaunal groups are molluscs, cirripedes (barnacles) and urochordates.

(2) Organisms previously mentioned, especially the examples fulfill the basic requirements for biological monitoring. Molluscs are a very successful and widespread group, inhabiting almost all marine environment, existing in large communities and serve as excellent concentrating mechanisms via their mode of feeding.

Polychaetes are often ecotypically related to fresh water forms such as Tubifex, a well-known indicator of sewage pollution. Barnacles are known to exist in even the foulest of harbors along our coastlines. The relationship of Urochordates to pollution is not well documented but because of their mode of feeding, research in this area should be extended.

(3) Future research should be in the direction of the relationships of groups of organisms to water and to sediments. Specifically associations should be understood in ecological terms, rather than previous surveys. These associations should be specifically understood as a direct result of interaction with water masses and sediment types. An excellent approach to research in this area is the use of a food chain bioassay.

Character of Coastal and Ocean Waters: Total Organic Carbon — Ocean waters contain total organic carbon on the order of magnitude of 2.4 mg/l with little vertical variation (in Atlantic). This has been confirmed by several investigators. The problem is to determine exactly how much organic carbon there is in the ocean and where and how it is distributed.

Polycarboxylic Acids — Oxalate and citrate (Cu salts) have been measured in

Antarctic bottom water. In this case, the citrate concentration was 0.14 mg/l.

Fatty Acids – The concentration of fatty acids has been found to vary from 0.05 to 0.5 mg/l in the ocean environment. These compounds were generally both saturated and unsaturated compounds of 10 to 18 carbons. Stearic and palmitic acids increased with depth but generally, the higher molecular weight fatty acids and more saturated fatty acids were found at deeper depths.

Coenzymes – It has been estimated that the coenzyme concentration in the ocean is 0.001 part per billion which is outside the capabilities of present chemical techniques. Biological assays or new developments in physical techniques will be necessary; however, many of these substances are only transient in the marine environment. The coenzyme Vitamin B_{12} has been found at a concentration of 5 mμg/l.

Estuaries: The coastal estuaries are subjected to tidal action and fresh water inflow. Because of this their inorganic salts content may vary from 1,000 to 30,000 mg/l. Obviously instruments designed to monitor nonspecific salinity (e.g., conductivity meters) will have no value as pollution monitors in open estuaries. However, such instruments may be used to indicate the degree and nature (type of stratification, for example) of seawater intrusion at a given location for all times. Such data accumulated over an extended period of time are necessary for estuarial modeling. Other instruments helpful in this regard are current meters and tidal depth sensors.

Temperature is also a necessary parameter in physical measurements, particularly where there is potential thermal pollution and where ecological studies are being conducted. The length of time required to gather data for estuarial modeling is indefinite. Certainly some daily patterns would be evident and seasonal cycles would be observable. It is important that inflow volume be known with some degree of certainty during the time.

Also meteorological data, such as wind velocity and direction, can have a bearing on turbulence and surface water movement. The above parameters are all significant with regard to physical definition of estuaries and it should be noted that all such parameters can be measured even though the estuary may be in a polluted condition during this time. The instrumentation for these measurements is within present capabilities for continuous and intermittent monitoring although design modifications may be necessary.

The monitoring of the pollutants in an estuary presents a far greater challenge. The complete spectrum of chemical substances, both natural and synthetic, and a multispecies ecosystem must be considered. As an example of the potential problems involved, specific chemicals may be active agents of extremely low concentration which require very complex analytical procedures for their detection and estimation.

In addition, such compounds may interact with other chemical constituents or with the existing ecosystem in ways that are not well understood. Some aspects of the pollution monitoring program which have minor importance in the physical monitoring system described above are listed below.

(1) Measurements may be made in a previously polluted system at the outset of the monitoring program.

(2) Measurements pertinent to one estuarial regime may not be applicable to another.

(3) Total flow into an estuary is not sufficient. Contributions due to waste outfalls and land drainage must be tabulated.

(4) The character of indwelling and bottom dwelling marine organisms must be determined. This would include study of their susceptibility to various stages and types of pollution, and their ability to concentrate certain pollutants.

Obviously continuous or semiroutine measurement of all possible pollutants is neither practical nor possible. Nevertheless, the capability for measurement of any type of contaminant as the need arises must not only be possible but practical, and some means of quickly locating its source must be available.

Effects of Solid Waste

This is the final report to the New England Regional Commission on a one-year research project on the biological effects of ocean disposal of solid waste. The problem area is extremely large, involving virtually all oceanographic disciplines. In many areas literature review or limited laboratory experiments had to serve where more detailed work may have been desirable.

The greatest value of this work may be the identification and definition of problem areas. Since resources to study the consequences of ocean disposal are limited, it is important to identify these aspects which are of greatest ecological concern and which are amenable to field and laboratory research.

Ocean dumping has been identified as an option for the disposal of both residential solid waste and incinerator residue for New England coastal cities (New England Regional Commission Solid Waste Proposal II).

A study which examined the economic aspects of ocean disposal of solid wastes showed that inland sanitary landfill would be little more expensive than disposal at sea. Incineration was found to be expensive whether land- or sea-based. On a basis of "non-ecological" costs ocean dumping was found to be an attractive alternative.

Two contrasting views concerning ocean dumping and the methods of disposal each engenders were discussed. The first is that the ocean is a link in the natural cycle of materials and that organic material should be dispersed in shallow waters where mineralization will be most rapid. The other is that wastes are potentially harmful and should be segregated from natural cycles by minimizing dispersion and dumping them in physically and biologically inactive areas. Segregation was assumed desirable in a report on deep ocean dumping of baled refuse prepared by the National Industrial Pollution Control Council (1971).

This research project was based on the assumption that containment would be a necessary part of any disposal plan. The problem of floating debris and uncertainty about the concentrations, biological effects, and rates of degradation of waste constituents all argue for containment. The ocean has been used as the ultimate disposal site for a variety of waste products. At first glance, solid waste would seem to be among the least desirable of these. The research discussed in this report attempts to answer the question of the ecological cost of ocean dumping of compressed residential solid waste. Solid waste in compact blocks has very different properties than either dispersed solid waste or other waste products. It is therefore necessary to study this form of waste to determine if it is indeed an undesirable material for ocean disposal.

During the course of these investigations, numerous reports were reviewed dealing with the management of sanitary landfills. It is clear that groundwater pollution by leachates will continue to be a problem in the eastern United States. Proposed collection and treatment of these leachates will make landfills more complex and expensive. An important feature of ocean dumping is that it circumvents this problem. Furthermore, total quantity of substances eluted from waste may be less in the ocean than in land disposal.

Both field and laboratory studies are necessary to determine the consequences of ocean disposal of solid waste. This report is based exclusively on laboratory studies, although type disposal locations were considered as a basis for the projection of results to the marine environment. The objectives of this project were to determine the following properties of compacted solid waste:

(1) The exudation of dissolved materials and their effects on the physiology and behavior of marine life.

(2) The loss of solid material and its dispersal characteristics.

(3) The interactions of fouling organisms and waste blocks.

(4) The interactions of motile animals (such as fish and lobsters) and waste blocks.

A variety of procedures were used in attempting to make these determinations. Scaling down waste blocks was an important problem in the design of these experiments. The heterogeneity and large size of the constituents of solid waste make it difficult to obtain small representative subsamples. Because of limited space and pump capacity, it was necessary to reach a compromise by using a few large tanks and a larger number of small ones. Both flow-through and batch tanks were used. A series of analyses of gases produced by waste were undertaken when it was observed that gas formation was occurring in all experimental tanks.

It was assumed that material to be disposed of would be residential solid waste and that shredding and compaction would precede dumping. The characteristics of three type disposal areas in the New England area were briefly considered as a basis for the design of experiments and for the projection of experimental results. The areas included a shallow shelf station off southern New England, a basin in the Gulf of Maine, and the continental slope and rise off New England.

Shallow areas are better known biologically and more easily monitored than deep areas, although they are close to fishing grounds. There is some evidence of active currents in the slope areas. Although the benthic environment of deep areas of the Gulf of Maine is not well known, this area merits further consideration on the basis of physical isolation.

Test waste was prepared containing milled paper, food, tin cans, aluminum, plastic, and glass. The percentage of paper (72%) was higher than in typical wastes. A number of different observations were carried out in a large flume (16' x 1' x 1') floored with test waste and a control flume floored with sand. A continuous flow of unfiltered seawater was provided in these tanks. Oxygen uptake was studied in smaller flumes (4' x 4" x 4") and waste degradation was monitored in 20-gallon tanks in a temperature controlled bath at 9°C.

Interstitial Solutions and Eluates: A variety of dissolved substances were monitored in water overlying waste and in interstitial water. It was possible to detect dissolved organic carbon, carbohydrates and organic nitrogen in the waste eluate for only a few days after placement of the deposit.

Interstitial solutions were monitored in the large flume and in closed tanks. The tanks contained: (1) high organic matter waste in salt water, (2) high organic waste in fresh water, (3) high organic waste in salt water poisoned with mercuric chloride and (4) low organic waste in salt water.

The same general pattern of chemical changes was found in the interstitial solutions of all biologically active saltwater deposits. There was an immediate drop in pH due to constituents in the paper. Oxygen was consumed in 6 to 8 days. Hydrogen sulfide (H_2S) was produced after about 50 days and continued to increase in concentration throughout the experiment. Dissolved iron concentration was initially high, but decreased to low values after H_2S production began.

Nitrate disappeared at the same time as oxygen. Ammonia concentrations were variable and reached values of 20 to 82 mg/l after 80 days in high organic matter waste. Phosphate reached a concentration of 22 to 32 mg/l after 50 to 60 days and then leveled off. Although carbohydrates decreased to low values, total dissolved organic carbon remained high at the end of 80 days. Turbidity of interstitial solutions increased through time. Variation in duplicate samples taken from single tanks indicated that rates of degradation and quality of degradation products were spatially heterogeneous.

Although temperature and pressure will limit the rate of degradation at an ocean disposal site, waste deposits will become anoxic and will contain H_2S. H_2S will remain high as long as degradable organic matter remains.

The only toxic substance other than H_2S which was measured was ammonia which reached toxic concentrations in interstitial waters. Volatile fermentation products which were not identified could have toxic effects on infauna or behavioral effects on motile animals.

Ammonia levels in the interstitial water of wastes were near those in highly organic natural sediments but phosphate levels were much higher, possibly due to relatively limited adsorptive surface. Diffusion of these nutrients would argue against disposal in shallow areas with restricted circulation.

In order to model the transfer of dissolved substances from waste deposits, it will be necessary to estimate diffusion coefficients. As a first attempt to describe the bulk properties of compressed waste, permeability was measured in test waste with varying densities. The waste was extremely impermeable at densities near those that would be used in actual dumping.

Solid waste contains many toxic metals. An important consideration in proposing ocean disposal is the mobility of these metals in anoxic environments. A review of the literature indicates that several toxic metals (including copper, chromium, mercury, and zinc) form insoluble sulfides in the presence of H_2S. Other experiments, however, have demonstrated the occurrence of these metals in soluble forms in anoxic organic sediments. There is a clear need for more laboratory research in this area.

Gas Production in Solid Waste: Model solid waste beds and slurries suspended in seawater were analyzed for gaseous content. At various stages of decomposition, H_2, O_2, N_2, CH_4, CO_2, and H_2S were detected. Initial gas production activity evolved large amounts of H_2; later CH_4 and H_2S were produced in lesser quantities.

In shallow water interstitial bubble formation is likely with associated buoyancy and ejection of pore water. In deeper water the possiblity of bubble formation, especially involving H_2 and CH_4, should not be discounted. Highly toxic levels of H_2S were characteristic of interstitial water. The possibility of rapid release of this water is a matter which should receive attention in baling considerations and dump site selection.

Oxygen Uptake by Solid Waste: Oxygen uptake was measured in both large and small flumes on waste which had been held in flowing seawater at ambient temperatures for periods of up to four months. During these experiments a volume of water was sealed in each flume, mixed by pumping, and the oxygen concentration measured with an oxygen electrode.

The rate of oxygen uptake was several times higher in the waste than in a control flume floored with sand. These rates were higher than those of most natural infaunal communities measured at similar temperatures. Uptake never exceeded 100 $ml/m^2/hr$, however. Examination of the literature suggests that diffusion rates of oxygen and reduced compounds limit oxygen uptake regardless of organic matter content.

Since these rates are orders of magnitude lower than rates of dispersion in aquatic environments there will be reduced likelihood of low oxygen levels or toxic H_2S levels in overlying water other than in the boundary layer in contact with a waste deposit.

Reactions of Organisms to Waste Deposits: A number of interrelated phenomena were observed in flowing seawater systems containing solid waste. A succession of microbial films developed on the waste surfaces. Eventually sulfide oxidizing genera dominated these films.

Fouling experiments were somewhat unsuccessful because fouling was limited in both waste and control flumes. The slides in the waste tank were fouled by filaments of sulfide bacteria. Nematodes, oligochaetes and harpacticoid copepods were found among these filaments. These groups are resistant to oxygen depletion and the presence of H_2S.

Only two large infaunal species successfully colonized waste deposits. *Capitella capitata,* an indicator species in organically polluted areas, was found in very high densities in a tank with measurable H_2S in the overlying water.

Nereis succinea colonized the large flume, burrowing into waste containing very high levels of H_2S (nearly 400 mg/l). Animals were screened for sensitivity to waste eluate by 72-hour bioassays in eluate flowing very slowly over a large surface area of waste. The most sensitive species tested was a marine fish. The only invertebrate suffering mortality was a shrimp. No mortality occurred in barnacles, hermit crabs, rock crabs, mussels, surf clams, or ocean quahogs. The crustacea were heavily fouled by bacteria and may eventually have died from restrictions of respiratory and feeding movements. The major source of toxicity in the eluate appeared to be H_2S.

The eluate did not stimulate feeding responses in three decapod species tested. It is believed that substances normally responsible for attracting scavengers to food were absent from the eluate. The following projections were made concerning the effect of waste on organisms in the marine environment:

(1) Only infaunal species tolerant to low oxygen levels and the presence of H_2S will colonize the waste.

(2) Both infauna and epifauna will be limited if the oxygen in the overlying water falls below 2 ml/l. This will be a function of rate of oxygen uptake; waste surface area; and the oxygen content, velocity and turbulence of overlying water.

(3) The effect of H_2S on epifauna and motile animals will depend on the dispersion rate in the overlying water. H_2S will be present in waters downstream from the dump site but little information is available on its attraction or repulsion of motile animals.

(4) Newly deposited waste will attract scavengers; however, this attraction may be diminished after a period of months.

(5) Fouling of the deposit may be restricted to species resistant to H_2S. Little fouling will occur in deep waters regardless of the deposit surface environment. It is not possible to state whether wood boring species will be attracted to waste.

(6) Bacteria on the waste surface may provide food for deep sea animals.

MECHANICS OF OCEAN DISPOSAL MONITORING

The following sources were used for the information provided in this chapter.

PB 198 032
PB 213 473
PB 224 793
PB 230 945
PB 233 019
AD 735 378
AD 757 599

The synthesis and evaluation of automatic regulatory dump monitoring systems requires a comprehensive understanding of operational dumping practices and detailed information and characteristics of electronic navigation techniques as well as of the dump vessels themselves, including vessel berth locations, speed, range, dump control specifics, and communication equipments. Additional factors to be considered include dump vessel traffic, type of dump material, existing shore facilities and personnel, and owner/captain cooperation.

An operational automatic vessel monitoring system must rely heavily on existing electronic navigation systems because it must integrate this position information with other events such as time of day, time of dump, duration of dump, and water depth.

OPERATIONAL ASPECTS

Monitoring

The monitoring of waste disposal operations at sea can be subdivided into two distinct, but often overlapping, categories:

(1) environmental monitoring, which deals primarily with the interaction of the wastes and the ocean environment, and
(2) regulatory monitoring, which is primarily concerned with the jurisdic-

tions, regulations, and operational matters involved in ocean disposal.

Environmental Monitoring: The EPA is the cognizant authority for granting permission to dump wastes in specified areas based upon the types and nature of the waste materials.

The objective of environmental monitoring, with respect to ocean disposal of wastes, is to determine both the short- and long-term effects of a given disposal operation. The monitoring results are essential for the generation and revision of water quality criteria applicable to the disposal of wastes, for assessment of trends and changes, and for identifying problem areas that require the attention of cognizant research and regulatory agencies.

Environmental monitoring includes periodic field observation and sampling, supported by laboratory testing and analysis. The objectives of environmental monitoring require a much more complex and costly program than regulatory monitoring, and require specialized personnel and equipment for both the at-sea phase and the shore-based laboratory facilities and staff.

In some cases, regulatory and environmental monitoring programs complement each other in that water quality data obtained as a regulatory requirement by the disposal operator is of value to the environmental monitoring phase of the waste disposal operation. However, this has been shown to be the exception rather than the rule.

The problem of environmental monitoring is complicated by the fact that the waste field developed, either at the ocean bottom or within the water column, is dynamic. It is subject to progressive modification by chemical, physical, and biological processes and, under the influence of currents, the effects of waste discharges in a given area will spread over a continually widening area, or be concentrated in undesired areas.

The problem of monitoring offshore disposal areas is further complicated by the fact that, with the exception of a few localities such as New York City, the wastes disposed at any one site commonly are dissimilar in character because they originate from more than one type of source. For instance, off Galveston, refinery wastes, pesticides, black liquor, and other industrial wastes are all disposed of in the same general area about 100 miles offshore. The frequency of disposal for a given operation in this area varies from weekly to once every two or three months. Thus, depending on the time that the area was sampled, the results could reflect the effects of all of the disposal operations or only that of the most recent discharge.

Another complicating factor is that the waste field developed either within the water column or on the bottom is not static, but is subject to progressive modifications by chemical, physical, and biological processes. Under the influence of currents, the effects of a waste discharged in a given area will extend over a much broader area. The acute effects within the waste field will depend on the toxicity of the wastes, dilution, etc.

An important consideration here is that a drifting waste field, inadequately diluted, can pose a possible barrier to migration and free movement of marine biota. Spawning migrations may be blocked or diverted to unfavorable areas. Similarly, if the waste field entered a heavily spawned area and mixed with waters containing large concentrations of eggs and larvae, conditions unfavorable for survival could be produced. Depending on the scale of the phenomena, this could seriously cripple the survival of a given year-class for valuable coastal fisheries.

Ongoing monitoring of present waste disposal operations (at sea and in the laboratory) for the assessment of environmental impact is clearly insufficient. Even in the case of the Galveston District of the Corps of Engineers, which probably had the most stringent requirements for environmental studies to be carried out prior to disposal authorization, lack of Federal and State funds for environmental control monitoring programs has limited the capability to verify original study results.

Regulatory Monitoring: In regard to regulatory monitoring, survey results and findings tend to indicate clearly that ocean dumping is occurring at locations other than those prescribed by licensing agencies, but the extent and exact nature of such violations are not precisely known because of the present lack of suitable regulatory monitoring systems. Some of the obvious factors producing violations of this type are rough seas, inclement weather, and faulty navigation, all of which contribute to premature dumpings or dumps in other than designated areas.

EPA is confronted with the problem of detecting such violations, whether brought about by somewhat accidental or uncontrollable conditions, or whether they are a result of purposeful neglect or willful disregard of regulations. In either event, the problem relating to detection, the exercising of legal restraints (invoking fine, revoking permits), and the defining of an effective correction action system are myriad. Some of the factors which influence the degree of the problem are:

Weather – Ocean dumping operations are performed 24 hours per day, 7 days per week, throughout the year except where severe weather conditions may jeopardize crew safety or result in vessel/dumper loss or damage. Dumping operations, however, can be expected to be performed over a gamut of weather conditions including fog, drizzle, snow, ice storms, and thunderstorms, as well as in fair weather, with corresponding conditions of temperatures, humidity, sea state, and wind. The weather environments may be expected to influence the operational performance of equipment and consequently become significant in evaluating oceanographic monitoring system approaches and in rating candidate regulatory monitoring systems.

Monitoring Records – From the standpoint of monitoring waste disposal methods and practices in the United States, the maintaining of records is a loosely applied function which falls into three general categories. The first of these is that used in the New York and Norfolk areas where disposal operators are required by the Corps of Engineers to submit regular (monthly, quarterly, etc.) summary data regarding the type and amount of wastes discharged at sea. Annual summaries of the operations are maintained and updated regularly by the Corps of Engineers. Similar data on disposal operations are obtained from disposal operations in the Philadelphia area, although strictly on a cooperative basis, because that particular Corps of Engineers District Office claims no jurisdiction over disposal operations beyond the three-mile limit.

The second type of record keeping situation was found in the cities of San Francisco, Los Angeles, San Diego, Boston, and New Orleans. In these cases, the submission of regular data on individual disposal operations is required by the regulatory agencies, but there is no evidence that the data is evaluated. As a consequence, the responsible regulatory agency in these cities is generally unable to provide an overall description of the disposal operations conducted under its jurisdiction.

The third type of practice is essentially one of no record keeping at all. After initial approval of a disposal operation, no additional records are required or maintained by the regulatory agencies.

New York City is the only major city in the United States in which waste disposals are subject to periodic surveillance while in transit to the disposal site. In order to ensure that operators disposing of sewage sludge, chemicals, dredge spoils, and construction and demolition debris have traveled the required distance to the disposal areas, surveillance is conducted by recording the times of departure and return; this is done by government vessels stationed in lower New York Harbor.

A similar situation existed in San Diego where staff members of the Regional Water Quality Control Board accompanied each load of barreled industrial wastes to sea.

In California and Louisiana, incidental surveillance from aircraft and patrol boats is sometimes conducted by State fishery agencies and the USCG during their normal activities.

Disposal Equipment

Dredges: Dredges fall into two basic categories—pipeline dredges, or hopper dredges. Either class may be equipped with various types of devices for scooping, sucking, cutting, or withdrawing by some other means, the dredge spoil material.

Pipeline dredges are essentially barge-like vehicles of varied sizes, equipped with dredging equipment on deck. The dredge spoil is piped from the dredging site to a disposal site through pipes as large as 24 inches in diameter and often longer than 10 miles. Pipeline dredges, whether self-powered or towed, are usually stationary platforms in operation, but may move at an extremely slow pace while dredging. Hopper dredges (all owned and operated by the Corps of Engineers) are large, self-propelled vessels which contain the dredge spoil on board in various types of hoppers for subsequent disposal elsewhere.

The Corps of Engineers relies heavily on the use of hopper dredges for maintenance dredging. In the continental United States hopper dredges are used for approximately 70,000,000 cubic yards of maintenance dredging annually. The hopper dredges are confined primarily to the coastal zone and the Great Lakes. Although hopper dredges operate on essentially the same principles as pipeline dredges, there are fundamental differences: (1) dredging is accomplished while the vessel is in motion, and (2) the dredged materials are contained on board in hoppers. Hopper dredges are unique in that provision is made for overflowing. This overflow capability concentrates the solids in the hoppers by overflowing finer particles through troughs located in the top of the hoppers. Overflowing results in return of the fine silt-like particles to the dredged site. (In some instances, this may not be desirable because of environmental impact or operating efficiency.)

Recent environmental concerns have resulted in the modification of several hopper dredges to permit direct pumpout for landfill or sidecasting for beach nourishment. Though technically a disposal operation, in the case of beach nourishment the dredge spoil is desirable. Pipeline dredges pumping from offshore areas are naturally subject to wind and wave limitations; submerged pipelines and flexible pipeline sections are occasionally employed.

Disposal Craft: Various types and sizes of disposal craft are used for particular operations and in different circumstances. The dump scows presently used for ocean dumping employ different dump actuating mechanisms and configurations. Some older and smaller scows generally contain six or eight pockets, each of which is equipped with double, gravity dump doors at the bottom. These doors are normally held closed by cables and a ratchet-and-pawl type mechanism. The pawl is released (for dumping) by hydraulic jacks operated by control valves. The scowman manually controls operation of the valves for each pocket dump mechanism.

Some scows are of a hinge-type configuration. These scows consist of a port and starboard section hinged topside (fore and aft) about which the two sections rotate during dumping operations. Large-diameter hydraulic pistons located beneath the fore and aft hinges cause the two sections to bottom separate, allowing the dredge spoils to gravity dump into the ocean. Dumping is activated by a scowman who operates the hydraulic control valves. Dump time is in the order of several minutes.

On some scows, the dumping is activated remotely, controlled from the towing vessel; a scowman need not, in this case, be aboard the scow itself. In some areas (New York, for example) self-propelled dumping scows are used. The self-propelled scows are equipped with manually operated gate valves and the sludge is either gravity dumped or pumped out. The number of pockets on the self-propelled vessels varies from two to six, depending on the type of vessel. Discharge times vary, but average approximately 15 minutes.

Hazardous chemicals and caustic wastes are hauled out to long-range dumping sites (approximately 100 miles) on scows towed by large ocean-going tugboats using, in most cases, Loran A for navigation. The scows are almost always unattended and dumping is performed by remote control from the tug.

On all types of disposal craft, whether self-propelled or towed, there is a significant difference in loaded and unloaded draft. The average difference in draft is 12 feet.

Power is minimal aboard the scows and is used only for dumping operations. On manned scows, the scowman's area is generally lighted by a gas lantern. On the other hand, the self-propelled dumping vessels and, of course, the tugs are equipped with both AC and DC power for radio, radar, and other equipment as well as for general lighting. The majority of the tugs and self-propelled vessels are equipped with various types of navigation equipment, including gyrocompass, radio direction finders, and radar. The long-range tugs have, in addition, Loran A receivers.

Navigation Aids

It is apparent that navigation ability is a significant factor in any comprehensive study of ocean waste disposal practices to insure that material is deposited in the designated disposal area. Some locations such as one shown in Figure 4.1 have visible landmarks that are used for taking bearings in good weather. Other sites, such as one shown in Figure 4.2 are far from shore and require the use of advanced navigation techniques.

The majority of the tugs and self-propelled vessels are equipped with various types of navigation equipment, including gyrocompass, radio direction finders, and radar. The long-range tugs have, in addition, Loran A receivers. An analysis was made of existing electronic navigation aids available on the U.S. coastline. A summary of characteristics and descriptions of these systems is presented in Table 4.1.

Loran A: Loran A is a hyperbolic system of radio navigation available throughout much of the ocean areas of the northern hemisphere. The system employs synchronized pairs of transmitting stations, master and slave, which transmit pulse signals with a constant time interval between them. Loran provides hyperbolic lines of position which are fixed relative to the earth's surface, as are latitude and longitude lines. These lines of position can be crossed with each other or with sun lines, star lines, or other broadcast stations. Special receivers, Loran charts, and tables are required for use of the Loran system. Both ground and skywave signals are used with Loran A.

For the purpose of nearshore work, only the ground wave signals will be considered. The range is 650 to 900 miles during the day, and the accuracy is, typically, 1.5 miles over 80% of the areas covered. Loran A is useful on the Atlantic Coast, Pacific Northwest, Hawaii, and Gulf of Mexico areas. The southwest coast of the United States is limited to a single line of position because of the distance in placement of the slave stations.

A second line of position must be taken using an ADF, console, radar, sun or star sightings. The Loran A system was scheduled for removal and replacement by Loran C and Omega by 1975; a definite date has not been established and service is likely to continue through 1980.

Loran C: The Loran C navigation system is a long baseline hyperbolic, area-coverage system, employing time difference measurements of signals received from three transmitters. It is basically a Loran A system with the following improvements:

(a) Low frequency transmission (100 kHz) for extended range.

(b) Pulse envelope measurement and phase measurement of the radio frequency signal provide a fine time-difference measurement.

(c) Automatic instrumentation aboard the ship can provide a continuous position indication.

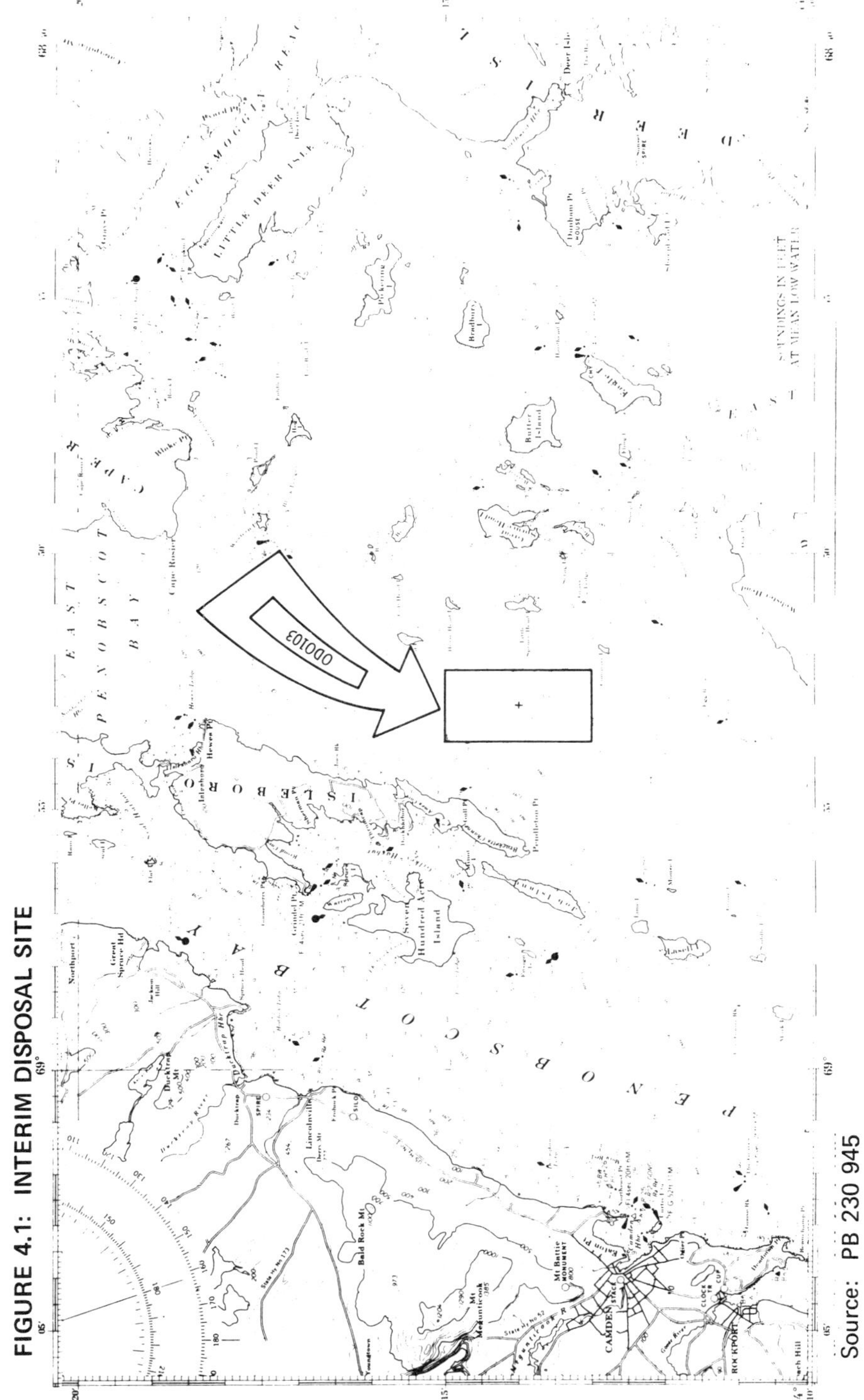

FIGURE 4.1: INTERIM DISPOSAL SITE

Source: PB 230 945

FIGURE 4.2: INTERIM DISPOSAL SITE

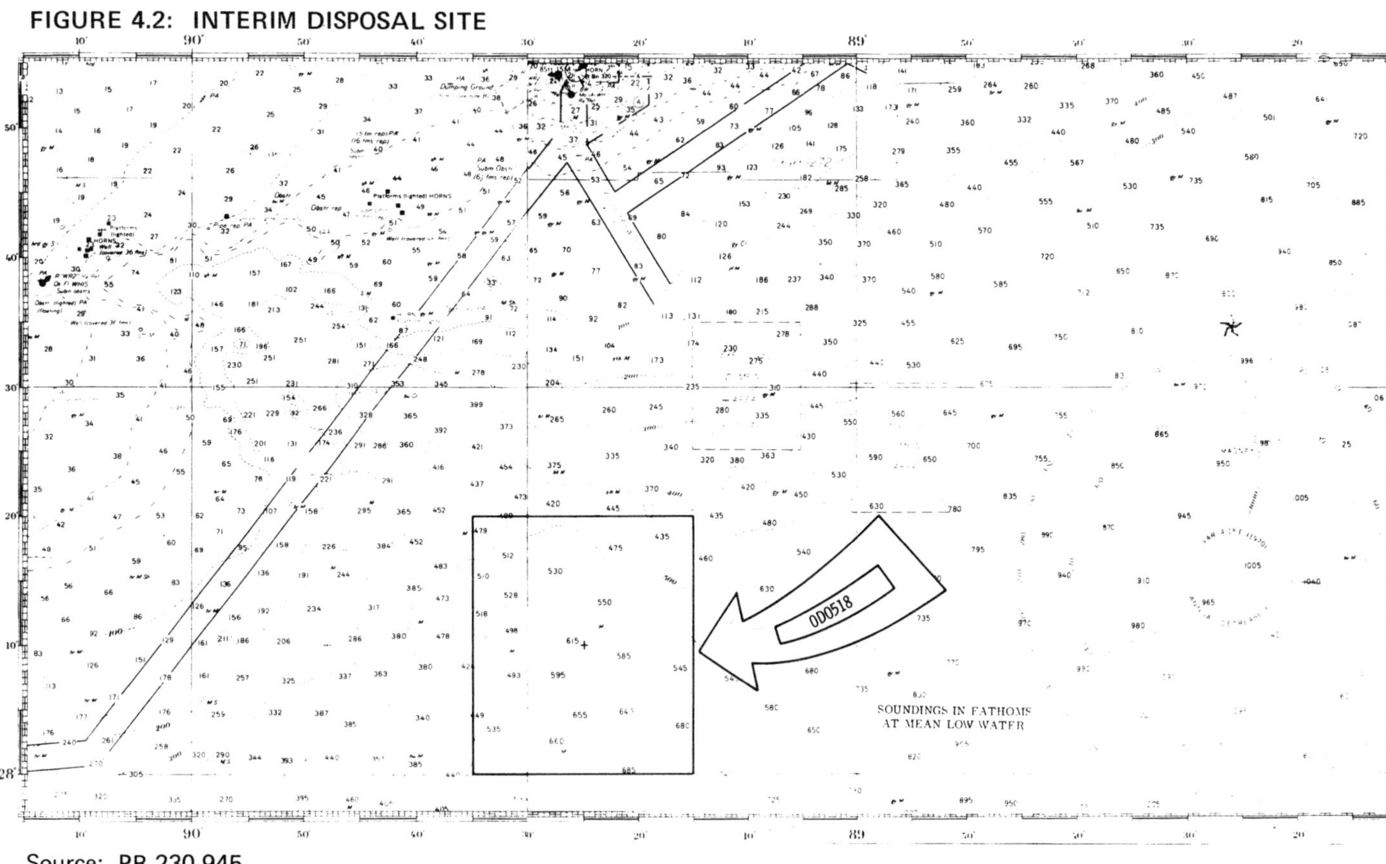

Source: PB 230 945

TABLE 4.1: CHARACTERISTICS OF ELECTRONIC POSITIONING SYSTEMS

System	Loran-A	Decca	Loran-C	Satellite	Omega	Raydist Passive	Cubic Auto Tape	Motorola RPS	Differential Omega	Ontrak II RNAV	VTS Radar
Principle	Pulse	Phase	Pulse & phase	Phase doppler	Phase	Phase	Phase	Phase	Phase	Phase	50 micro-sec pulse
Freq band	1900 kHz	100 kHz	100 kHz	150 MHz & 400 MHz	10.2 kHz	1600 - 4000 kHz	3000 MHz	8000 MHz	10.2 kHz	10 - 20 kHz	9.4 GC
Range (nm)	600	250	1,200	World-wide (hourly)	World-wide '75	350 day 175 night	Radio line of sight 50 mi	Radio line of sight 50 mi	3 - 500 miles of a shore-based rec'ver	World-wide	To 40 mi
Time for fix	1 - 2 min	Instan-taneously	30 sec indexing	10 min (hourly)	Instan-taneously	Instan-taneously	Instan-taneously	Instan-taneously	Instan-taneously	Instan-taneously	Instan-taneously
Sensitivity	500'	50'	50'	60'	400'	1 - ½'	0.1 meter	0.1 meter	100'	60'	30'
Accuracy	1 - 2 nm	¼ nm	¼ nm	100' to 0.1 nm	1 - 2 nm	10	50 cm + range x 10 -5	Soft		1200	0.3°
Freedom from dilution problems	Medium	Medium	Good	Good	Good	Fair	Good	Good	Good	Best	Good
Probable maximum terrain distortion	40' x dilution	800' x dilution	800' x dilution	Velocity & orbital errors 0.1 - 1 nm	10,000 x dilution	40'	Negligible	Negligible	10,000 x 40' dilution	--	Negligible
Skywave error	Negligible	Up to 1 nm (subtle & not easily forecast)	Negligible (if careful)	Negligible	10 nm (but generally pre-dictable to 10%)	Negligible (unless obvious)	None	None	X2.5 miles	Variations stored in memory-aut correction	NA
Percentage of world coverage	15%	5%	15%	100% (hourly)	40% now 100% '72	Special radio location only	Line of sight	Line of sight	40% now 100% '75	100%	Line of sight

(d) Higher average transmitting power and phase coding of the multi-pulse groups allows station identification and discrimination between ground and skywaves.

The ground wave coverage of Loran C extends to approximately 1,200 nautical miles. The accuracy is, typically, 1,500 feet over 95% of the coverage area. Loran C may be considered useful on the Atlantic and Gulf Coasts, but not on the West Coast of the United States. The Coast Guard presently has a budget request of more than $5 million to replace the approximately 30 Loran A stations with 11 new Loran C stations and modify six others to cover the continental coastline and southern Alaska. The existing Loran C system has 8 chains containing a total of 31 transmitting stations. Several of the existing Loran C stations are located in Southeast Asia.

Omega: Omega is a long-range radio navigation system utilizing phase difference measurements of 10.2 kHz carrier frequencies received from each of two stations whose transmissions are phase-synchronized. Hyperbolic lines of postion of constant phase difference with the stations lying at the foci of the hyperbolas provide position fix at the intersection of the two lines of position. The accuracy of the fix is proportional to the LOP angles of intersections, 90° being optimum, characteristic of CW phase measuring systems. Cyclic ambiguity causes isophase LOPs or lanes every eight nautical miles.

To increase the lane ambiguity these stations cyclically transmit the CW waves at several frequencies, i.e., 610.2 kHz, 13.6 kHz, and 11.33 kHz with a 0.2 second off-period between each transmission. With a two-frequency receiver, the resolution or lane ambiguity increases to 24 nautical miles, and with a three frequency receiver improves to 72 nautical miles.

Generally, with some minimal dead reckoning navigation equipment aboard the vessel, lane ambiguity is easily resolved. The propagation of Omega signals conforms to the earth/ionosphere wave guide which is a diagonally varying dimension along the propagation path. This variation in ionospheric height produces an effective variation in propagation velocity, which must be compensated for as a function of time and approximate position to assure predictable phase comparison.

The variation predictions known as skywave corrections have been tabulated based on a prediction model as a function of time and day for a specific location. Results of measurement programs have shown an operational accuracy of one to two nautical miles rms depending on time of day. Improved accuracy is possible using a differential Omega approach.

Differential Omega: In the differential Omega concept, a remote Omega receiver at a known geographic location is utilized to correct certain unpredictable propagation anomalies thereby resulting in improved fix accuracy. It is assumed that the Omega receiver used for position fixing is experiencing the same unpredictable variations as the remote Omega receiver at the known location and, hence, suitable corrections may be determined and applied to correct the data received by the actual navigation receiver. This approach removes time-dependent errors and increases accuracy repeatability to approach the relative accuracy of the two receivers operating in a simultaneous environment. Experimental data obtained with differential Omega shows an improvement of 4 to 1 over a conventional Omega system with an average LOP error of 4 to 7 centicycles at night and 1 to 3 centicycles during the day (1 centicycle equals approximately 1 microsecond, which is 150 meters on the baseline).

Decca: The British-developed Decca system is a hyperbolic radio navigation system which utilizes low-frequency (70 to 100 kHz) CW transmission signals from a master and three slave stations to provide a position fix. Each station transmits a stable CW frequency signal with a fixed relationship to the frequencies of the other three stations. Phase comparison of the signals produces hyperbolic LOP where the phases are equal. Typical frequencies transmitted would be as follows:

(a) Master station 85 (6F)
(b) Red slave station 113.333 (8F)
(c) Green slave station 127.500 (9F)
(d) Purple slave station 70.833 (5F)

These frequencies are multiples of frequency F which in this case is 14.166 kHz. The receiver incoming frequency signals are multiplied by factors to produce frequency differences for the stations which are either 30F (purple), 18F (green), or 24F (red). These differences are measured by a phase meter of the continuously integrating type (deccameter) which indicates total and fractions of cycle that the receiver passes through.

Instrument accuracy is on the order of 1 to 50 of a lane corresponding to five yards along the baseline. The Decca system utilizes a lane identification technique for solution of the lane ambiguity problem. Each station transmits, in addition to its fine fixing signal, a lane identification signal by a second transmission at specified intervals. This technique, coupled with a comparison of the F frequency for each of the three phase comparison systems for half a second, reduces lane ambiguity to 1/100 of a lane.

Practical coverage for Decca is limited to about 200 nautical miles because of continuous wave propagation and skywave contamination. At this time, a Decca system is in operation on the East Coast of the United States. California has proposed the installation of a Decca system on the West Coast; however, this has not been finalized. The Decca system requires special receivers leased from the British-owned Decca company.

Radio Direction Finding: The use of ground-based radio direction finders for fixed locations has been utilized for many years. In this system, transmissions from the vessels are received at two shore RDF stations from which bearings to the vessel are measured. The two bearings uniquely fix vessel location. The basic principle of direction finding (DF) is the measurement of differential distance to the transmitter using a loop or Adcock type antenna.

Currents are generated in each vertical segment of the loop, induced by vertically polarized transmissions, when the loop is 90° to the direction of the arrived signal. Many types of RDF antennas are produced, but the Adcock type is perhaps most attractive for a shorebased RDF station. In its simplest form, this antenna consists of two vertical antennas connected to a receiver. Operation is similar to the loop antenna, the null indicating signal direction. Because of the size of antennas utilized in the 400 kHz to 3 MHz range, physical rotation of the antenna is not practical and a goniometer in conjunction with four or eight antenna towers is used. The goniometer is an instrument consisting of two sets of windings at right angles to each other with a central rotor which, in effect, translates the received radio field at the antennas into a miniature magnetic field in which the rotor operates.

The angle output of the goniometer rotor then provides the direction of the transmitted signal. Accuracy of an RDF system depends not only on instrumentation errors, but also on external error factors such as phase interference effects, polarization errors, tilt of the ionospheric layer, and site irregularities. In a modern RDF system, bearing accuracies of $\pm 1^\circ$ with calibration corrections are possible. At night with skywave contamination the accuracy may vary 2° at 100 nautical miles, and as much as 4° at 500 nautical miles.

Radar: Vessel location using a shorebased radar is determined from the time elapsed between transmission and reception of a radar signal (range) and the radar beam antenna directivity (bearing). The operation principles of a radar in its simplest form utilize a transmitter which generates high-power, short-duration pulses which are radiated in a narrow beam by a parabolic reflector which is rotated mechanically or electrically in azimuth. When the pulse strikes the target a small amount of power is radiated back to the antenna and amplified in a receiver. The receiver output is displayed in a pulse position indicator (oscilloscope).

The radial scan is generated in synchronism with the transmitted pulse rate and a rotary scan with the azimuth rotational rate. This causes a spot to be illuminated on the PPI scope in which the distance and azimuth are proportional to the true position of the target. When the target is cooperative, a secondary radar (radar beacon transponder) can be used which reduces power requirements of the radar transmitter and reduces clutter by utilizing different frequencies.

Modulation techniques can be incorporated on the beacon to provide identification and other coded data. The frequency of radar operation varies depending on range, environment, and accuracy required. Generally, radar range accuracy, which is primarily a function of pulse duration and display resolution, is on the order of 1,000 feet, and bearing accuracy, which depends on azimuth beam width, is less than 1°. Because of the high operating frequencies of radar systems, line-of-sight limits range coverage.

Variations of the radar system are the Cubic Auto Tape and the Motorola range positioning system (RPS). Both of these systems use a shipboard interrogator and two shorebased transponders. The accuracy of RPS is 50 feet at 50 miles, and the auto tape claims accuracies of 6.4 feet at 30 miles. These systems are limited to line-of-sight, and range in cost between $40,000 and $90,000.

Vessel Traffic Radar: San Francisco Vessel Traffic System (VTS) – The Ports and Waterways Safety Act of 1972 (PL 92-340) gives the Department of Transportation the authority for the development, administration, and operation of Vessel Traffic Systems in U.S. ports and harbors. The U.S. Coast Guard is the agency responsible for carrying out this function. The San Francisco Vessel Traffic System is one of the first such systems to be put into service, with eventual coverage of the deep draft waterway system of the San Francisco Port complex, including the bay tributaries extending north and east to Sacramento and

Stockton, south to Redwood City, and seaward approximately 20 miles. The San Francisco system incorporates the functions of the Coast Guard experimental facility formerly known as Harbor Advisory Radar (HAR). The Vessel Traffic Center, operated continuously by Coast Guard personnel, maintains communications with vessels via vhf fm radiotelephone and monitors the position and movements of larger vessels by shorebased radars and position reports.

A traffic separation scheme is now being implemented to separate vessels traveling in opposite or nearly opposite directions. Future developments will include a Coast Guard operated Vessel Movement Reporting System for the Sacramento and San Joaquin Rivers, and a navigational safety summary broadcast, similar to present weather reports, primarily for the benefit of small vessels.

The Vessel Traffic System is a voluntary system of vhf voice communications used in conjunction with a high-resolution radar surveillance system and computer to track traffic in and out of the San Francisco Bay. The system is used out to 8.8 miles, and operation is similar to an air traffic control system. The Coast Guard operator identifies a target on the radar through voice contact; he then enters the data into a computer which stores the information and tracks the vessel through the bay. The primary purpose of the system is safety.

It is intended as a pilot program for several nationwide systems. Presently, Puget Sound maintains a voice-only operation while New York, New Orleans, Houston, Chesapeake, and Delaware are in the planning stages for voice and radar systems. The San Francisco system will use two radars, one at Yorba Buena Island, which would be the Vessel Traffic Center, the other at Point Bonita. VHF radios will be located at Point Bonita, Concord, and Yorba Buena Island. Radios will operate on the vhf fm Channels 13, 16, 18, and 21.

An existing Coast Guard computer will be utilized with special programming. The radars are of a special design for the Coast Guard and feature horizontal, vertical, or circular polarization and digital output. The computer is designed to store seven basic charts of the Bay area, which will be displayed to the operator on 17-inch CRTs for easy identification and tracking. When a ship is logged into the system, it is identified by pilot name, name of the ship, ship registration number, Coast Guard assigned number, and destination.

Information is then stored and the ship is tracked to its destination, in port or to sea, automatically. Provision has been made for automatic handoff from radar to radar to the computer. The radar display is photographed every three minutes for permanent records. This system is meant for use with large ships; pleasure craft are not to be included although the system has sufficient resolution.

This system appears to be an excellent candidate for control of offshore dumping out to 20 or 40 miles depending on the elevation of the radar transmitters. The addition of a radar transponder, with auxiliary sensor inputs to indicate the time of dump, would make up a complete monitoring system. This system is to be implemented in New York, New Orleans, Houston, Chesapeake, and Delaware so that the majority of the major dump sites will be covered.

Satellite – The Navy Navigation Satellite System (NNSS) is a worldwide all-weather system from which accurate navigational position fixes can be obtained using the data transmitted from five orbiting satellites, four tracking stations, two injection stations, the U.S. Naval Observatory, and a computing center. Any number of user navigational installations can exist with no interference between them. The Navigation Satellites are placed in circular polar orbits about earth at an altitude of approximately 600 nautical miles. The orbital planes of the satellites have a common point along the earth's rotational axis. Each satellite orbits the earth approximately every 107 minutes.

The geometrical placement of the orbiting satellite allows an earth-bound observer to cross directly under the satellite twice daily. Typically, the observer receives data from the satellite twice each time he is near the orbit, because the satellites appear to traverse longitudinally as the earth rotates. The earth rotates 27° longitudinally per satellite pass. At the equator, about 20 daily fixes are possible. Realistically, about 15 daily fixes can be realized. In Los Angeles (34° latitude) an average of about 29 passes is observed daily, of which approximately 20 provide usable fixes.

The utilization of the satellite data to compute a position fix is similar in concept to any hyperbolic positioning system where the satellite simulates the multiple transmitting stations by its inherent motion relative to the user. The shipboard equipment required to use the NNSS consists of an antenna, preamplifier, satellite receiver, computer, and teleprinter. This equipment must be installed in an environmentally controlled area.

A position fix using this system permits latitude and longitude on a worldwide, all-weather basis to an accuracy of 40 meters rms from a single satellite pass. Because satellite passes are available on a 1 to 2 hour interval, dead reckoning between satellite passes, using manual or electrical inputs from the ships' speed and heading sensors, is required. A typical shipboard installation of this system will cost between $30,000 and $65,000.

Raydist – Two basic Raydist systems are the DR-S which is an active system utilizing a mobile or shipboard transmitter, and the Raydist T which is a passive system utilizing four shorebased systems. The Raydist system generates dual hyperbolic/hallop coordinate geometries which provide a high degree of operational flexibility and simplicity. A sensitivity of 1.5 feet at a geographic position accuracy of better than 10 feet make Raydist suited to applications ranging from general navigation to precision electronics survey.

The Raydist system can be used out to 150 miles without losing its accuracy. Operationally, a CW signal at approximately 3 MHz is transmitted from the vessel. A reference signal is simultaneously generated at the shore station at a frequency equal to one-half the mobile transmitter's carrier frequency, plus or minus 200 Hz. The shore station reference frequency is doubled and heterodyned with the received signal from the mobile transmitter (ship) to obtain an audio beat note of approximately 400 Hz.

To obtain red range, the audio tone generated at the red base station is returned to the mobile installation together with the base station reference signal. This is done with minimum use of frequency spectrum by incorporating the audio tone as single sideband modulation on the base station reference carrier. The audio tone information is extracted from the received signal on the vessel, and the base station reference is again doubled and heterodyned with the mobile

CW signal within the navigator. This locally generated audio tone has precisely the same frequency as the one derived at the base station, and the two tones exhibit a phase relationship proportional to distance between the vessel and the base station. The two audio tones are then applied to a precision electromechanical phase meter to obtain red range. This process is repeated with the green shore station to obtain the green range coordinate.

In the Raydist-T configuration, the CW mobile transmitter is placed ashore, establishing a baseline with respect to the red shore transmitter. For optimum coordinate geometry, a second CW transmitter is positioned to form a baseline with the green base station so the four stations form an approximate rectangle. The resulting independent hyperbolic baselines provide an easy-to-interpret hyperbolic geometry.

Three independent lines of position are available from the four-station arrangement, providing a convenient means for automatic or manual lane identification. This feature can be used for lane determination by ships approaching outside the coverage area. The Raydist signals are completely continuous and tracking response can be made extremely fast permitting the systems use in high performance aircraft and rapidly maneuvering vessels with negligible accuracy degradation. The major drawback of this system is the limited number of installations in the U.S. This system, however, is available for lease anywhere in the world.

VLF Area Navigation – Ontrac II is a receiving/computing (RNAV) system which uses existing vlf navigation and communication transmitters providing worldwide navigation. The basic principle of operation is that all vlf stations are phase stable and can be used to generate hyperbolic lines of position, which the built-in computer converts to latitude/longitude. Unlike Omega, which utilizes phase-synchronized transmitters, this system measures the phase difference at the beginning of a run and stores the information in memory. Six receivers are operated simultaneously, allowing automatic selection of the best three signals. A built-in minicomputer looks at the phase difference (arrival times) and displays the changes in latitude/longitude and speed as the vessel is underway.

To use the system, the operator must enter (through a small keyboard) his point of departure latitude/longitude in degrees, minutes, and seconds; date; time; destination latitude/longitude; and any way-points he may desire. The unit then displays latitude/longitude of his present position; heading and distance to destinations (five way points); speed and time to destination; time (GMT); and left/right track. All of these displays are available as external outputs. Special tables and charts are not required to operate this system to its stated accuracy of 1,200 feet.

Another feature of this device is a dead-reckoning mode. Because the device has its own computer and memory, it is able to remember its last position and, in the event of a complete loss of signal, will continue to compute heading, position, and speed based on its last known inputs. The price of the system is estimated to be under $20K. The unit is small in size, operates from 24 Vdc, and requires no special installation. This system is a strong candidate for navigation/control of the dumping vessels.

The level of documentary monitoring of operational practices of ocean disposal seems to be inadequate for the number and scope of disposal operations. In connection with regulatory monitoring, inspection of disposal operations is inadequate. There should be monitoring by means of an automatic, tamper-proof vessel log similar to that used by airlines and trucking firms.

MONITORING SYSTEMS

Establishing an Information System

The establishment of an Ocean Disposal Information System (ODIS) is of paramount importance. A well-designed ODIS will provide effective communications; control and management; storage and dissemination of data; and generation of reports.

A consolidation and grouping of several agencies from the six geographical areas surveyed during the ocean disposal study would be the first logical step in a plan.

Several categories of information are essential to the ocean disposal program. These categories contain data and information pertaining to:

Permits
Disposal Sites
Regulations, Criteria, and Administration
Enforcement
Research and Monitoring

Permit Data: Pertinent ocean disposal permit data should be filed and maintained using the National Pollutant Discharge Elimination System (NPDES) presently under development by the EPA. Permit data is typically a mixture of administrative and technical information pertaining to individual dischargers. The primary sources of this data are the EPA regional offices. Other sources are concerned federal and state regional agencies.

Site Data: Data on individual ocean disposal sites will become quite voluminous and detailed in content, but an anticipated small number of dumping permits to be issued tends to minimize the applicability of an ADP system for storage and retrieval. Thus, a manual or site description catalog becomes useful. Major essential catalog elements for each site are:

(1) A chart and a detailed description of the site location, its surroundings, history, present use, future plans, and type of materials dumped.
(2) A detailed technical description covering physical, chemical, biological, geological, topographical and bathymetrical features.

Sources of these types of information are field studies; scientific literature; marine data banks (such as NOAA's National Oceanographic Data Center), and other similar public and private systems; and research surveys. Users of this data are EPA's national headquarters personnel and the EPA regions.

Regulations, Criteria, and Administration: Ocean dumping rules, regulations and criteria are published in the Federal Register, and various administrative procedures exist throughout the regions.

Enforcement Data: Enforcement data is an important requirement to assure fair and uniform application of the law.

Research and Monitoring Data: Most of the pure monitoring data can be made available through various existing automatic data processing systems such as EPA's STORET or NOAA's NODC systems.

Establishing a Pollution Control System

Quantitative and interpretable data are needed on both municipal and industrial waste discharge characteristics. Waste streams should be metered, sampled, and analyzed to yield statistical descriptions ($\overline{X}$ and S_x) of the significant pollutant mass emission rates (i.e., lb BOD_5/day). The pollutant mass emission rates, hereafter referred to as MER provide the basis for the most important and interpretable waste discharge characteristic, the unit mass emission rate, UMER. The UMER can be defined as the MER divided by the rate of waste generating activity.

The UMER adds significantly to the important pollutant mass emission rate, MER, and it can be computed easily provided accurate information is obtained on the level of waste generating activity. For municipal waste streams the waste generating activity unit is the individual; that is the UMER is the MER divided by the tributary (connected) population.

With industrial wastes the waste generating activity is some convenient unit of production, one unit of which is 10^3 lb of specific product. It should be noted that the product should be identified and should be the direct result of the waste generating activity. A typical form of expression for UMER for both municipal and industrial wastes is shown below:

	Primary Effluent	Settled Refinery Waste
$UMER_{phenol}$	1×10^{-4} lb/capita-day	5×10^{-3} lb/10 gal gasoline

UMER data are of particular value for all waste discharge descriptions and especially for industrial wastes. A few of the significant reasons for obtaining such data can be summarized as follows:

1. It is possible to compare the performance of various types of municipal waste treatment systems and to make reasonably accurate cost-effectiveness analyses of the treatment system.
2. The magnitude and variation in the UMER provides the best practical estimate (short of an expensive industrial waste survey of the municipal sewer system) of the relative contribution of industrial wastes in a combined, municipal-industrial sewer system (the usual metropolitan sewerage situation).
3. The magnitude and variation in the UMER and MER provide the only meaningful check on the level and uniformity in operational control of waste treatment systems.
4. For discrete industrial waste discharges, the UMER provides the only meaningful information available on the level or degree of in-house waste elimination and treatment practices.

5. Data on industrial UMER for comparable or similar waste producing processes in different plants provides a basis for estimating or comparing the relative waste elimination practices in the different plants and should provide a rational basis for focusing needed waste control measures to achieve the lowest cost benefit ratio for waste reduction.

6. Statistically valid data on UMER from an industrial plant with several waste producing processes permits a first-cut evaluation and identification of the principal process sources of significant pollutant emissions using multiple regression analyses with the aid of digital computers. The alternative to this would be to make a MER analysis of each and every process waste stream which may be an expensive project.

Methods: Characterization of waste streams requires that attention be given to three independent but important aspects of the problem. To develop reliable data on pollutant mass emission rates, MER, and unit emission rates, UMER, one must have accurate data on the volume flow rates, F; representative samples of the waste stream; and accurate analytical methods for determination of the pollutant concentrations in the collected waste sample.

As the objective, generally, is to determine the mean pollutant mass emission rate, MER, it is apparent that the mean MER is the product of the mean flow rate, F, times the mean concentration of the pollutant in the waste stream. Determination of the mean pollutant concentration in the waste stream requires collection of a representative sample for analysis by withdrawing continuously a sample from the waste stream (or collection of an aliquot or volume of sample in direct proportion to the work flow rate or volume during the interval of sampling).

The general problem of composite sampling in proportion to flow has not been given adequate attention. In most waste treatment plants the question of flow metering and composite sampling is inadequately monitored. Some installations use only grab samples for analysis. Because of the manner of sampling as well as the time holdup through process for a variable concentration feed stream, the results of such grab samples may be as much as 30 to 100% in error.

An additional complicating factor is the fact that waste streams are rarely, if ever, at steady state; the flow and pollutant concentrations are quite variable, especially with industrial wastes which are most frequently the results of batch type process discharge rather than from continuous flow. Consequently the problem of proportional to flow sampling to obtain mean concentrations of the contaminant in the samples collected is worthy of serious consideration.

The relatively uniform domestic waste stream is not as uniform as some think and its variation must be considered if one is to obtain representative estimates of MER. The normal daily variation in domestic sewage flow rates is from $F_{avg}/2$ to $2F_{avg}$. In addition there is an organic pollutant concentration variation from $C_{avg}/2$ to $2C_{avg}$. Surprisingly the minimum pollutant concentration occurs at the same time that minimum flow occurs and the maximum pollutant concentration occurs during the daily maximum flow period (i.e., $F = 2F_{avg}$). Thus it is obvious that the MER data obtained from grab samples could vary a factor of 4 depending upon the time of sampling and flow measurement.

Metering – To permit collection of samples for analysis on a composite basis, it is imperative that accurate and reliable flow meters be used. Meters with mechanical rotors such as propellor meters often offer substantial maintenance problems. Venturi meters, venturi flumes (PB Meters) and magnetic meters are most acceptable. However venturi meters need to be equipped with pressure cleanout taps for the pressure transmitting lines.

In most situations the meter should be connected to a continuous recorder and totalizer or flow integrator. Integrators that can be set to activate an external circuit or transmit an electrical pulse at any predetermined increment of flow integration (500 gal, 5,000 gal, 20,000 gal, etc) are the preferred recorder integration combinations. As is usual for flow meters an accuracy of measurement of ±3% is preferred with ±5% the outside limit.

Sampling — One of the most neglected phases of waste monitoring has been the attention given to the type, mechanics, accuracy, and precision of waste stream sampling methods. Moreover the representative sampling of waste streams is not a simple task. Waste constituents of concern include not only the substances in solution but also the suspended solids fraction consisting of both fine and coarse material, the major portion of which settles to the bottom and a somewhat smaller fraction that rises to the surface.

The latter fraction commonly referred to as floatable matter consists of oil and grease as well as buoyant particulates of waste origin. Problems associated with the physical and chemical properties of the pollutants add to the general problem of obtaining appropriate volumes of samples in proportion to the flow rate so that the concentration of pollutants in the sample volume collected is representative of the flow weight concentration in the waste stream.

Many near-continuous proportional type sampling systems are available for open channel flow conditions. Most of these samplers employ variable speed pumps or dippers geared in proportion to waste flow rates. The samplers are often followed by some type of flow splitting device to reduce the sample volume to manageable proportions. Most of these samplers have limitations of sample flow rates that affect the representative character of the selected sample.

One of the vexing problems associated with a waste monitoring program is selection of the significant and necessary analyses to be included. The final selection of parameters monitored obviously must depend on the general character of the waste stream, the type of waste treatment, and the character and use of the receiving water. In other words, the exact mix of analyses to be conducted should depend upon the local problem and circumstances.

Nonetheless, it would appear that these should be a near-minimum set of analyses that should be conducted on essentially all waste streams. However, the final selection of parameters or analyses to be included in any monitoring program will be a compromise between what is wanted or needed and what can be included within the limits of available funds. In selecting the parameters to be included in a waste inventory program, a few guiding principles can be stated as follows:

(1) The central core of parameters to be included should be those that are considered absolutely essential in either the waste treat-

ment and disposal planning and design function or are of known or presumed practical significance in interpreting or evaluating the effect on the receiving waters.

(2) Analyses of second level significance can be included considering both the cost of the analyses as well as the interpretability of results.

(3) Serious consideration must be given to the routine addition of newer or less conventional waste parameters at the expense of dropping some of the classic or traditional but seemingly uninterpretable parameters—at least insofar as the design or evaluation function is concerned. Specific examples of some analyses that might be added as well as those that might be deleted from routine waste monitoring programs are as follows:

Likely Additions	Likely Deletions
Toxicity (48 or 96 hr TL_{ms})	Total Solids
Biostimulosity*	Complete mineral analyses**
Floatable matter	Alkalinity
Pesticides (hard, i.e., half life—6 months)	Turbidity
Total heavy metals (gross)	Settleable solids
Threshold odor number	

*Analogous to TL_m. The concentration required to increase selected algal growth rate in chemostat by 50%.

**Most ions such as Ca, Mg, K, Na, SO_4, etc.

(4) Consideration should be given to reducing the number of analyses that measure essentially the same thing or eliminating those analyses that studies have shown for particular types of waste streams that are highly correlated with the selected analyses.

(5) For every parameter or analysis included in the monitoring program, information should be available or developed on the accuracy and precision of both the sampling method as well as the laboratory analyses.

Establishing a Surveillance System

The synthesis and evaluation of candidate systems for the Dump Monitoring System requires a comprehensive understanding of present operational dumping practices and detailed information and characteristics of the dump vessels themselves including vessel berth locations, speed, range, dump control specifics (actuation mechanisms and control), and navigation and communication equipments. Additional factors to be considered include dump vessel traffic, type and composition of dump material, existing shore facilities which may be utilized for the DMS and vessel owner/captain cooperation.

With the above information, system requirements including operational requirements, mission requirements and subsystem performance requirements can be formulated and utilized for evaluation of system approaches and rating of candidate systems.

Operational Requirements: Fleet Composition — The dump fleet for the New York Bight is comprised of both self-propelled dump vessels and towed barges or scows operating from berths in Manhattan, Westchester, Long Island and New Jersey.

Dump scows used for ocean dumping of dredge spoils are of several basic types employing different dump actuating mechanisms and configurations. Older and smaller scows generally contain 6 or 8 pockets, each of which contain double, gravity dump, bottom doors normally held closed by cables and a ratchet and pawl type mechanism. Release of the pawl for dumping is provided by hydraulic jacks operated by control valves located within the scow bridge; the scowman manually controls operation of the valves for each pocket dump mechanism.

Several of the dump scows are of the hinge type configuration; the scow is comprised of a port and a starboard section which are hinged topside (fore and aft) about which the two sections rotate during dump operation. Large diameter hydraulic pistons located beneath the fore and aft hinges cause the two sections to bottom separate thus allowing the dredge spoils to gravity dump into the ocean. Dumping is actuated by a scowman activating hydraulic control valves. Dump time is on the order of several minutes.

On some barges the dumping is remotely activated and controlled from the towing vessel and a scowman need not be aboard the barge. Self-propelled dumping vessels are primarily used for sewer sludge disposal in New York and Long Island. Sewer sludge from New Jersey, however, is carried out by barges. The self-propelled vessels are outfitted with manually operated gate valves and the sludge is either gravity dumped or pumped out. The number of pockets on the self-propelled vessels varies from 2 to 6, depending on the vessel. Discharge time varies but is on the order of 15 minutes.

Hazardous chemicals and caustic wastes are carried out to the long range dump site (approximately 100 miles) on scows towed by larger tugs using, in most cases, Loran A for navigation. The scows are unattended for the most part and dumping is performed by remote control from the tug. Dumping time ranges from 30 minutes to 1½ hours.

Significant draft change from loaded to unloaded states is evident on both the self-propelled dumping vessels and towed barges. On the average a change in draft of 12 feet may be expected.

Navigation equipments found on the majority of the tugs and self-propelled vessels include gyrocompass, radio direction finder and radar. Tugs used for the 100 mile dumps in addition have Loran A receivers.

Power aboard the dump scows is minimal and only used during dump. Scowmen aboard the barge generally use a gas lantern for lighting their quarters. The self-propelled dump vessels and tugs, on the other hand, have both DC and AC power available for radio, radar and other equipments as well as general lighting.

Operational Procedures – Operational procedures utilized by licensed ocean dumping companies are similar but vary somewhat due to the waste material to be dumped, the type and capacity of dumper utilized and vessel berth location. Permits allowing ocean dumping are issued for a specified period of time, depending on job circumstance, and may vary from 1 day to as long as 1 year.

The dumping fleet berths are located on Long Island, New Jersey coastal and river locations and along the East and Hudson Rivers of New York. Operating procedures while enroute to the dump sites depend on the berth location since navigable inland bays and rivers, due to heavy traffic, will necessitate a closer towing distance between barge and tug as well as slower vessel speeds. In addition, on some routes low bridges will require the tug or vessel captain to phone ahead to have the bridge opened upon his arrival at that point.

For the return leg of the dump mission to inland water bays and rivers, the captain may have to schedule his dump mission and time his arrival so that the tide is outgoing to permit control of the barge/tug in the event speed must be seriously curtailed due to heavy traffic. Otherwise the barge may become uncontrolled due to tides and currents.

Present operational procedure necessitates the vessel captain enroute to the dump site to communicate to the harbor supervisor when in the vicinity of New York Harbor. Dumping is generally performed on the move; the permit, however, usually spells out the dump method to be used. Upon reaching the dump site, the captain initiates commencement of dumping to the scowman by tug whistle.

Weather conditions sometimes prohibit dumping vessels from going out to the dump sites. This may occur an average of 8 to 10 times during the year. With sewer sludge, large storage tanks, located at the sewer plant, are typically used for storage until the weather permits dumping operations to resume.

Environmental Requirements – Ocean dumping operations are performed 24 hours per day, 7 days per week throughout the year except for severe weather conditions which may jeopardize crew safety or result in vessel/dumper loss or damage. In the New York offshore area, severe weather conditions which prohibit dumping include hurricanes, severe blizzards or strong Northeasterly winds and gales. Dumping seem to be curtailed on the average between 8 to 10 days per year. Dumping operations, therefore, may be expected to be performed over a gamut of weather environments including fog, drizzle, snowstorms, ice storms, thunderstorms as well as in fair weather with corresponding conditions of temperatures, humidity, sea states and winds.

Mission Requirements: Dump Sites – Within the New York Bight, five dumping sites are utilized:

Acid waste dumping ground
Sewer sludge dumping ground
Cellar dirt dump ground
Mud and one man stone dumping ground
Wreck dumping ground

In addition to the above, a hazardous material dumping ground is provided for the dumping of caustic wastes and other chemicals considered hazardous if dumped near shore. The location of these dump grounds relative to Ambrose Light is given in Table 4.2.

TABLE 4.2: LOCATION OF DUMP SITES

Dump Ground	True Bearing from Ambrose Light	Distance from Ambrose Light (NM)
Acid – Summer	135°	10.7
Acid – Winter	145°	9.2
Sewer sludge	124° 30'	4.5 (10.0 NM to point of nearest land)
Cellar dirt	170°	4.7
Mud and one man stone	190°	4.0
Wreck	168° 30'	14.3
Hazardous material	--	Approx 100NM

Time and Range of Missions – Since loading piers and docks are located at various points in New York, New Jersey and Long Island, the range to present dump sites from these locations varies.

Total mission time, including dump time, varies and depends on vessel berth location, dump material carried, vessel type and characteristics, traffic, and weather conditions. For the nearby dump sites, total mission time may vary between 5 to 8 hours whereas the long range dump mission is on the order of 40 hours. Generally the return trip time is somewhat less than the outbound trip, possibly due to the change in vessel draft or ocean currents.

Locating Dump Sites – The dump vessels leaving N.Y. Harbor and New Jersey navigate along either Ambrose or Sandy Hook channels to the nearshore dumping grounds. The captain generally utilizes dead reckoning navigation (gyrocompass, tachometer and clock) to reach the dump sites and then takes several radar or radio direction finder fixes to locate himself accurately relative to the site. On the long range dump missions, loran is utilized as a navigation aid to and from the dump site. For this mission a N.Y.D.C.E. inspector usually boards the dump vessel; it is required that the N.Y.D.C.E. be notified at least 48 hours in advance of the departure time for the long range dump mission.

A suitable Dump Monitoring System (DMS) would be a basic system embodying loran for navigation and position fixing, an events unit for entering start and end of dump and other significant events, and a printer to provide a written record of dump-related activities. This basic system, called LEPS for Loran Events Printer System, would be augmented by positive dump sensing when appropriate.

In operation, ship position is continuously recorded (every 6 minutes) by printing two lines of position (LOP) from two on-board automatic-tracking loran receivers. Also, the two LOP's are presented to the captain for use, at his discretion, as a navigation aid. The captain presses a button on the events unit at the start of dump and another button to indicate the completion of dump. The events unit is also used to enter other significant events such as "passing Ambrose now," etc. The printer is a 21-channel paper tape alpha-numeric printer.

Appropriate fusing and loss-of-function alarms are provided. The equipment for the basic system LEPS is housed in a single equipment rack which can be table or deck mounted, requiring very little shipboard space. Only four electrical connections to the rack are required. Two for primary electric power, and two for r-f (one antenna and one ground). The simple installation and low weight packaging is a decided advantage of the LEPS, since it permits use of a portability concept for use aboard ships that only occasionally dump and do not justify the investment of a permanently installed system.

An additional feature of the recommended system, SEPS, is that it requires no equipment aboard a towed dump scow. Thus in an operation where any of numerous tugs may tow one or more scows to the dump area where the trip is frequently made, the LEPS, when not in use, could be kept at the scow-loading area and, using the LEPS portability feature, placed aboard the selected tug at the time the tug picks up the scow. Furthermore, the fact that LEPS does provide two loran LOPS for use by the captain as navigation aids (at his discretion) is another decided advantage.

The LEPS includes no equipment for sensing the occurrence of dump and, instead, relies upon the captain to enter the start of dump and end of dump via the events units. This system has many attractive features but certain situations require a more positive dump detection. To accommodate these applications, a draft sensing sub-system would be added to the basic system. When contained on one vessel the LEPS plus draft sensor is called DELPS (Draft Events Printer System).

Examples of such installations are self-propelled dumpers or sophisticated barges with larger on-board crews and significant power generation capability. When the application involves a barge or scow which has crew or power limitations, the towing tug would carry the basic LEPS and the barge would carry a Scow Indicating Draft System (called SIDS).

The problem of monitoring sea-dump operations can be solved in a practical way by employing one of three systems, DELPS or SIDS and LEPS. The recommended systems satisfy all requirements while representing low-cost approaches which are immediately applicable for the present dump sites and require no modification should the dump sites move offshore 100 or 150 miles. Furthermore, the recommended systems can be used in any area covered by loran (for practical purposes, all of continental USA), for the monitoring of ocean dumping of waste material.

Physical Description of Systems: LEPS – When installed there are only two apparent elements to the LEPS, namely the equipment rack and the whip antenna. For convenience in handling during installation and maintenance, there are four plug-in units packaged in equipment rack. The LEPS thus comprises the six units with Unit Designation (UD) 101 through 106 as shown in Table 4.3. The equipment rack, housing in one case the complete LEPS (less antenna), is to be mounted in the wheelhouse. Two different size equipment racks are offered as shown in Table 4.3 to accommodate different wheelhouse space available on different vessels.

TABLE 4.3: UNITS FOR LEPS & DELPS

UD	Component	Qty/System	Approx Size
101	Antenna	1	15 ft whip
102, 103	Loran receiver	2	14" x 9" x 12"

(continued)

TABLE 4.3: (continued)

UD	Component	Qty/System	Approx Size
104	Printer unit	1	9" x 6" x 18"
106	Equipment rack	1	26" x 19" x 20"
	or, Option 1 (for use when less floor space is available):		
106-A	Equipment rack	1	16" x 32" x 20"
	plus Option 2 (added when positive dump sensing is needed):		
107	Draft sensing unit	1	10" x 17" x 5"

DELPS – When positive dump sensing is desired, the DELPS may be obtained from the basic system, LEPS, by the simple addition of the draft sensing unit, UD 107, as shown in Table 4.3. The draft sensing unit comprises eight pressure switches, preset for each installation to switch in sequence at the following fractional parts of full load: ¼, ½, ⅝, ¾, $^{13}/_{16}$, ⅞, $^{15}/_{16}$ and Full.

These switch outputs provide discretes for appropriate recording on one channel of the digital printer. The unit is connected to sea pressure via a ¼" pipe line, which can be made up shipboard of an existing sea cock or to a simple sea chest. No difficulty from sea waves is anticipated, but if necessary a damping pot could be added.

SIDS – When the positive dump sensing is to be obtained from a towed scow, the tug is equipped with the basic system (LEPS) and the scow is equipped with a scow indicating dump system, SIDS. The SIDS comprises the two units as shown in Table 4.4. The draft sensing unit, UD 201, is the same as the DELPS UD 107. The recorder unit, UD 202, is a unit comprising a resistor bank, an analog printer, a "mark now" button and "on-off" switch.

The resistor bank is connected to the switches of the draft sensing unit so that a stepped current proportional to the preset step changes in draft is recorded on the analog printer. The SIDS requires less than 50 ma at 12 v DC nominal and would normally operate from the scow's batteries. However, if necessary, this low power demand can be provided by a rechargeable external battery pack good for at least 100 hours between charges.

TABLE 4.4: UNITS FOR SIDS

UD	Component	Qty/System	Approx Size
201	Draft sensing unit	1	10" x 17" x 15"
202	Recorder unit	1	8" x 14" x 10"

Functional Description: LEPS – A functional flow diagram for LEPS, showing also the electrical interface signals is presented in Figure 4.3. The two loran receivers, UD 102 and UD 103 provide two hyperbolic lines-of-position (lop), to establish CEP vessel position to better than 0.25 nmi. The two loran receivers automatically track the preselected loran A station pairs after the station masters are manually acquired by the Captain (a very simple procedure).

The two lop's are displayed on nixie tubes on the receiver front panels. The two lop's are also transmitted to the printer and are automatically printed every six minutes and every fifteen seconds for two minutes immediately following each event.

FIGURE 4.3: LEPS FUNCTIONAL FLOW DIAGRAM

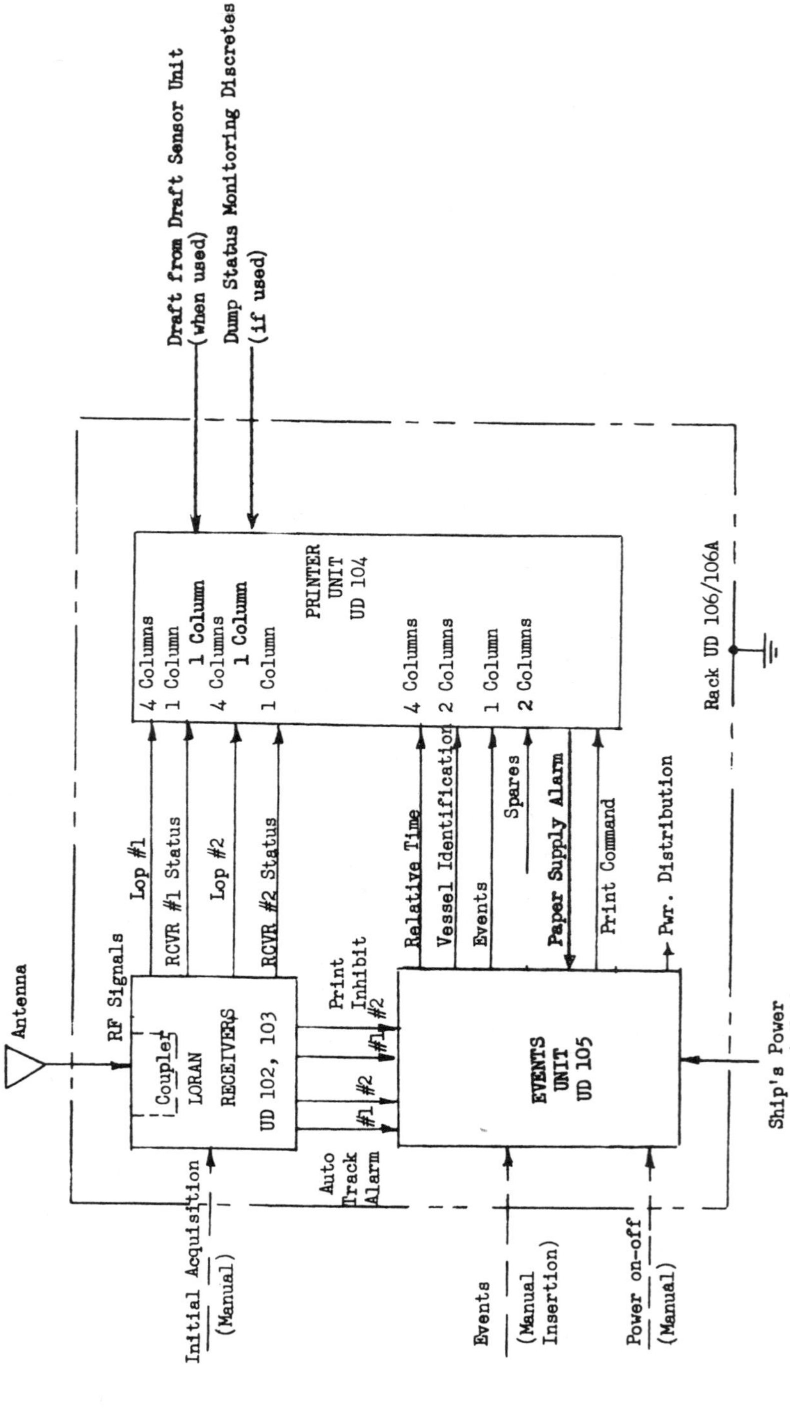

Antenna
RF Signals
Coupler
LORAN
RECEIVERS
UD 102, 103
Initial Acquisition
(Manual)
Lop #1
RCVR #1 Status
Lop #2
RCVR #2 Status
4 Columns
1 Column
1 Column
4 Columns
1 Column
1 Column
PRINTER
UNIT
UD 104
Draft from Draft Sensor Unit
(when used)
Dump Status Monitoring Discretes
(if used)
Auto
Track
Alarm
#1 #2
Print
Inhibit
#1 #2
Relative Time
Vessel Identification
Events
Spares
Paper Supply Alarm
Print Command
4 Columns
2 Columns
1 Column
2 Columns
Events
(Manual
Insertion)
EVENTS
UNIT
UD 105
Power on-off
(Manual)
Pwr. Distribution
Rack UD 106/106A
Ship's Power
11 to 65V DC

Source: AD 735-378

This is timed by an electronic clock in the events unit. The clock is used to transmit relative time to the printer. The events unit also contains ten buttons which are depressed to indicate, and record in coded form, the occurrence of a particular event, such as shown in Table 4.5. Five of the buttons are changeable, preset to accommodate different vessels, routes, etc., and that one button is used to synchronize recorded data of SIDS with LEPS.

TABLE 4.5: EVENTS UNIT BUTTONS

Button No.	Event
1	Leaving Dock Now
2	Starting Dump Now
3	Completing Dump Now
4	Return to Dock Now
5	Passing Fixpoint No. 1 (e.g. Buoy "xx")
6	Passing Fixpoint No. 2 (e.g., Ambrose)
7-9	Additional Customized Events
10	"Mark Now" (for synchronizing with SIDS)

Signals showing status of the two loran receivers are also printed. In addition, loss of automatic track in the loran receivers will actuate a visible and audible alarm in the Events Unit. Appropriate print-inhibit signals are used to prevent printing during any interval when the lop registers are being updated. A visible and audible alarm is also to be provided to indicate an approaching need for replacement of printer paper. The printer is a 21-channel alpha-numeric printer using 3.5" paper tape. The channel allocation is shown in Table 4.6.

TABLE 4.6: PRINTER COLUMN ALLOCATIONS

Column	Function
1 and 2	Vessel identification
3	Events
4 and 5	Spares
6 through 9	Relative time
10	Draft (when used)
11	Status, Loran Rcvr No. 2
12 through 15	LOP No. 2
16	Dump Status (if used)
17	Status, Loran Rcvr No. 1
18 through 21	LOP No. 1

At the end of each dump trip the Captain signs the recorded data certifying its validity and delivers it to NYDCE by Courier or posts it in U.S. mails within 12 hours after return to port, using preaddressed envelopes provided by NYDCE. The delivered data is quickly reviewed for place of dump and for total trip elapsed time to identify suspect dumps for further examination.

DELPS – As previously described, DELPS is formed by adding positive dump sensing to the basic system, LEPS. This is accomplished by connecting the draft sensing unit to sense external water pressure and thus vessel draft. The draft sensor unit comprises eight preset pressure switches, adjusted to throw at preset pressures representing the following fractional parts of full load: ¼, ½,

$^5/_8$, $^3/_4$, $^{13}/_{16}$, $^7/_8$, $^{15}/_{16}$, Full. Note that the size of the steps are less nearer to full load. The outputs of these switches are draft discretes connected electrically to be recorded on one channel of the digital printer, as shown in Figure 4.3.

SIDS – SIDS comprises two units, namely a draft sensor unit (same unit as for DELPS) and a recorder unit. In this case, the draft is recorded aboard the scow by SIDS (this approach permits operation without a data link between the tug and the towed scow). The draft recorded data must be time-coordinated with the LEPS recording.

This is accomplished by simultaneously depressing "mark now" buttons on SIDS and on LEPS at the start and the end of each trip (conceivably at the time the tug hawser is made fast and is released). An additional checkpoint exists at the time of dump when the tug captain depresses the start and completion of dump buttons and a corresponding change in draft should occur on the SIDS recording. (The strip paper drive for the SIDS recording is at a fixed speed within 2%).

Legal Effectivity – The printed data of LEPS provides a complete timed history of the entire dump mission. The continuous timed record of vessel position with indications of specific locations at specific times (events) would be very difficult to fabricate or to manipulate or to tamper with. In addition, the print-out is in English language (with only a simple coding for events) so that the Captain can make an intelligent review, thereby making his signature more meaningful. Accordingly, the data usually will be a true representation of the dump trip.

It may be noted that, when only LEPS is used, the occurrence of dump is indicated only by the Captain depressing the corresponding events button for start of dump and completion of dump. Accordingly, the integrity of the Captain is relied upon and, of course, this always can be challenged. He might purposely misrepresent or, making a human error, he may forget to depress these events buttons at the correct time if at all. The review of such data would then simply look for the farthest traveled point, (generally the turnaround point) as determined by the two LOPs.

It is then assumed that the dump occurred at that point. Certainly the dump did not occur at a further point, since the vessel travelled no further. The dump may have occurred earlier, but this is not very likely since there is normally nothing to be gained by dumping early and then travelling further. However, in some situations early dumps are anticipated (such as a trickling dump made for comfort in heavy weather with the hope of escaping detection). In fact, an early dump might be attempted at times by almost any Captain. For these cases positive detection of occurrence of dump should be made, using DELPS or SIDS.

Should experience show that additional independent sensing of the occurrence of dump is necessary, recording the status of dump control valves, actuators, scow doors, etc., can be added. Printer spare column 16 (Table 4.6) has been reserved for this use to give a separate indication of the time of dump. However, it is felt that for many situations the LEPS alone or with SIDS or DELPS will be adequate without such addition. It is realized that the major purpose of a DMS is not to increase the exchequer by fines won in court cases, but rather it is to control ocean dumping practices. In this regard, the value of the DMS as a deterrent to illegal or improper practices is important. The systems defined are believed to provide as strong a deterrent as would be provided by more sophisticated, more costly systems. Of course, an empty black-box will initially provide a strong deterrent too (that is, until the Captain learns that the box is ineffective in catching violations). The defined preferred systems provide a rather high confidence that violations will be detected and thus should retain their value as deterrents.

It is expected that suspect dumps will be resolved more often out of court than in court. It is likely, however, that the Captain involved will be given strong warnings, such as might occur, for example, if a particular Captain delivered data establishing a pattern of abnormally frequent loss of loran auto track (and thus no valid data for such trips). The NYDCE could add weight to the impact of their warnings by barring the renegade Captain from dump vessels upon the second warning; barring the vessel/owner from dump activities upon the third warning, etc.

Finally, it is suggested that the fact that not only is the relatively high-cost monitoring of ocean dumping required to help protect the environment, but that the ocean may be the only logical place to dispose of much of the waste be kept sight of and, accordingly, that any monitoring activity should not stifle ocean dumping but rather should facilitate the proper disposal of waste in the ocean.

Using Satellite Techniques (ERTS)

In 1972, the National Aeronautical and Space Administration launched the first of a series of experimental satellites, the Earth Resources Technology Satellite (ERTS). This program is a major effort in the application of remote sensing technologies for practical and efficient managment of earth resources. The sensor system installed in ERTS A incorporates a four-channel multispectral scanner system. It has provided some rather spectacular results that can be applied to the monitoring of waste disposal operations from point source and nonpoint source origins in lakes, rivers and oceans.

Figure 4.4 illustrates the overall ERTS system. Systematic repeating ERTS coverage under nearly constant observation conditions has been obtained. The ERTS observatory operates in a circular, sun synchronous, near polar orbit at an altitude of 494 nautical miles (900 kilometers). It circles the earth every 103 minutes completing 14 orbits per day. Complete coverage of any given point of the earth is achieved every 18 days. This provides accurate registration of any point within 20 nautical miles.

A typical one-day ground coverage sweep is shown in Figure 4.5 for the daylight portion of each orbital revolution. The multispectral scanner is a line scanning device which uses an oscillating mirror to continuously scan as the observatory orbits. Six lines within the same band-pass are scanned simultaneously on each of the four spectral bands with each mirror sweep. The track of the spacecraft provides progression of the scanning lines. Optical energy is received simultaneously by an array of detectors in four visible spectral bands from 0.5 to 1.1 microns. In the ERTS B (1973 launch) a fifth band in the near-infrared (thermal) from 10.4 to 12.6 microns is provided.

FIGURE 4.4: OVERALL ERTS SYSTEM

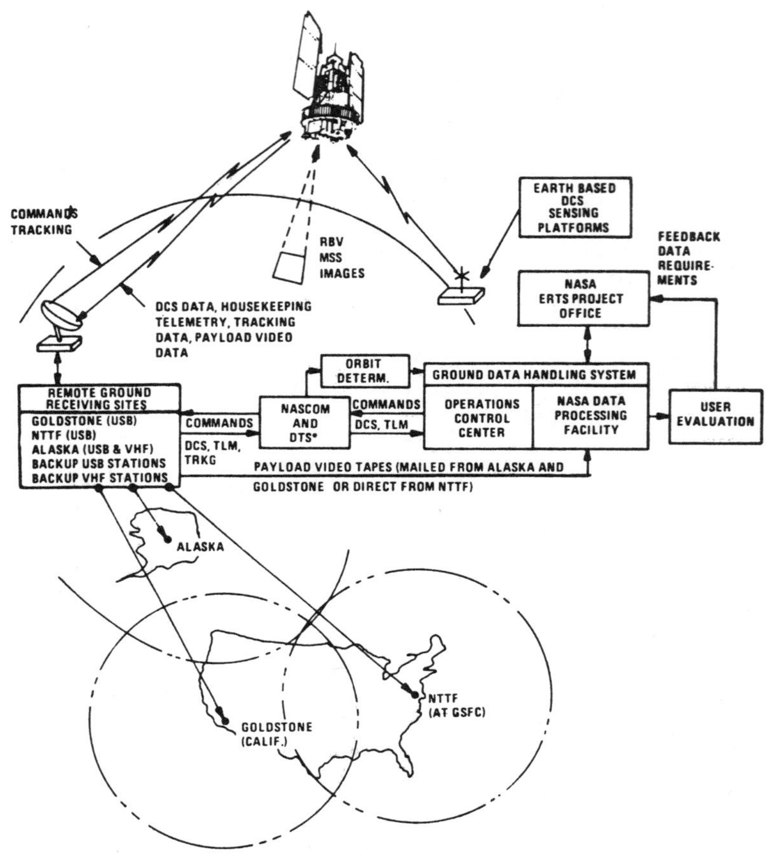

Source: PB 233 019

The outputs of the detector array are digitally encoded and transmitted to earth on a high speed data stream operating at 15 megabits per second. In addition to the multispectral scanner, precision color photography is available.

Using an actual ERTS photograph showing an ocean waste disposal operation in the New York Bight and plotting a grid of latitude and longitude to provide an accurate ground location reference, a low cost management tool may be derived from data routinely provided by orbiting satellites. It is recommended that the Ocean Disposal Program establish a means of utilizing ERTS data to provide a spot audit of disposal operations.

FIGURE 4.5: TYPICAL ONE-DAY GROUND COVERAGE SWEEP

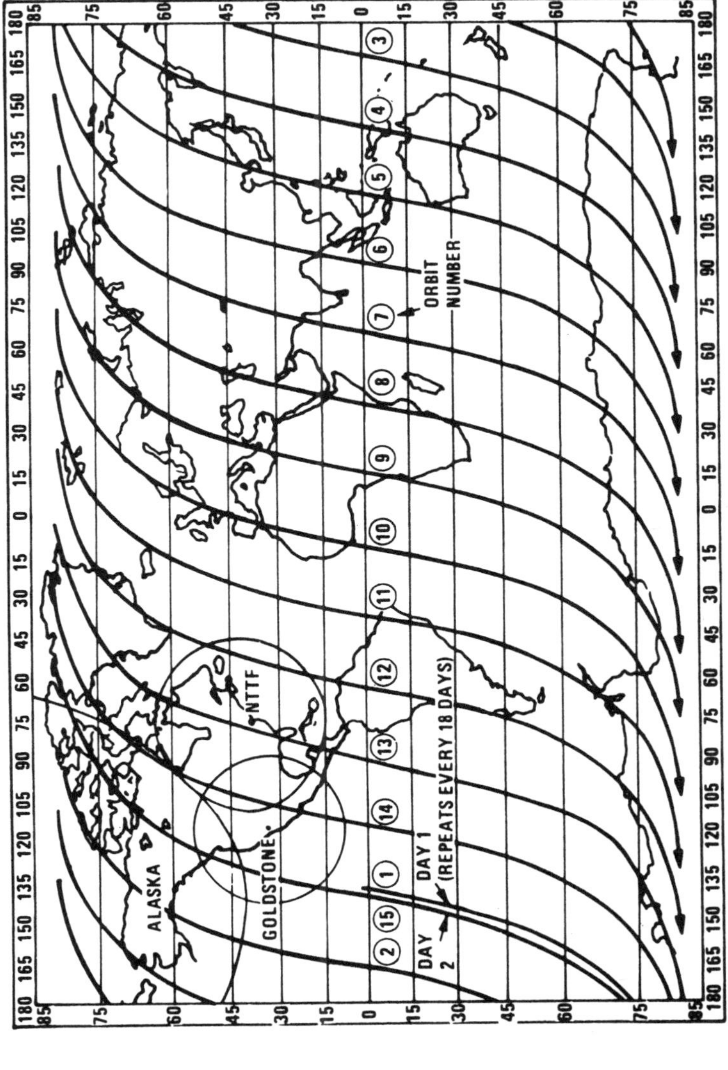

Source: PB 233 019

This would be done by checking the coordinates illustrated in the satellite photograph in a daytime group against the permits issued authorizing disposal that day. These tools would be particularly useful for spotting unauthorized dumping operations, dumping short of the authorized disposal area, and observing phenomena such as international disposal or disposal in the broad ocean area by foreign vessels that may reflect on U.S. disposal operations.

CONTROL CRITERIA

Coordinating Material/Method/Site

Implementation of the national ocean dumping policy requires the enactment of laws and executive decrees that are effective in minimizing pollution and are legally enforceable. Adherence to these regulations must be technically, socially, and economically possible. The regulations must be sufficient and necessary, but not be capricious nor vague. They may contain quantitative standards and limits which are set arbitrarily, but if the values are scientifically derived, these standards are defensible and constitute justifiable criteria.

The following are a distillation of a very large mass of data and provide information on the types of waste materials, dumping methods, disposal sites, compatible combinations, and quantification considerations respectively.

Classification of Waste Materials: Table 4.7 lists material types that are or have been dumped at sea, and includes generalized comparative descriptions of their usual attributes and most common transport and discharge modes. The materials are grouped according to whether they affect the bottom, the water column, or both. The column headings are considered to be the factors which are most influential in determining the fate of a dumped waste and its environmental effect.

Solubility is also an important factor but it is determinable only for specific, relatively homogenous wastes, or for discrete elements of a heterogenous mixture. Thus, it cannot be indicated on a listing of general waste types. The phase and persistence of the bulk materials are partial indicators of solubility. Phase, density and mode are indicators of the horizontal and vertical dispersivity of a dumped waste. Chemical activity, persistence, toxicity, and discharge rate are the prime determinants of effects on the biota of a dump area.

Other combinations and comparisons of these factors are applicable for specific areas, depth zones and/or habitats. There are many local exceptions, especially in the type of transport vessel and some materials may be dumped in bulk or in drum at different times. Chemical activity and biological toxicity refer to their immediate or short-term effects and not necessarily to the effects over the period of persistence.

Disposal Methods: Vessel Types – The vessel type and method of disposal are major factors in determining the initial dispersion, aesthetics, ultimate fate of some wastes and the relative economic advantages of ocean waste disposal. (See Table 4.7 for typical material and vessel combinations.)

Solid Wastes: There are three normal methods used for the disposal of solid

TABLE 4.7: MATERIAL TYPES

Material	Phase	Density	Activity	Persistence	Toxicity	Mode	Vessel	Discharge	Rate
Rock	solid	>2.0	Inert	Permanent	Nil	Bulk	Deck Barge	Overside	Tons/sec
Dredge spoil (sand)	slurry	2.0±	Inert	Permanent	Nil	Bulk	Hopper	Dump	Tons/min
Construction debris	solid	2.0±	Inert	Permanent	Nil	Bulk	Deck Barge	Overside	Tons/sec
Wrecks	solid	>1.0	Inert?	>20 years	Low	Bulk	Tow Barge	Overside	Tons/sec
Conventional munitions	solid	>4.0	Low	Low	Low	Container	Ship	Scuttle	Tons/sec
Radioactives	solid-liquid	>1.0	Low	>20 years	High	Drum	Deck Barge	Overside	Drums/hr
Garbage	slurry	1.0±	BOD	<1 mo.	Nil	Bulk	Dump Scow	Dump	Tons/min
Cannery Wastes	slurry	1.0±	BOD	<1 mo.	Nil	Bulk	Tank Barge	Pump	Gals/sec
Acids/Bases	liquid	1.0+	High	<1 day	High	Bulk	Tank Barge	Pump	Gals/sec
Common salts	slurry	>1.0	Moderate	1 day-1 year	Low	Bulk	Tank Barge	Pump	Gals/sec
Sodium-Calcium	slurry	>1.0	Extreme	<1 hour	High	Bulk	Tank Barge	Pump	Gals/min
Petroleum	liquid	1.0-	Moderate	<1 year	Moderate	Bulk	Tanker	Pump	Gals/hr
Petrochemicals	slurry	>1.0	Low	>1 year	High	Bulk	Tank Barge	Pump	Gals/min
Insecticides, PCB's	liquid	1.0+	Low	>1 year	Extreme	Drum	Deck Barge	Overside	Drums/min
Dredge spoil (mud)	slurry	2.0-	Inert?	Permanent	Nil?	Bulk	Hopper	Dump	Tons/min
Sewage sludge	solid	1.0+	High BOD	<1 mo.	Moderate	Bulk	Dump Scow	Dump	Tons/min
Trash	semi-solid	<1.0	Inert?	1 year	Nil?	Bulk	Dump Scow	Dump	Tons/min

wastes: deck barges with over-the-side dumping; dumping barges with clamshell bottoms for disposal; and hopper barges used in the dredging industry, with valves in the bottom of the barge for dumping.

Deck barges are the most frequently used type of solid waste disposal vehicle, excluding dredge spoil, and for some materials (rock, construction debris, etc.) the most effective means of dumping at sea. This is because the disposal time is long by comparison to other methods, and allows the spreading of materials over a larger area and minimizes buildup in one small area of a dump site. Most dumping is performed by pushing the materials over-the-side with hydraulic jets or a bulldozer. Some barges are available that tip down on one side to allow dumping.

Semisolid Wastes: Dump barges are commonly used for the disposal of semisolid materials, such as sewage sludge, dewatered dredge spoil and other materials with high water content or good flow characteristics. Materials are dumped by the opening of hinged bottom doors. One variation of dump barge has the hinging mechanism at deck level and the entire barge opens like a giant clam shell. This type of barge can dispose of large volumes of material in a short time and is the most economical of the three solid waste barges per unit volume of material.

Dredge Spoil: Almost all ocean disposal of dredge spoil is done by hopper dredges. These are ships with built-in hoppers which have smaller gates at the bottom than the doors of the dump barges. The U.S. Army Corps of Engineers uses hopper dredges to clear entrance channels through offshore bars by pumping sand into the hoppers through suction pipes. When full, they move to the side or to deeper water to dump. The hoppers are cleaned out by hydraulic jets. A few dredges can pump spoil out of the hoppers into a pipeline or a barge brought alongside. Others have sidecasters. The spoil is pumped from the bottom and discharged out the side without using the hoppers.

Liquid Wastes: There are two methods commonly used for the disposal of liquid wastes – by tanker barges; and by containerization and dumping over-the-side. Tanker barges are the most economical method for the disposal of liquid wastes. The removal of several manual steps by pumping results in both smaller crews and fewer vessels required for a given disposal operation. Pumping of the material gives a greater degree of control over disposal rates and by proper plumbing allows dispersion over a large area or into the water column at depth.

Miscellaneous: There are variations on each of the methods for disposal of liquid and solid wastes. In one case the disposal of liquid wastes was performed by breaking individual containers and pouring the contents over-the-side. Hulks of decommissioned ships are frequently scuttled as a method of disposal. And in some cases containerization is used for transport of materials to a site with the container being ruptured by hydrostatic pressure while sinking or by physical methods to promote sinking of the container.

Containerization is usually used for the disposal of materials that are immediately toxic to surface fauna or react with water and could cause harm to the personnel dumping the material. Radioactive materials have been disposed of by this method in the past because of their long-term effects and slow degradation. The objective was to contain the materials over their toxic life span.

Discharge Techniques – The preferred method of waste disposal is generally that which allows the greatest control of distribution. This applies to wastes consisting of large discrete objects such as wrecks, hulks, rocks and containers, as well as liquids, slurries, and fine to coarse granular wastes. Dispersal over a large area or through a large volume of water is usually preferred for liquids and some slurries, whereas concentration over a small area or even spreading or spacing is desired for the particulate wastes.

In all cases, accurate and reliable navigation or plotting is required. The following paragraphs describe the various material/vessel combinations, control capabilities and preferred distribution schemes.

Liquids: For nontoxic and readily miscible liquids or slurries, such as common brines and comminuted cannery wastes, discharge rate control is not critical. They may be dumped from hoppers, tilt tanks or pumped at maximum speed from tanker barges. For fluids which are toxic and degradable, have toxic-like physical effects, or discolor the water, tanker barges with controllable pumps and/or nozzle systems should be employed. Initial discoloration or toxic concentrations may be controlled in the receiving waters on the surface, at depth, or through a given depth range by selection of vessel speed, pumping rate and the number, size and placement pattern of orifices.

If aeration is advantageous, liquids may be sprayed downwind through fog nozzles. Pre-discharge dilution may be obtained using venturi suction intakes. Fluids (or solids) which are persistently toxic should not be dumped in the ocean unless they are: containerized in weighted, extended-life vessels; offloaded from deck barges in a spaced pattern that will ensure adequate dilution volume if rupture or leakage occurs; and into an area where natural or artificial deposition of sediments is occurring at a rate that will ensure permanent burial within the lifetime of the containers. Water depth must be adequate to avoid interference with navigation and fishing and the area must not be a benthic breeding, spawning or nursery habitat.

Large Objects: Relatively inert and dense material such as wrecks, hulks, rocks, and construction debris rapidly sinks to the bottom with no effect on the water column and very rarely has any prolonged effect upon bottom water quality. The size of the individual pieces usually does not allow use of hopper or dump barges except those with very large doors or those of clamshell construction. Deck barges with a small bulldozer or sweep-arm are the most convenient vessels to use because the waste may be distributed as desired, whereas tilting deck barges can only dump the entire load at one spot.

A few benthic animals may be killed during a dump, but the rate of repopulation is sometimes spectacular. This class of waste is frequently used for construction of fishing reefs and the net result is habitat enhancement. The only disposal requirements are adequate water depth for navigation and noncontamination by grease, oil, garbage, or other floating or soluble debris.

Munition filled hulks or containers are a special case. The primary requirement is placement in a deep, out-of-the-way area with warnings on the navigation charts. This automatically places them in an essentially barren habitat. Sealing of the hulks or containers is not required because most ordnance is already in heavy metal casings and even the bare explosive chemicals are usually insoluble.

Mixed Particulates: Garbage and trash are heterogenous, affecting the entire water column and the bottom. Disposal techniques are either the use of a deck for transport and pushing or washing over the side, or by bottom dump barges. Major problems with this waste are the aesthetic unpleasantness associated with the materials that float, and the nondegrading materials now becoming more evident in the common household waste can. Except for the problem of ingestion of physically harmful materials such as zip-top tabs or plastic bag remnants this waste is not generally considered toxic or harmful.

In the study of U.S. coastal areas only one instance of this waste was determined to be disposed of at sea, in small amounts off the coast of Southern California. The present method of disposal is probably acceptable for the volume being disposed, but there is no control over the exclusion of persistent components. Some sorting of materials should be called out in restrictions on this operation. High pressure baling might be used to eliminate any flotsam. The bales could possibly be used in reef construction.

Fine Granulates: The hopper barge or dredge is best used for materials that are in a fluid state. Dump scows and deck barges may be needed for materials that do not flow as readily or have been dewatered. For most areas, the preferential ranking of discharge techniques by vessel type would be:

(1) Hopper barge (or dredge) – allows the control of discharge rate, extent of area and partial control of depth of release.
(2) Dump scow – although dumping is performed in a short time period and not readily varied, it does allow release below the surface of the water.
(3) Deck barge – the time honored method, but causes the most surface turbidity and is the least variable and flexible of the common methods.

The finely divided particulate wastes, principally dredge spoil and sewage sludge, affect the bottom and the entire water column. The major problems with disposal of unpolluted dredge spoil are the burial of benthic biota and the temporary increase of turbidity. Vessels used for dredge spoil disposal offer little advancement in reducing turbidity associated with the disposal process There have been some experiments in the use of sheet plastic barriers and agglutinating chemicals to reduce turbidity, but further research is needed.

Burial is unavoidable, however in some areas it may prove more practical to spread spoil in thin layers over large areas at a rate that allows plants and benthic animals to survive. Only hopper dredges with their speed and gate valves have adequate control to accomplish this discharge technique. Hopper barges would be too slow to cover the area in a reasonable time. Studies have shown that biomass recovery of a buried area occurs in approximately one season and that productivity is even greater afterwards because of the added nutrients. Biotic diversity is reduced by burial and at present the time required to return to a natural community is unknown.

If the wastes are sewage sludge or contaminated dredge spoil, the effects on water quality and benthic habitat are much greater and more persistent. There is need of a trade-off study to determine whether these materials should be handled like degradable toxic liquids, i.e., diluted and dispersed, or if they should be confined to small areas that are simply written off for near term biologic productivity. This decision must be based upon the level and persistence of the toxic effects, the relative economics and habitat value as compared to the cost of alternate disposal methods.

Classification of Disposal Sites: At most places around the margins of the continents, beyond the gently sloping continental shelves, the steep continental slopes descend to the oceanic basins. Figure 4.6 shows the commonly accepted nomenclature for marine environmental zones, both in the water column (pelagic) and on the bottom (benthic). Also shown are the usual relationships of these zones to the major geomorphic divisions.

Table 4.8 is a list of environmental parameters which are deemed important in describing the characteristics of an ocean area in which waste disposal is to be performed. Many of these parameters have been reported using descriptive terms as well as numerical values. Every existing or potential dump site is unique and the values of the parameters listed in Table 4.8 will not only change from one area to another, but many will vary with time, especially seasonally, at the same location.

Valid comparisons between the marine environmental zones may be made only within a specific climatic region for a given time of year. Table 4.9 is an example of an interzone comparison chart for a north temperate region (Cape Cod to Cape Hatteras) in October. By examining a quarterly or monthly series of such charts, the following generalizations for this region may be made:

(1) Width, relief and sediment type do not vary seasonally in any benthic zone.

(2) Temperature, salinity, DO, turbidity, and biota are static in the lower bathypelagic, abyssopelagic, bathyal rise, abyssal and hadal benthic zones.

(3) Bathypelagic waters are cooler and less saline than shallower waters.

(4) Minor seasonal variations occur in the mesopelagic and bathyal slope zones. Water is warmest and saltiest in winter.

(5) Greater seasonal variations occur in the neritic and epipelagic zones. Water is warmest in summer, but still saltiest in winter.

(6) A thin layer of fresher water usually overlies the saline waters of the neritic zone near river mouths.

(7) Neritic waters are freshest and most turbid shortly after the maximum stream runoff period (late spring).

(8) Benthic biota is most varied in the littoral zone, most abundant in the silty, outer sublittoral zone, and both diversity and biomass decrease sharply seaward.

(9) Pelagic biota is most varied and abundant in the neritic zone with a rapid decrease in the oceanic zones where it is almost restricted to the epipelagic and near-bottom waters. Seasonal maximum is in the spring.

(10) Dissolved oxygen decreases with depth and towards land. Seasonal maximum is in the spring.

FIGURE 4.6: MARINE ENVIRONMENTAL ZONES

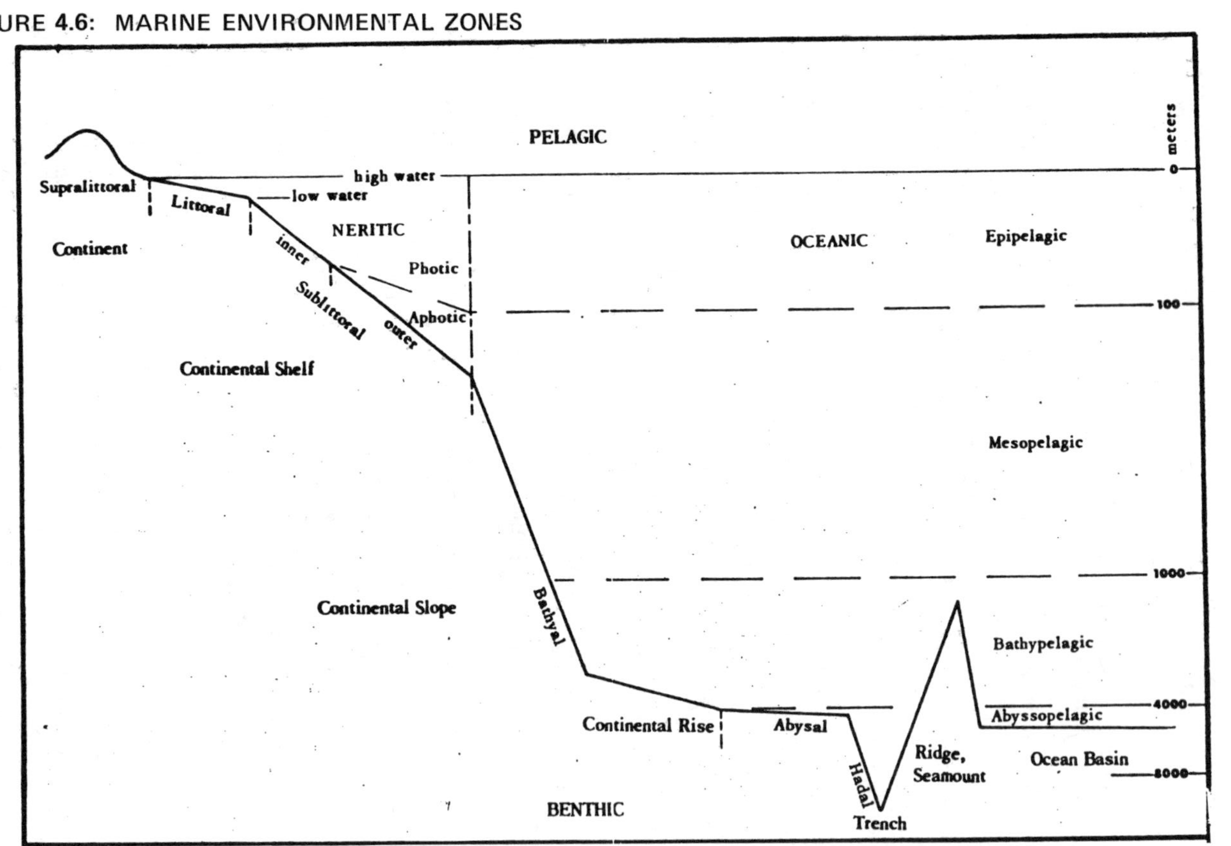

Source: PB 233 019

TABLE 4.8: DESCRIPTIVE AND NUMERICAL DIVISIONS OF ENVIRONMENTAL PARAMETERS

Parameter	Divisions
1. Depth (Meters):	a. Shelf 0-200, b. Slope 200-3000, c. Rise 3000-4000, d. Basin 4000-6000, e. Trench >6000.
2. Relief (Degrees):	a. Flat <0.1, b. Shelving - Undulating 0.1 - 0.5, c. Sloping - Incised 0.5 - 2, d. Steep - Irregular 2 - 10, e. Precipitous - Rough >10.
3. Sediment Type:	a. Lithogenous, b. Biogenous, c. Hydrogenous, d. Meteoritic
4. Sediment Size ($\phi = -\log_2$mm.):	a. Clay >8, b. Silt 8 to 4, c. Sand 4 to -1, d. Granules -1 to -2, e. Pebbles -2 to -6, f. Cobbles -6 to -8, g. Boulders >-8.
5. Sediment Sorting ($\phi = \frac{1}{2}$ (ϕ 84 - ϕ 16)):	a. Well Sorted <1, b. Moderately Sorted 1 - 2, c. Poorly Sorted >2.
6. Sediment Thickness (Meters):	a. Bare 0, b. Veneer 0-1, c. Thin 1-10, d. Medium 10-100, e. Thick >100.
7. Sedimentation Rate (cm./yr.):	a. Very Slow 0 -0.1, b. Slow 0.1 - 1, c. Moderate 1-10, d. Fast 10-100, e. Very Fast >100.
8. Sediment COD (mg/L):	a. Low <250, b. Moderate 250-500, c. High 500-750, d. Very High >750.
9. Sediment Eh (Volts):	a. Oxidizing +0.4 to 0.0, b. Neutral 0.0 to -0.2, c. Reducing >-0.2.
10. Average Annual Precipitation (Inches):	a. Arid 0-20, b. Light 20-40, c. Moderate 40-60, d. Heavy >60.
11. Average Daily Solar Radiation (Langleys = gm-cal./cm^2):	a. Very Low <100, b. Low 100-200, c. Moderately Low 200-300, d. Moderately High 300-400, e. High 400-500, f. Very High >500.

(continued)

TABLE 4.8: (continued)

12. Operational Wind Speed (Knots):	a. Calm <1, b. Light Air 1-3, c. Light Breeze 4-6, d. Gentle Breeze 7-10, e. Moderate Breeze 11-16, f. Fresh Breeze 17-21, g. Strong Breeze 22-27.
13. Wind Direction:	a. Onshore, b. Offshore, c. Up Coast (Nörthward), d. Down Coast.
14. Surface Turbidity (JTU):	a. Very Low 0-10, b. Low 10-20, c. Moderate 20-30, d. High 30-40, e. Very High >40.
15. Bottom Turbidity (JTU):	Same.
16. Net Surface Current Speed (Knots):	a. None, b. Slow <0.2, c. Moderate 0.2-1.0, d. Fast >1.0.
17. Net Surface Current Direction:	a. Onshore, b. Offshore, c. Up Coast, d. Down Coast.
18. Net Bottom Current Speed (Knots):	a. None, b. Slow <0.2, c. Moderate 0.2-1.0, d. Fast >1.0.
19. Net Bottom Current Direction:	a. Onshore, b. Offshore, c. Up Coast, d. Down Coast.
20. Net Vertical Current Speed (Knots):	a. None, b. Slow <0.2, c. Moderate 0.2-1.0, d. Fast >1.0.
21. Net Vertical Current Direction:	a. Upwelling, b. Downwelling.
22. Tidal Current Speed (Knots):	a. None, b. Slow <0.5, c. Moderate 0.5-3.0, d. Fast >3.0.
23. Tidal Current Orbital Eccentricity (Asix L/Axis S):	a. Rotary <2, b. Elliptical 2-10, c. Linear >10.
24. Surface Water Temperature (^{o}C.):	a. Cold -2 to 5, b. Cool 5-15, c. Warm 15-25, d. Hot >25.
25. Bottom Water Temperature (^{o}C.):	Same

(continued)

TABLE 4.8: (continued)

26. Surface Salinity (^{o}/oo = ppt = g/Kg.):	a. Brackish <30, b. Hyposaline 30-34, c. Normal 34-36, d. Hypersaline 36-40, e. Brine >40.
27. Bottom Salinity (^{o}/oo):	Same
28. Surface Water Density (g/cm^3):	a. Very Light 1.00-1.02, b. Light 1.02-1.03, c. Moderately Light 1.03-1.04, d. Moderate 1.04-1.05, e. Moderately Dense 1.05-1.06, f. Dense 1.06-1.07, g. Very Dense >1.07.
29. Bottom Water Density (g/cm^3):	Same
30. Surface pH:	a. Acid <6.0, b. Slightly Acid 6.0-6.5, c. Neutral 6.5-7.5, d. Slightly Alkaline 7.5-8.0, e. Normally Alkaline 8.0-8.5, f. Very Alkaline 8.5-9.0, g. Extremely Alkaline >9.0.
31. Bottom pH:	Same
32. Surface Dissolved Exygen (DO, ml/L):	a. Extremely Low <2.0, b. Very Low 2-4, c. Low 4-6, d. Moderate 6-7, e. High 7-9; f. Very High 9-11, g. Extremely High >11.0.
33. Bottom DO (ml/L):	a. Extremely Low <1.0, b. Very Low 1-3, c. Low 3-4, d. Moderate 4-5, e. High 5-6, f. Very High 6-8, g. Extremely High >8.0.
34. Pelagic Biomass (g/m^3):	a. Barren <0.01, b. Very Low .01-.1, c. Low .1-1, d. Moderate 1-10, e. High 10-100, f. Very High >100.
35. Benthic Biomass (g/m^2):	a. Barren <0.1, b. Very Low .1-1, c. Low 1-10, d. Moderate 10-100, e. High 100-1000, f. Very High >1000.
36. Pelagic Diversity (No. of Species):	a. Monospecific 1, b. Restricted 2-10, c. Varied 10-100, d. Diverse 100-500, e. Highly Diverse >500.
37. Benthic Diversity (No. of Species):	Same

TABLE 4.9: INTERZONE COMPARISON CHART

Mid-Atlantic Coast Profile - October

Zone	Width (n. mi)	Relief (°)	Currents (Knots)	Temp. (°C.)	Salinity (°/oo)	DO (M^l/l)	Biomass (G/M^3)	Sediment Type
Photic Neritic	30		0.2	18°	32	5	1.0	
Aphotic Neritic	25		0.1	13°	35	3	.3	
Epipelagic	--		0.1	16°	34	6	.5	
Mesopelagic	--		0.1	8°	35	4	<.2	
Bathypelagic	--		0.2	4°	35	3	--	
Abyssopelagic	--		0.1	2°		2	--	
Littoral	0.1	0.9					4	sand
Inner Sublittoral	30	0.07		18°			6	sand
Outer Sublittoral	25	0.2		13°			10	silt
Bathyal Slope	40	0.8		8°			2	silt
Bathyal Rise	80	0.2		4°			-	clay
Abyssal	--			2°			-	clay

(11) Except for temporary, wind-driven surface currents up to two knots and the rotary tidal current of 0.5 knot, permanent currents average about 0.1 knot through all zones. Neritic currents flow seaward (SE) on the surface and toward estuaries near bottom. Oceanic currents trend SSW, representing the western side of a large eddy from the Gulf Stream. Velocities of 0.2 knot are common in the bathypelagic zone (countercurrent).

Material/Method/Site Compatibility

Present ocean disposal is of three general types. Inert or nontoxic materials are dumped at shallow, nearshore sites, conveniently located near the waste source, to reduce hauling costs. Dangerous or toxic materials are dumped in deeper water at larger, offshore sites. An intermediate class consists of garbage, cannery wastes, and sewage sludge which have an oxygen demand but are at least partially degradable.

Much dredge spoil, especially from industrial ports, also has a high BOD and like most sewage sludge, has an appreciable quantity of heavy metals and other persistent toxic material. This intermediate class of material has been dumped in both shallow and deep water, mostly dependent upon the required hauling distance. On the Gulf and East coasts, sludge and contaminated spoil have been dumped in relatively shallow water (less than 30 meters). Off California, garbage has been dumped in water over 800 meters deep, but this site was located only six miles offshore.

Depth and/or distance from shore is a good basis for determining whether a particular site is suitable for the disposal of a given waste. Nearly all other location factors, socio-economic as well as environmental, are determined by the depth/distance factor. Assuming that ocean disposal has the lowest priority in competitive usage of a marine area, Table 4.10 is a matrix which combines the material types of Table 4.7 with the environmental zones of Table 4.9 and gives a compatibility rating of good (A), fair (B), or poor (C) for the acceptable combinations.

Ungraded combinations are considered unacceptable because of either exclusive competing uses or needless hauling distance. It must be further assumed that the available area and/or water volume is large enough to assimilate the amounts of wastes to be disposed and that preferred operational practices are employed for each material type.

Based upon this information, it would seem to be a simple task to develop criteria stipulating which wastes may not be dumped into the ocean at all and where and how to dump those wastes that are acceptable. Controversy occurs when one begins to apply numerical answers to such questions as: What is the dividing line, in years (months?) between degradable and persistent? When does turbidity, in Jackson Turbidity Units, become aesthetically offensive? What color should water be? What concentration of "X" is acceptable? What volume should be allowed for a mixing zone (where higher concentrations of "X" are accepted)? Some of these subjective values can be partially obtained by objective methods. For example, bioassays can determine mortality rate for a given organism at various concentrations of "X". This only compounds the question to: Is the death of 10% (5%, 1%, a single individual) unacceptable and was the best organism selected for the test?

TABLE 4.10: MATRIX OF COMPATIBILITY

Material	Photic Neritic	Aphotic Neritic	Epi-pelagic	Meso-pelagic	Bathy-pelagic	Abysso-pelagic	Littoral	Sub-littoral	Sub-littoral	Bathyal	Abyssal
Rock							*	*	A		
Dredge spoil-sand							*	A	B		
Construction debris								*	A		
Wrecks								*	A		
Conventional munitions					C	B				C	B
Radioactives											
Halogenated organics											
Dredge spoil-mud	C	B	A	A				C	B	A	
Dredge spoil-toxic			C	C	B	B				C	B
Sewage sludge											
Sodium-Calcium			C	C	C	B				C	B
Organic chemicals					C	B				C	B
Trash			C	C	B	B				C	B
Acids/Bases			C	B	B	B					
Common salts			C	B	B	A				B	
Petroleum			C								
Garbage		C	A	A					C	A	
Cannery wastes		C	A	A					C	A	

ECONOMIC ASPECTS OF OCEAN DISPOSAL

The following sources were used for the information provided in this chapter.

PB 195 225
PB 204 868
PB 221 684

BARGED OCEAN DISPOSAL

Barge Characteristics

Wastes are discharged from barges either in bulk or containerized form. Containerized methods have been used for toxic, radioactive, and a variety of industrial wastes. Generally the most popular waste container is the 55-gallon steel drum which can be carried to the site and simply dropped overboard. Reclaimed drums have an expected life of 10 years, but this can be extended by filling with a concrete mixture containing the contaminants. Scientists have indicated that the voids in the concrete may result in implosion of the drum and fracture may occur at depths between 100 and 1,000 meters.

Two types of barges, towed or self-propelled, employing either pumped or gravity discharges are usually used for bulk disposal. Until recently the self-propelled barge had been limited to the hopper dredge, however, the "Glen Avon," an automated sewage disposal vessel was purchased by the city of Bristol, England. It has the following features: 900 ton load capacity, maximum speed of 12 knots, discharge time of 15 minutes and low pressure air-gravity discharge system.

These characteristics may be compared with those of the Hopper Dredge (Table 5.1) which operates in the following manner. During the dredging process the bottom material is pumped in a diluted state into hoppers equipped with overflows. The hoppers can be emptied in 3 to 15 minutes dependent upon the volume and nature of the material. Generally, relatively fluid materials are dumped quickly whereas sticky clays and certain granular materials may require the washing of the hoppers with large volumes of water, a process known as monitoring.

FIGURE 5.1: BASIC BARGE CONFIGURATIONS

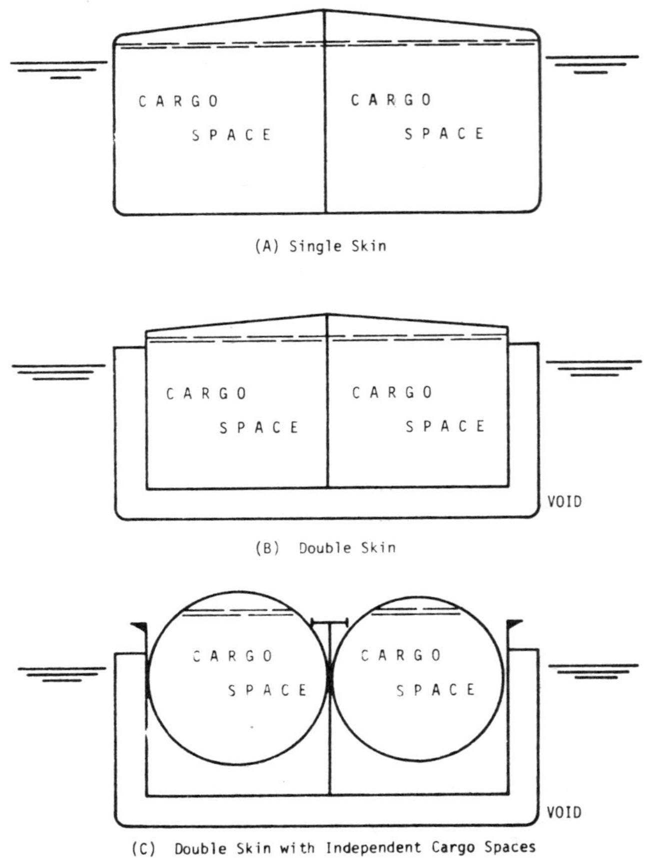

Source: PB 204 868

Towed barges are by far the most numerous and are quite variable in characteristics. They range from the simple bottom release mud scow used in small dredging operations to specially designed, automated tank barges for sewage and industrial sludges, toxic liquids and gases, and pressurized liquids. Three basic barge configurations are illustrated in Figure 5.1. They are (a) single skin, (b) double skin, (c) double skin with independent containment vessel.

Most petroleum products are carried in single skin barges (a) with poisons, acids and cargos requiring heat or insulation utilizing double skinned (b) vessels. The cylindrical tank barges (c) are generally used to carry liquids under pressure, however, it is not uncommon for pressurized liquids to be transported in double skinned vessels. Barges can carry large waste loads and discharge the contents quickly.

TABLE 5.1: SPECIFICATIONS OF CORPS OF ENGINEERS HOPPER DREDGES

Name	Length, beam, and depth	Maximum hopper capacity	Maximum draft loaded	Dredge pumps		
				Number	Size	Horsepower
	Feet	Cu. yd.	Ft. in.		Inches	
Essayons ...	525x72x40	8,270	31-0	2	32	1,850
Goethals ...	476x68x36	6,442	29-0	2	30	1,300
Biddle ...	352x60x30	3.060	24-3$^{3}/_{4}$	2	28	1,150
Comber ...	352x60x30	3,422	24-3$^{3}/_{4}$	2	28	1,150
Gerig ...	352x60x30	3,060	24-3$^{3}/_{4}$	2	28	1,150
Langfit ...	352x60x30	3,060	24-3$^{3}/_{4}$	2	28	1,150
Harding ...	308x56x30	2,682	20-3	2	20	650
Markham ...	339x62x28	2,681	20-0	2	23	650
Mackenzie...	268x46x22	1,656	21-0	1	26	900
Hains ...	216x40x15	885	13-0	1	20	410
Hoffman ...	216x40x15	920	13-0	1	20	410
Hyde ...	216x40x15	720	13-0	1	20	410
Lyman ...	216x40x15	920	13-0	1	20	410
Davison ...	216x40x15	720	13-0	1	20	410
Pacific ...	180x38x14	500	11-3	1	18	340

Table 5.2 summarizes towed barge characteristics readily available on this subject.

TABLE 5.2: BARGE CHARACTERISTICS

(Tons) Capacity	Type of Waste	Average Depth of Discharge	Type	Discharge Characteristics			
				Pipe Size	Number Spacing	Towing Speed	Discharge Rate
		Feet		Inches		Knots	Tons/min.
5,400	Iron-acid	10	Pumped	12	2 @ 50 ft.	8.5	78
	Ore washing mud	10	Gravity	-	2 @ 50 ft	8.5	
3,200	Iron-acid	10	Pumped	12	2 @ 43 ft	6.0	16 - 39
5,000	Chem-Insecticides	10	Gravity	-	-	-	-
1,200	Chlorinated Hydrocarbons	12 below deck	Pumped	8	1	6.0	5
1,100	Chlorinated Hydrocarbons	8	Pumped	4	1	6.0	4
8,000	Philadelphia Digested Sludge	-	Gravity	24	8		267
6,300	NYC Digested Sludge	-	-				210
350	Dredge Spoil	17 - 20	Gravity		Bottom dump	-	100 - 200

Barging Economics

Many factors have been shown to influence the dispersion and dilution of wastes dumped into the ocean. The physical parameters over which some control can be exercised to minimize the effects of waste disposal on the ocean environment are the same parameters which influence the associated costs. These include discharge rate, water, depth, barge capacity and distance to the disposal area. The reported costs and average disposal costs on a $/wet ton basis are shown in Table 5.3, and are representative of the following geographic areas: Philadelphia, New York City, Elizabeth, N.J., Baltimore and Washington, D.C. The costs cited in this section on barging economics are based on 1971 dollars.

TABLE 5.3: REPORTED COSTS OF BARGING OPERATIONS IN $/WET TON

Waste	Total	Pacific	Atlantic	Gulf
Industrial (a) bulk	1.70	1.00	1.80	2.30
(b) containerized	24.00	53.00	7.73	28.00
Refuse and garbage	15.00	15.00	----	----
Sewage sludge	1.00	----	(0.8-1.2)	----

To explore and present a method that analyzes the towing costs for various barge sizes, sludge quantities and towing distances the following assumptions are made. It is assumed that:

(1) The community owns its own barges and contracts for tug services thus removing the hidden costs of profit and overhead margins.
(2) A standby barge, similar in design to the primary barge, is required. The capacity of this barge is set at 1,000 tons.
(3) The availability of tug services is limited to 260 days per year allowing for holidays, strikes, repairs, etc.
(4) Loading costs are included in the operating cost of the facility.
(5) There are three major cost categories, namely:
   (a) Capital Costs
   (b) Maintenance Costs
   (c) Towing Costs

Capital Costs: The capital costs used are based on a figure of $170.00 per ton and represent the purchase of new, specially constructed barges with radio controlled rapid discharge systems. The average discharge time will be taken at 90 minutes. Service life of ocean going barges varies between ten and twenty years. For this example a barge life of ten years will be assumed for the primary barge and 20 years for the standby barge. Annual costs will be calculated using an eight percent interest rate, equal replacement cost and no salvage value or benefits from depreciation.

Maintenance Costs: A maintenance cost of $800.00 per trip for a 1,500 ton barge making six trips per year or a total yearly cost of $4,800.00 was estimated. This is approximately 12% of the annual capital costs of a barge, as described for this example, using the capital cost format itemized. This practice of calculating maintenance costs as a percentage of the annual capital cost is used in the

Portland, Oregon area by several firms. This approach, where maintenance is independent of both frequency and duration of use, is felt justifiable as salt water corrosion accounts for the major portion of required maintenance and repairs, usually accomplished during one annual dry docking. The capital and maintenance are combined to provide one fixed yearly operating cost. Table 5.4 shows the distribution of these costs for this example.

TABLE 5.4: CALCULATED ANNUAL FIXED COST FOR BARGING OPERATIONS

Capacity (tons)	Capital Costs Primary	Secondary	Maintenance	Annual Fixed Cost (Nearest $1000)
1000	25,340	17,330	5,120	48,000
2000	50,680	17,330	8,161	76,000
3000	76,020	17,330	11,202	105,000
5000	126,700	17,330	17,284	161,000

Towing Costs: The greatest amount of variability is expected to occur in the towing costs. This cost can be influenced by discharge rate, tug speed, distance traveled and the existing environmental conditions (i.e., wind, sea state, etc.). Costs in the Oregon, Alaska and Washington areas were found to range between $75.00/hr and $85.00/hr and on this basis an average cost of $80.00/hr was selected as representative of the cost of a tug capable of handling barges up to 5,000 ton capacity at an average net speed of six knots. The towing costs on a per trip basis can be calculated from a simple relationship of the following form:

$$C_t = Tc\left(\frac{x}{v} + t\right)$$

C_t = total towing cost in dollars/trip

Tc = towing charge in dollars per hour

x = round trip distance traveled in miles (point to point)

v = tug speed in miles per hour

t = unloading time in hours.

From this relationship the total costs per trip can be shown to increase directly with travel time for constant discharge times. Solving the above equation for various values of (x) reveals a cost/trip mile relationship as given in Figure 5.2. Assuming a loading time of 3 hr/ton the limiting number of possible trips for a single barge operation can be determined for the stated tug availability of 260 days per year. These limits are shown for this example in Figure 5.3.

The total annual cost (fixed and towing) can be presented by a series of graphs where total annual cost is plotted against the number of trips made in a year for given barge sizes, round trip haul distances, and the total annual tonnage to be disposed of.

FIGURE 5.2: COST PER TRIP MILE AS A FUNCTION OF ROUND TRIP HAUL DISTANCE

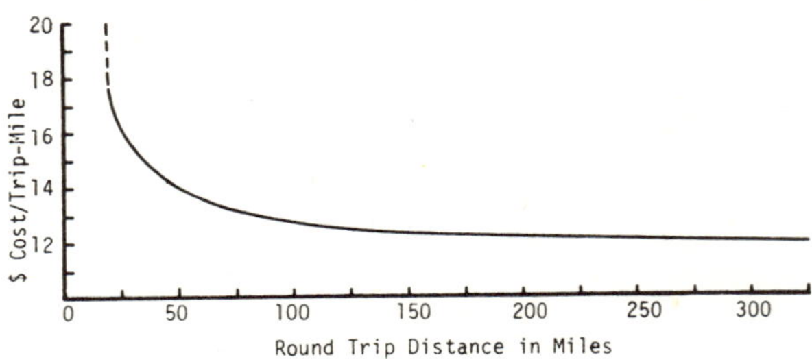

FIGURE 5.3: LIMITING NUMBER OF TRIPS PER YEAR FOR PRESET BARGE SIZES

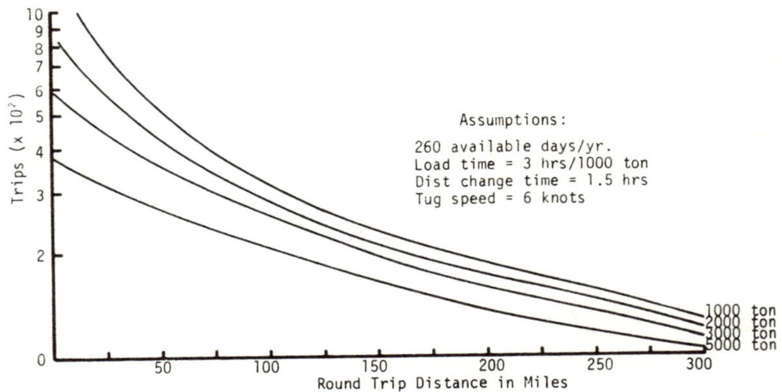

Source: PB 204 868

This method of display (Figures 5.4, 5.5 and 5.6) can provide a method that will aid in the evaluation of alternatives as well as predict the barge size that will accrue the least cost.

The relationships and figures presented in this example can be used to determine many factors, for example Figure 5.4 for x equal to 20 miles shows a 1,000 ton barge to be most economic size for yearly sludge loads that do not exceed 100,000 tons with the 2,000 ton barge economically feasible between 100,000 and 300,000 tons and the 3,000-ton barge becoming the cheapest to operate at 600,000 tons per year. These tonnage figures can be related to population of the community if the percent solids and the percent reduction in solids of the process are known.

Using 250 mg/l SS, 250 gpd per capita, 10% solids and a 33% reduction by digestion it can be shown that a population of one million will produce 365,000 tons of sludge per year.

If the haul distance were increased to 100 miles round trip a 5,000 ton barge would be necessary to minimize the costs. The additional operating costs incurred when the standby barge is needed are also given in Figures 5.4, 5.5 and 5.6.

The average cost shown in Table 5.3 for sewage sludges ranged between 0.8 and 1.2 dollars per wet ton. If an average yearly sludge load of 100,000 tons is assumed with a round trip haul distance of 50 miles, Figure 5.6 can be used to calculate the estimated minimum yearly cost and barge size. This results in a barge of 1,610 ton capacity making 64 trips per year at total cost of $117,000 or $1.17 per wet ton.

FIGURE 5.4: ANNUAL OPERATING COSTS FOR ROUND TRIP HAUL DISTANCE OF 20 MILES

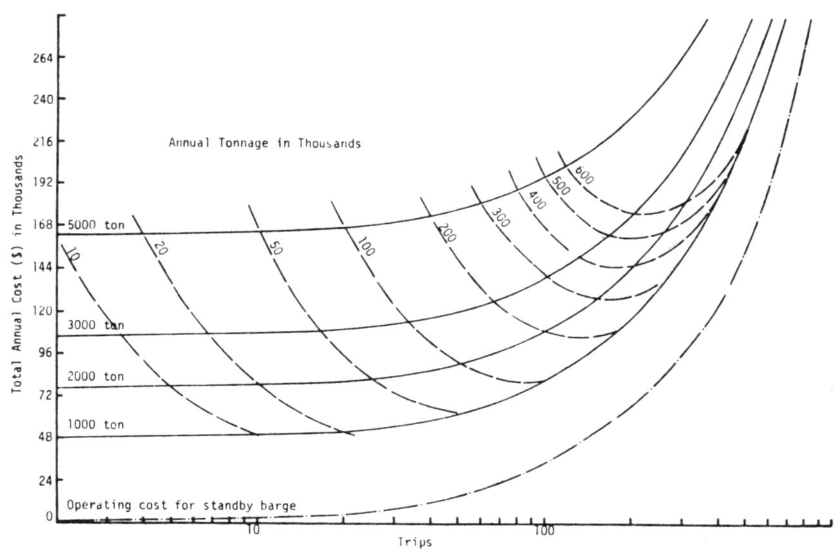

Source: PB 204 868

FIGURE 5.5: ANNUAL OPERATING COSTS FOR ROUND TRIP HAUL DISTANCE OF 50 MILES

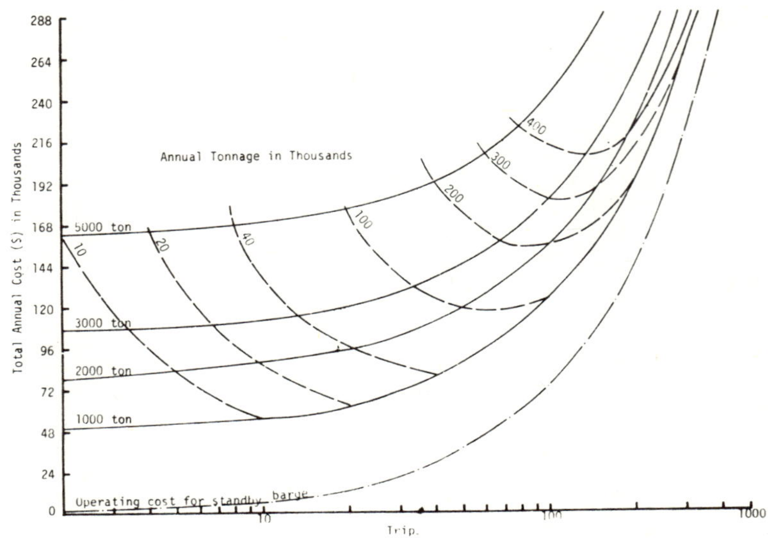

FIGURE 5.6: ANNUAL OPERATING COSTS FOR ROUND TRIP HAUL DISTANCE OF 100 MILES

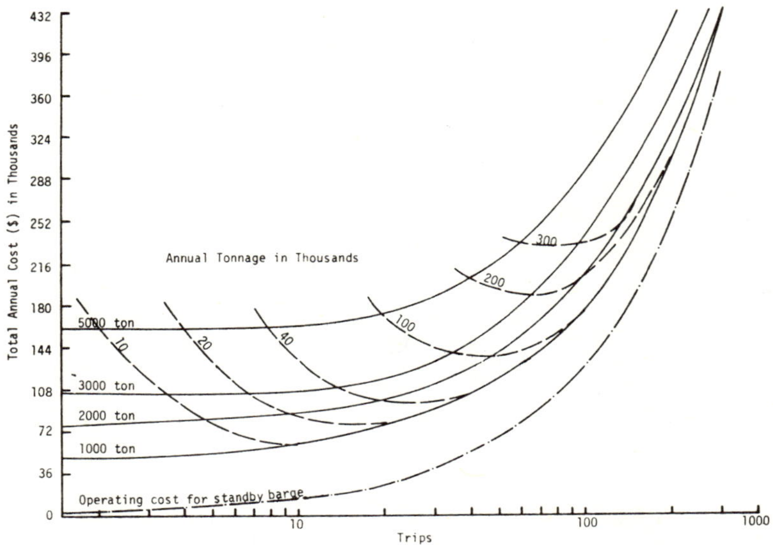

Source: PB 204 868

SOLID WASTE DISPOSAL AT SEA

Solid waste has been defined to include all "material which is normally solid, and which arises from animal or human life and activities and is discarded as useless or unwanted." This report however concentrates almost exclusively on the problems facing the large coastal cities and, in particular, on the problem of disposing of the solids normally found in the refuse collected regularly by these cities.

Given this restricted definition of solid waste, the purpose of this study is to determine under what conditions disposal of these solids at sea becomes economic, where the term economic is interpreted in a sense wide enough to include all the costs and benefits associated with this activity, and not merely those which are reflected in the market. With respect to solid waste disposal at sea, the main nonmarket economic variable is the ecological effect which the introduction of solid waste will produce in the marine environment. The investigations have shown that given the present state of knowledge in this area, it is impossible either to predict reliably the effects of a given dose of solid waste on the marine environment or estimate the values which the public places on these effects.

The report concentrates on a comparison of the market costs of various disposal alternatives and derives through present value analysis unit market disposal costs for sanitary land fill via rail haul, incineration on land, dumping of compacted bales at sea and incineration at sea in a number of situations, pointing out the potential ecological problems inherent in each system, and the relevant available information. A deterministic answer to the question of under what conditions solid waste disposal at sea becomes economic is not provided by the report; it serves to define, however, the critical unknowns and point out the areas where further study is necessary.

In order to provide a realistic picture of the potential for solid waste disposal at sea, the problems faced by a large coastal city will be considered. In the past, these cities have typically relied on coastal landfill. However, they are rapidly running out of politically feasible shoreline sites. New York City is taken as the prototypical situation. The best estimates of the 1970 unit costs of disposal for the New York situation are given in Table 5.5. These estimates assume close-in landfill sites are not available. They purport to cover all the market costs incurred from the time garbage leaves the collection truck to its ultimate disposal. They do not include differentials in collection truck haul distances implied by the different alternative.

These differentials can be quite significant. Total collection costs, cost to the point at which the refuse leaves the collection truck, average $28 a ton in New York City in 1968. And, in general, incineration and barge haul imply longer collection truck distances than rail haul because more rail haul collection stations can be supplied than incinerators or sea transfer stations. All these unit figures are based on operations of approximately the same scale and it is assumed that scale is large enough so that no further economies of scale are likely. These are assumed to be the best of the set of alternatives which they represent.

The most important limitation is that it takes no account of the ecological effects implied by the different alternatives other than through the fact that the incineration on land figure includes pollution control devices sufficient to meet present Federal standards. Hence, the data presented in Table 5.5 is not in itself an argument for or against any of the alternatives, but rather a listing of the premiums that will have to be paid to avoid an undesirable ecological effect. For example, it will cost 56 cents per ton ($7.34 to $6.78) to avoid dumping at sea.

TABLE 5.5: UNIT COST OF VARIOUS DISPOSAL METHODS (Based on NYC Labor, Land, and Construction Costs)

	Cost in $/tons for interest rate i: i=5%	i=8%
A. LAND-BASED		
1. Rail Haul-Sanitary Land Fill*	$ 7.34	$ 7.62
2. Incineration**	10.50	11.00
B. SEA-BASED		
1. Dumping of Compacted Bales		
a. Coastal City***	6.78	7.09
b. City 50 miles inland +	10.61	11.02
c. City 100 miles inland +	10.97	11.37
d. City 150 miles inland +	11.42	11.82
2. Incineration at Sea		
a. Inland Incinerator - Sea Dump	11.46	11.96
b. Water-borne Incinerator	10.89	12.00

* Based on 50 mile railhaul

** Includes pollution control equipment sufficient to meet present federal standards.

*** Based on Westchester to Hudson Canyon; bailing but no packaging. (80 mile ocean tow)

\+ Bailing at inland city, railhaul to coast, and 80 mile ocean tow.

Source: PB 195 225

Ocean Dumping of Refuse

The nonmarket effects of disposal at sea can usefully be grouped under two headings: ecological and sociological, where ecological refers to the effect on fauna and flora that would be measured by the biologist and sociological refers to effects on the ocean which will have a direct impact on humans. The need for this distinction arises from the fact that the effects which, by any biological measure, are beneficial may be valued negatively by humans. For example, distribution of chopped paper in nutrient-poor surface waters may have a very favorable effect on the ambient fauna but be regarded as quite offensive by the human users of the area.

Consideration of the ecological effects as used here implies that one must be able to guarantee that the value of any deleterious effects on plant and animal life due to dumping at sea will be smaller than the savings achieved by ocean dumping (the differential in costs between ocean dumping and the cheapest feasible alternative).

This requires keeping the following two facts in mind. The ecology of the oceans is in a very dynamic and extremely delicate state of equilibrium. This system may be very insensitive to certain perturbations but extremely unstable with respect to others. Second one must consider the long-term effects of the disposal of solid wastes at sea. Even a disturbance which appears to have a beneficial short-term effect may result in harmful changes in the long-term patterns of growth and distribution.

Consideration of sociological effects implies that the value of the detrimental effects of ocean dumping on recreational opportunities and enjoyment, on navigation and industry must also be weighed in any complete analysis of disposal at sea. In this context, it is important to remember that rather large volumes of waste are being considered. New York City alone generates some 20,000 tons of municipal refuse per day.

The ecological and sociological effects of ocean disposal can be controlled by one of two differing philosophies.

(1) An area of small ecological-sociological value is selected for a disposal site. Disposal and containment are handled in such a way that biological effects are as small as possible and are guaranteed to be restricted to the boundaries of the site. In this way, a site of low value is essentially "written off" and hopefully the ecology of the surrounding areas is not affected. In general, this philosophy points to disposal in deep waters well offshore, careful packaging of the wastes, and careful control of the dumping operation to insure dumping at the prescribed site. This is the philosophy which has governed most dumping to date.

(2) The entire open ocean within range of the refuse generation area is divided into classified sectors, with a classification based on sociological and ecological value (e.g., prime fishing area, recreational area, spawning area, harbor area, ecologically rich area, ecologically barren area, area already used for dumping, etc). For each area classification, an acceptable concentration of floating, suspended, and sinking solid waste would be computed. Dumping would then occur in each sector throughout the entire open ocean until the limiting concentration in a sector had been reached. Dumping operations in such a sector would then cease until the ecology had absorbed the waste and thereby lowered the concentration again.

This philosophy points toward dispersed dumping in shallow or surface waters of loosely packaged or ground refuse. It almost certainly implies some segregation of the waste stream to avoid particularly toxic or nondegradable materials. It attempts to take advantage of the ocean's biological activity to return the refuse to the life cycle quickly. In so doing, it will often find ecological and sociological values in conflict.

Description of Alternative Systems: A review of the logistics of several alternative systems which employ dumping at sea follows.

Dumping of Loose Refuse (A1) — This system consists of: trash being delivered by primary collection vehicles to dockside transfer station; loose refuse being loaded into barges and barges being transported to sectors designated for that day's dumping and refuse being discharged into water.

Such a scheme could be conducted only under the "concentration" philosophy i.e., a particular area of the ocean is classified according to its sociological and ecological value, and a correspondingly acceptable concentration of refuse is computed. Refuse is then dumped in the area until the limiting concentration is reached. However, a preliminary estimate of the composition of the incoming refuse is approximately 50% (by weight) paper which brings two opposite situations into play.

(a) Since almost all of the paper products will either float or remain suspended near the surface, a volumetric concentration level becomes meaningless. Only the volume very near the surface will be utilized. Conceding that only the upper 100 ft of the ocean shall be considered for the volumetric concentration, the area required to dispose of metropolitan solid waste becomes enormous.

(b) On the other hand, paper products are generally quite easily decomposed by the physical and biological ocean environment, and rather high temporary concentrations may be permissable.

Thus the conclusion that consideration (a) combined with the fact that the percentage of nonbiodegradable (within reasonable time limits) solid waste is significant, outweighs consideration (b), and that this method is unworthy of further consideration can be reached.

Dumping of Bales Compacted Dockside (A2) — This system consists of: trash being delivered by primary collection vehicles to dockside transfer stations and compacted into bales; bales being loaded onto barges, the compacted bales being dumped by monorail into a barge, the barge loading facility being entirely enclosed within the station; barges being transported to disposal sites and bales being deposited on the sea bed.

A 1969 report prepared for the New York State Pure Waters Authority detailed the specific system necessary to dispose of the solid wastes of Westchester County, New York, however, the economics of the system analysis presented there seem relevant for general application. The operating cycle and costs are representative of the design of a system capable of handling 1,000 tons per day (based on a 16 hr day).

For the study a towing company quoted a flat rate of $2.60 per ton upon the specification of a minimum annual quantity of 264,000 tons from each of three stations. These costs are also contingent upon approval of a dumping site approximately 80 miles off Sandy Hook at the mouth of the Hudson Canyon. A detailed study of barge haul costs together with their effect on overall system costs is presented below.

This analysis is based on a system which has the capability of handling 1,500,000 tons of waste per year. The costs were calculated for a set of three ranges, namely 20, 50 and 100 miles.

It was assumed that 1,500,000 tons of garbage would be generated over a year-long period. This rate would increase at 5% per year. The tug-barge system was assumed capable of operating at least 300 days per year. One tug is capable of towing two barges out to sea. These barges are of the bottom dump type used in dredging and dump automatically. Nontransit time per round trip was as follows:

	3 hours part-time for loading, etc.
	1 hour at sea-dumping time
Total	4 hours nontransit time per round trip

The calculations for each of the given distances follow.

20 MILES: Towing speed = 5 knots

Transit time required =	4 hours out
	4 hours in
Total transit time	8 hours
	+ 4 hours nontransit time
	12 hours per round trip

Each barge has a capacity of 1,300 tons. Therefore, a 2-barge and one tug combination can make 2 trips per day with capacity 5,200 tons per day. The tug is capable of producing 1,800 bhp and towing at 5 knots.

Capital Costs:

1 tug	$ 500,000
2 barges	1,000,000
Total	$1,500,000

Operating Costs:

Tug Operating Costs	$2,000 per day	
Barges: 2 deck hands	240 per day	
Maintenance and repairs	200 per day	(60,000 yearly) / (300 per year)
System Daily Operating Costs	$2,440 per day	

System Annual Operating Costs:

(Daily Costs x 300 days)	=	$732,000 per year
Interest Rate	=	8%
Life of Equipment	=	20 years

Present Value of 20-year Operating Costs	=	$\$7.28 \times 10^6$
Capital Cost	=	$\$1.5 \times 10^6$
20-year System Cost		$\$8.78 \times 10^6$

Discounted Solid Waste Flow at 5% over 20 years = 18.7×10^6 tons

Cost per ton = $0.47 per ton

50 MILES: Towing speed = 5 knots

Transit time	10 hours out
	10 hours in
	20 hours
	+ 4 hours nontransit time
	24 hours per round trip

To meet the goals of a capability of 5,000 tons per day 2 tugs and 4 barges are needed.

Daily Operating Costs:

$2,400 per tug x 2 tugs	=	$4,800
$220 per barge x 4 barges	=	880
Total		$5,680 per day

Capital Costs:

2 tugs (1,800 bhp)	$1,000,000
4 barges (1,300 tons)	2,000,000
Total	$3,000,000

Operating Costs Discounted at 8%:

For 20 years	$\$16.7 \times 10^6$
Capital Costs	3.0×10^6
	$\$19.7 \times 10^6$

Discounted Garbage Stream at 5% = 16.7×10^6 tons

Cost per ton = $1.05 per ton

100 MILES: Towing speed = 7 knots requiring larger tugs

Transit time	14.5 hours out
	14.5 hours in
	+ 4 hours nontransit time
	33 hours per round trip

Operating 300 days per year = 7,200 hours per year.

Therefore, one tug-two barge system can make 219 round trips a year. Assuming that on each round trip, it carries 2,500 tons to be dumped, one such system can carry 550,000 tons per year. Therefore, three such systems are need to meet the requirements of 1,500,000 tons per year.

Capital Cost of System:

3 tugs	$3,000,000
6 barges	3,600,000
Total	$6,600,000

Round Trip Operating Costs for One Tug-Two Barge System:

Tug costs per round trip	\$3,300
Barge costs per round trip	660
	\$3,960 per round trip
	x 220 per round trips per year
	\$8.7 x 10^5 per year
	+ 0.8 x 10^5 per year
	\$9.5 x 10^5 per year
	x 3 such systems
System Yearly Operation Costs	\$28.5 x 10^5

Annual Operating Costs

Discounted at 8% over 20 years	=	\$48.0 x 10^6
Capital Costs	=	6.6 x 10^6
Total System Cost over 20 years	=	\$54.6 x 10^6

This yields \$1.85 per ton to dump 100 miles out.

Summary of Calculations — for Barge Transportation Costs only (no shore-handling or transfer costs included):

Basic Assumption:

(1) 1.5 x 10^6 tons to be dumped per year: this comes to 5,000 tons per day for 300 days.
(2) One tug tows two barges each with a capacity of 1,300 tons. Thus, one tug barge train has a capacity of 2,600 tons.
(3) System operates 300 days per year.
(4) Nontransit time per round trip:

1 hour for dumping at sea
3 hours loading in port
4 hours nontransit time per round trip

Summary of Costs:

Towing Distance in Miles	Towing Speed in Knots	Cost per Tons
20	5	\$0.47
50	5	\$1.05
100	7	\$1.85

The cost of barge transportation on inland waterways of the United States for 1969 was \$0.0033 per ton-mile. Towing at sea is more expensive with costs substantially rising with the distance from shore. In these cost calculations, the system was assumed to be able to operate reliably only 300 days per year, since rough weather will inhibit ocean dumping. These costs will vary, of course, with the interest rate used in the present value calculations. Assuming an interest rate of 5% without an additional inflationary factor, the following costs are obtained.

Summary of Costs:

Towing Distance in Miles	Towing Speed in Knots	Cost per Tons
20	5	$0.57
50	5	$1.30
100	7	$2.25

The economic analysis of System A2 (reported in detail below) leads to the unit costs of $6.78 ($7.09) per ton for interest rates of 5% and 8% respectively. The figures quoted are those reached in the design of a particular station with a capacity of 1,000 tons per day for 260 days per year = 260,000 tons per year.

Initial Capital:

(a) Land (3 to 4 acres)		$ 120,000
(b) Transfer Station Building Cost		
(1) Pile Foundations and Sheeting	$ 200,000	
(2) Main Building	1,190,670	
(3) Barge House	229,512	$1,620,182
(c) Equipment Costs		

Item	Quantity	Unit Cost	Total Material Cost
Scale and Recorder	1	10,000	10,000
Bridge Crane, 10-ton	3	135,000	405,000
Apron Conveyor	2	24,000	48,000
Shredder	2	102,000	204,000
Shredder Motor and Accessories	2	37,000	74,000
Shredder Refuse Conveyor	2	15,000	30,000
Baler	4	48,000	192,000
Baler Motor and Accessories	4	3,125	12,500
Strapping Unit	4	20,000	80,000
Monorail	1	27,800	27,800
Traveling Crane for Bulky Wastes	1	14,600	14,600
Miscellaneous Hoppers		15,000	15,000
Transformers		26,000	26,000
Total			$1,138,900

Operating Costs:

(a) Overhead	
Shredder Power	$ 65,800
Bailer Power	19,100
Crane Power	17,000
Lighting and Ventilation	12,600
Heating	6,500

Strapping Bales	91,000
Hogger Blade Replacement	130,000
Maintenance on Equipment	45,000
Maintenance on Building	21,300
Total	$408,300
Contingencies 15%	61,200
Total Annual Cost	$469,500

(b) Labor	Number per Day	Yearly Salary	Total
Weight Master	1	6,700	6,700
Mechanic	3	9,000	27,000
Crane Operator, 10-ton	4	10,000	40,000
Superintendent	2	10,500	21,000
Laborer	5	6,700	33,500
Baler Operator	2	8,000	16,000
Monorail Operator	2	8,000	16,000
Bulky Waste Loader	1	8,000	8,000
(Pension, Taxes, Payroll, Office, etc.) 50%			92,100
Total			$260,300

Transportation and Disposal Cost:

Towing Company Quote = $2.60 per ton
For 260,000 tons per year = $676,000 total.

Present Value Determination:

Capital Costs

Item	Cost	Life	Total Cost
Land	$ 160,000	20 yr	$ 160,000
Transfer Station Building	1,620,182	20 yr	
Roads, Dredging etc.	170,000	20 yr	
	$1,790,182		
ENR correction	x 1.575		
	$2,819,537		$2,819,537
Equipment	1,138,900	20 yr	
30% Installation	341,670		
	$1,480,570		$1,480,570
			$4,460,107

Operating Costs (Annual):

Transfer Station	469,500
Labor	276,300
Disposal	676,000
	$1,421,800

Present value price determination yields a price per ton charged to the community of:

for i = 5%	P_t = \$6.78
for i = 8%	P_t = \$7.09

where i = interest rate and P_t = price per ton on annual basis.

Dumping of Bales Compacted Inland (A-3) – This system is almost identical to system A1, the only difference being in the location of the transfer station. System A1 located the transfer station at the barge loading facility, which seems reasonable for coastal municipalities. However, if the system of compaction and dumping at sea becomes economically and/or socially superior to rail haul and sanitary landfill for inland municipalities, this system would provide a feasible solution.

Using the same study as a source of cost figures, a system capable of handling 1,000 tons per day for a transfer station located 50 to 150 miles inland is considered. The distance selected affects only the railhaul rate. This system consists of:

(a) Trash being delivered to inland transfer stations and being compacted into bales.

(b) Bales being transported by rail to dockside facilities and loaded onto barges. The refuse bales are then transported in specially designed sideloading rail cars to the barge loading site. Refuse would be removed from the rail cars at the unloading area by a high capacity fork lift truck, and placed in the barges by means of monorail loader.

(c) Barges being transported to disposal site and bales are deposited on sea bed. (A towing company quoted a flat rate per ton and minimum desired quantity which are given.)

The economic analysis for this system is reported below and leads to unit costs (dollars per ton) for interest rates of 5% and 8%, respectively, of:

Inland Transfer Station	**i = 5%**	**i = 8%**
50 miles	\$10.61	\$11.02
100 miles	10.97	11.37
150 miles	11.42	11.82

The figures quoted are an estimate in the design of a system with a capacity of 1,000 tons per day for 260 days per year = 260,000 tons per year.

Initial Capital:

A. Inland Rail Transfer Station*

(a) Land	450,000
(b) Transfer Station Building Cost	1,128,000
(c) Equipment Cost	1,125,500
	\$2,703,500

B. Barge Transfer

(a)	Land (5 to 6 acres)		$150,000
(b)	Building Cost		
	Pile Foundation and Sheeting	140,000	
	Barge House	230,000	
			$370,000
(c)	Unloading Area Cost		
	Apron	16,000	
	Rail Spur	150,000	
			$166,000
(d)	Equipment Costs		

Item	Quantity	Unit Cost	Total Material Cost
20-Ton Fork Lift	2	210,000	420,000
Close-Coupled Engine	1	35,000	35,000
Conveyor	2	24,000	48,000
Monorail	1	27,800	27,800
Total			$530,800

Operating Costs:

A. Inland Rail Transfer Station*

(a)	Overhead	461,500
(b)	Labor	276,300
		$737,800

B. Barge Transfer Station

(a) Overhead

Conveyor and Monorail Power	6,000
Heating, Lighting, Ventilation	13,000
Fuel	6,000
Maintenance on Equipment	20,000
Maintenance on Building	11,000
Total	$56,000
Contingencies 15%	8,400
Total Annual Cost	$64,400

(b) Labor	Number per Day	Yearly Salary	Total
Supervisor	2	10,500	21,000
Mechanic	2	9,000	18,000
Fork Lift Operator	2	8,000	16,000
Monorail Operator	2	8,000	16,000
Laborer	4	6,700	26,800
			$97,800
(Pension, Taxes, Payroll Office, etc.) 50%			48,900
Total			$146,700

(c) Transportation

These figures are based on 100 tons per car loading density in cars attached to regular trains and railroad-owned cars.

Shipping Distance	Dollars per Ton	Yearly Cost **
50 miles	2.65	689,000
100 miles	3.00	780,000
150 miles	3.45	897,000

*Summary Figures from Rail Haul-Sanitary Land Fill study.
**Yearly cost for 260,000 tons per year in dollars.

(d) Disposal

(Towing company quote)	$2.60 per ton
Total Cost for 260,000 tons per year	$676,000

Present Value Determination:

Capital Costs:

Item	Cost	Life	Total Cost
Land—RTS	450,000	20 yr	$450,000
Transfer Station Building—RTS	1,228,000	20 yr	
Roads, Sidings, etc.	200,000		
	1,428,000		
ENR Correction	x 1.575		
	$2,249,100		$2,249,100
Equipment—RTS	1,125,500	20 yr	
30% Installation	337,650		
	$1,463,150		$1,463,150
Land—BTS		20 yr	$150,000
Transfer Station Building—BTS	370,000	20 yr	
Roads, Dredging, etc.	170,000	20 yr	
	540,000		
ENR Correction	x 1.575		
	$850,500		$850,500
Equipment	530,800	20 yr	
30% Installation	159,240		
	$690,040		$690,040
Total Capital Cost			$5,852,790

Operating Costs (Annual):

Rail Transfer Station	461,500
RTS Labor	276,300
Barge Transfer Station	64,400
BTS Labor	146,700
Disposal	676,000
	$1,624,900

Rail Haul Distance	Charge	Total Operating Costs
50 miles	$689,000	$2,313,900
100 miles	780,000	2,404,900
150 miles	897,000	2,521,900

Value price determination yields a price per ton charged to the community of the following amounts:

Transfer Station Inland	P_t	
	i = 5%	i = 8%
50 miles	$10.61	$11.02
100 miles	10.97	11.37
150 miles	11.42	11.82

Barge Transportation Costs: In all of the ocean dumping systems, the cost of the barge-haul and dump procedure constitutes about 25% of the total disposal costs. The available proposals of solid waste disposal systems utilizing barge transportation have all recommended placing responsibility for this phase of the operation in the hands of a private contractor for a fixed fee. If considerations of the economies of scale of barge transportation were made part of the initial design parameters, instead of considering the barge haul operation as an external "black box", the scale of proposed sea-based solid waste disposal systems might be significantly increased.

The analysis carried out with respect to system A2 illustrates the economy of scale available in a barge system capable of handling 1,500,000 tons of refuse a year and leads to the following results. These costs will vary with the interest rate used in the present value calculations. Assuming an interest rate of 5% without an additional inflationary factor, the following costs are obtained.

Summary of Barge Transportation Costs:

Towing Distance in Miles	Towing Speed in Knots	Cost per Ton
20	5	$0.57
50	5	1.30
100	7	2.25

Dispersal Effects of Dumping of Compacted Bales: Both systems A2 and A3 employ the same ultimate disposal technique, the placing of compacted bales in a selected disposal site. This is an example of the "containment" philosophy, where an area of minimal ecological-sociological value, and hopefully minimal circulation is restricted as a dumping area. This concept places five constraints on the operation of the system.

(1) The system must guarantee that each and every bale sinks to the bottom, without reaching a condition of neutral buoyancy. This means the compaction system must produce bales of a density greater than 64 lb/ft^3. High density compaction can be attained by the use of converted scrap metal presses or extrusion-type bales

similar to those used in the baling of paper and cotton. One-, two- and three-stroke compactors, both with and without preshredding, were investigated. The baler selected for preliminary design studies is a one-stroke baler (American Baler Company) which also requires a preshredding unit.

The one-stroke baler was selected on the basis that neither two- nor three-stroke balers appeared capable of producing the required capacity without excessively expensive and cumbersome modifications. With preshredding, the baler unit can guarantee an output of 25 tons per hour. The bale size is roughly 30" by 40" by variable length of from 40" to 80".

Bales prepared for tests by the Sandy Hook Marine Laboratory were baled at about one-third normal operating pressure, and had a density of about 68 lb/ft^3. Densities of 70 lb/ft^3 under better operating conditions seem attainable. However, due to the variation in composition and moisture content of the refuse delivered to the baler, even with preshredding, it was felt that the baler should have the capability of guaranteeing a minimum density of 70 lb/ft^3 on every bale produced.

(2) The system must guarantee that the bale maintains its integrity during descent and upon impact on the ocean floor. The system designed around the proposed mechanism includes a unit to provide automatic wire bale strapping.

The Sandy Hook Marine Laboratory gave indications that the quality of the bales presently being produced may not be high enough to prevent extensive disintegration of the bale both in transit to the site and during the dumping operation. Inadequate preshredding allows many bales to contain whole bottles, magazines, cans, and other whole objects which would easily slough off under the influence of currents or wave action, and during handling procedures.

While the presently available shredding, baling, and strapping mechanisms appear to have the capability of producing bales of adequate integrity, further design improvement and intensive quality control procedures are clearly necessary.

(3) The system must guarantee that the bales are deposited within the prescribed site limits. This will require suitable navigation equipment and aids to assure positive location of the barges above the dumping site. It also requires some knowledge of the descent trajectories of the bales under various sea-state conditions. This information is not yet available, but some sort of fathometric trace may provide adequate instrumentation to roughly determine descent trajectories.

(4) The system must guarantee that the bales do not disintegrate in the ocean environment. While it appears that the bale will remain

intact in its descent and impact phases, quick disintegration due to biological and chemical decomposition may still be a problem. At present no data on this problem is available.

(5) The system must guarantee that the bales themselves will not be transported beyond the dumping site limits by current and tidal action. This problem must be solved by a suitable site selection criteria. Experience with incinerator residue indicates that a depth of 200 feet will ensure negligible ocean transport while fifty feet will not. A Summary Cruise Report of the R.V. *Challenger* research vessel is presented below to illustrate the type of information available.

SANDY HOOK MARINE LABORATORY
U.S. Bureau of Sport Fisheries and Wildlife
Highlands, New Jersey

Summary Cruise Report R.V. *Challenger*

Dates: October 16, 17 and 25; November 25, 1968.

Purpose: To investigate the possibility of utilizing compressed solid wastes and garbage as reef material.

Procedure: On October 16, a test bale of compressed solid waste and garbage was obtained from the towing company. The bale had been prepared, specifically for this test at sea.

The bale, which was loaded on the *Challenger* was 30½" by 40½" by 48" and weighed 2,225 lb. This bale occupied 34.3 ft^3 and had an indicated density of 67.75/ft^3 on the day it was baled.

On October 17, divers located and buoyed off a relatively flat area of bottom about 100 ft west of the Shrewsbury Rocks' bell buoy in 45 ft of water. The bale was dropped on this spot from the stern of the *Challenger*. It floated at the surface for several minutes, then sank to the bottom. The bale was then tied to the bell buoy anchor with a tag line to facilitate finding the package on future inspection dives.

Results: Five to ten minutes after the bale reached bottom, several areas on the bale which appeared somewhat loose were observed. However, most of the bale remained intact and soon attracted a few small cunner, *Tautogolabrus adspersus*. The cunner did not appear to be feeding on the compressed garbage but merely attracted to the unusually high relief offered by the bale on the relatively flat bottom.

On October 25, the bale was again inspected by diver biologists. They indicated that the bale appeared very much the same as the day it was put down with the exception of a slight rounding of the edges. There were a few pieces of material scattered around

Results: (Continued) the bale and the surface area was slightly soft after the eight days of submergence. However, the bale was still compact and felt quite solid when a diver attempted to stick his knife into it.

The next attempt to observe the condition of the test bale was delayed by weather until November 25. This dive came shortly after an intense northeast storm which had developed a sizable storm surge. Divers were unable to locate the bale and it was assumed that the storm surge moved the compressed garbage some distance from the test site.

At present, there is insufficient information to make a statement concerning the possibility of utilizing such bales as reef material. However, it is believed that another attempt to study the fate of similar bales in deeper water where there would be less chance of damage or loss by storm surge should be made. Since ocean disposal is becoming increasingly more popular, efforts should be made to comment intelligently on the feasibility of using compressed solid wastes and garbage as reef materials. The effect of these wastes on the ecology of the marine environment needs to be understood.

The baler company indicated they used only one-third of the possible operating pressure in preparing this test bale. In view of ocean disposal it may be better to make the bales denser. A bale produced by utilizing two-thirds to full operating pressure of the compressing unit would probably sink quickly in salt water, and reduce the chance of bales dispersing away from a dump site by drifting.

Incineration at Sea

The concept of incineration and/or the disposal of incinerator residue at sea has been proposed as a sea-based alternative to existing solid waste disposal methods. As sites for the land-filling of incinerator residue become more costly and scarce, and as the need for the construction of new efficient incinerators appears to become more acute, a system for burning and disposal at sea may gather a great deal of backing in both the public and private sector.

Three main system alternatives present themselves for disposal of incinerator residue. System (1) involves: incineration at existing inland incinerators and proposed regional incinerators; transport of incinerator residue to dock facilities and loading onto sea-transport mechanism, and transport of incinerator residue from dockside to dumping site and placement on sea floor.

System (2) involves: transport of raw garbage in primary collection vehicles to dock-side transfer station; incineration at new incinerator facilities constructed at waterfront sites; and transport of incinerator residue from dock-side incinerator to dumping site and placement on sea floor.

System (3) involves transport of raw garbage in primary collection vehicles to dock-side transfer station, loading of raw garbage onto a waterborne incinerator, and transport of raw garbage to dumping site; and incineration while en route and/or at dumping site and placement of incinerator residue on sea floor at dumping site.

These systems present alternatives regarding the amount of material transported and distance transported; the use of already existing facilities; the type and cost of initial capital equipment expenditures; and system reliability. Before these alternatives are considered, it is necessary to review the biological effects and the question of dispersion of the residue of incineration.

Any system which includes the ultimate deposition of incinerator residue on the sea floor, whether in a state of static or dynamic equilibrium, must guarantee a controlled minimal effect on the marine life there, both plant and animal, free-swimming and bottom dwelling. It must be realized that:

(1) Many perturbations will appear to have little or no short-term deleterious effects but may have very serious and irreversible long-term (5 to 25 years) effects on the ecology of the sea bed.

(2) Since an extremely delicate equilibrium is in operation in the food chain of the sea floor, the very slightest perturbation, even one which seems to have beneficial short-term effects, may cause violent, irreparable damage.

(3) Since the residue of the present state-of-the-art incineration is essentially nonbiodegradable, the self-cleansing and renewing mechanisms of the sea ecology will have a long and difficult task in assimilating this residue. (See Municipal Waste Disposal by Shipborne Incineration and Disposal of Residues.)

The second constraint common to all systems depositing incinerator residue on the ocean floor is the need for suitable site selection. Once minimal deleterious effects on marine life outside the immediate dumping area can be guaranteed, the dispersion of the residue and gradual growth of the disposal site must be able to be controlled.

Based on the Harvard School of Public Health Study, it has been concluded that incinerator residues should be deposited in waters at least 200 ft deep to avoid significant migration from the assigned site under the combined effects of wave action plus tidal and bottom currents and that depressions in the ocean floor, where bottom currents are minimal, would be ideal. It seems necessary that certain logistic consideration be added to this conclusion.

Going to deeper disposal sites (and therefore sites further from shore), the problem of locating the site under poor operating conditions (storm seas, darkness, fog, etc.) becomes extremely difficult. Once the dumping site has been located, the problem of positioning the unloading facility to assure that the trajectories of the falling residue carries it near to the center of a one or two square mile site also becomes exceedingly difficult. There appears to have been no study of the descent rate of the residue.

Some migration rates are allowable in trying to select a site on the basis of existing current data. Most of the more easily transported residue components, such as tin cans, are also those that are most susceptible to disintegration from the abrasive action of the transport process. Observations of incinerated cans suspended in seawater indicate that within one year the metal disintegrates into a mass of small flat crumbly scale particles.

While the 200-foot depth does appear to be a safe minimum depth limit, the costs of surveying prospective sites at this depth, and the costs of navigational aids to ensure dumping within the subscribed site once it has been selected, must be figured into the cost of any sea disposal of incinerator residue system. These costs will increase rapidly with increase in depth.

Analysis of Alternative Systems: Analysis of System (1) — Inland incinerators as of September 1970 handle approximately 10% of the total municipal refuse accumulation. Construction of large-scale regional incineration systems could guarantee an 80 to 90% reduction by volume of solid waste and incinerate the solid waste at an average cost of $10.50 per ton. (All costs cited are 1970 dollars.)

However, in order for this residue to be suitable for sea disposal, it must contain no floatables. Efficient high-temperature incineration (as distinguished from existing local incinerator systems) claims to guarantee that no combustible materials come through unburned. For such a furnace, the only necessary sinkage treatment would be a crush process to get rid of noncombustible floatables such as semiclosed cans.

The poor quality of incineration available from existing local medium-temperature incinerators would not suffice for guaranteeing the sinking of the residue, even with a crusher treatment added. Too many combustible paper and wood products, whose buoyancy would be essentially unaffected by crushing, are found in the residue of these incinerators. More exotic treatment of the residue, such as weighted packaging, or compacting, or tarring may be necessary.

For the transport of incinerator residue to dock facilities the residue is (under normal operating procedures) quenched by a water spray or both as the residue leaves the incinerator. Sufficient water is absorbed or retained to increase the weight of the residue from 40 to 70% above its dry weight. Records of refuse disposal operations in Washington, D.C., Arlington County, and Alexandria show these general relationships of incinerator residue to raw refuse for existing incinerator plants: the wet bulk density of residue averages about 1,000 lb per cubic yard; and approximately 0.35 to 0.5 tons of wet residue results from burning one ton of raw refuse.

In the solid waste disposal study for the Washington Metropolitan Region, the following cost figures for a haul involving 20 miles of expressway travel and 5 miles of nonexpressway travel are presented.

10-Cubic Yard Vehicle	35-Cubic Yard Vehicle
$3.50 per ton of residue	$1.25 per ton of residue

Since the larger 35-cubic yard vehicle will be more compatible with the proposed large regional incineration plants, the cost of $1.25 per ton has been chosen for use in the subsequent analysis. Conversion of this cost per ton of residue to a cost per ton of untreated refuse yields a price of $0.53 per ton based on the calculation below.

$$\frac{\$1.25}{\text{1 ton residue (wet)}} \times \frac{\text{0.425 ton residue (wet)}}{\text{1 ton solid waste}}$$

The transfer of the residue from the trucks to the barges should be accomplished directly at a paved loading dock area. The relatively high density (and therefore small volume) of the residue should minimize congestion problems. Using a barge transportation cost analysis as presented above the following costs can be established.

Miles to Dump Site	Cost per Ton Residue	Cost per Ton Solid Waste
20	$0.65	$0.28
50	$1.00	$0.43
100	$1.27	$0.54

For the sake of comparison with the other systems involving a sea dump operation, the 50-mile distance to the ocean dump site is used in determining an overall price for disposal. This leads to an overall cost figure for System (1) as follows:

(a)	Incineration	$10.50 per ton	$11.00 per ton
(b)	Transport to dock	0.53 per ton	0.53 per ton
(c)	Sea transport and dump	0.43 per ton	0.43 per ton
	Total	$11.46	$11.96

Analysis of System (2) — This system concept differs from System (1) in only one essential area—the location of the incinerators. In this one area it is inferior on two counts:

> Regional inland location is designed to minimize the distance traveled by (and thereby the costs of) the primary collection vehicles. By locating the incinerator in the port complex, both the distances and costs of primary collection are increased which forces the collection vehicles to operate in the already congested and usually inadequate dock transfer area; and
>
> Regional inland location allows some latitude in finding suitable land at a reasonable cost, and minor deviations to capitalize on local geographic and economic conditions. Dockside location forces the purchase of easily accessible (both by land and sea) waterfront property, which is at somewhat of a premium.

System (2) is therefore both economically and logistically inferior to System (1), and appears unworthy of further consideration.

Analysis of System (3) — This system represents a radical departure from any existing solid waste incineration techniques, both in overall system concept, and in a large part of the necessary equipment. The only completed study on this area at this time consists of a 1968 report for the City of New York City Planning Commission. In the preliminary design considerations, the main system functions were characterized by the following mode of operation.

Transfer stations are designed and constructed to accept garbage from existing primary collection vehicles. The refuse is compacted into storage containers, and these containers are then placed in special racks in the ship's hold. These containers are designed to allow ram discharge into the incinerator. The size and rated capacity of the transfer system is a function of the cycle time.

The filled containers are unloaded using a bridge crane and track extension to remove the filled container from the dock and place it in its assigned rack position. The garbage is then transported and burned on the basis of either a 2- or 3-day cycle. The cycle time determines ship size and incinerator capacity as well as transfer station size.

Incineration may occur while en route provided that all burning takes place at least 10 miles offshore (or at the dumping site) to ensure adequate air pollution control. The following general components were formulated for the incinerators: a continuous feed-water cooled chute; a hung and tied-back refractory lined furnace; a separate refractory lined combustion chamber for complete gas burn-out; a traveling or reciprocating feed grate; combustion air fans; an insulated metal stack; a crusher-type, wet quench residue removal system; and the inclusion of no separate air-pollution control devices.

Residue leaving the incinerator will pass through crushers into a water-quench hopper. It is then pushed out of the hopper onto a conveyor belt, carried to the top deck, and dumped into an open well located amidships. This will provide for the separation of flotables and nonflotables in the residue.

The nonflotables sink through the well to the ocean bottom. Any flotables remaining on the surface will be periodically skimmed off and recycled through the incinerators. The equipment and operating costs of such a system have been computed for three different operating scale and/or cycle time arrangements given as follows:

(A) *Two-Day Cycle System* – All refuse received in a two-day period is initially stored at a transfer station, then loaded on board and transported to the burning and dumping site. The capacity of this system is approximately 650 tons per day, (two 500 tons per day furnaces used) operating 6 days per week.

(B) *Three-Day Cycle System* – All refuse received in a three-day period is initially stored at a transfer station, then loaded on board and transported to the burning and dumping site. The capacity of this system is approximately 780 tons per day, (two 500 tons per day furnaces used), operating 6 days per week.

(C) *Liberty-Ship Conversion System* – Due to ship size constraints, the Liberty-ship incinerator will operate on a two day cycle. The capacity of this system is approximately 433 tons per day, (two tons per day furnaces used) operating 6 days per week.

The details of these analyses follow.

Economic Analysis of Incineration at Sea:

(A) Two-day Cycle System
(Calculations based on six-day working week)

Initial Capital Costs:

(1) Vessel

Division	Material	Labor
Hull Steel	840,000	1,800,000
Hull Outfits	145,000	170,000
Incinerators and Housing	4,000,000	Included
Hull Engineering Items	525,000	465,000
Machinery and Propulsion	1,235,000	445,000
Total	$6,745,000	$2,880,000

Total Material	=	6,745,000
Total Labor	=	2,880,000
Total Cost (1) Ship		$9,625,000

(2) Transfer Station

Building Structure	300,000
Ramp	200,000
Compactors—7 @ $40,000 (installed)	280,000
Crane	100,000
Storage Racks, Winches, etc.	250,000
Scale	20,000
Total Cost (1) Transfer Station	$1,150,000

(3) Container

Containers—110 @ $8,000	$880,000

Total Initial Cost:

Vessel (1)	=	9,625,000
Transfer Station (1)	=	1,150,000
Containers	=	880,000
		$11,655,000

Operating Costs (Annual):

(1) Vessel

Wages	562,000
Sustenance	34,000
Stores and Supplies	15,000
Insurance on Crew	38,000
Fuel @ 340 days	164,000
Maintenance and Repair	85,000

Insurance on Ship	54,800
Overhead and Misc.	87,000
	$1,039,800

(2) Transfer Station

Total Personnel Cost	188,000
Heat, Power, Water, etc.	9,400
Maintenance	40,600
	$238,000

Estimated Unit Cost:

It is estimated that the waterborne incinerator will be out of service for maintenance work two weeks out of each year. Therefore, total annual tonnage would be 650 tons per day by 6 days per week by 50 weeks per year equals 195,000 tons per year.

Present Value Determination:

Capital Costs:

Item	Life	Total Cost
Vessel	20 yr	$ 9,625,000
Transfer Station	20 yr	1,150,000
		$10,775,000
Containers	10 yr	$ 880,000

Operating Costs (Annual):

Vessel	$ 1,039,800
Transfer Station	238,000
	$ 1,277,800

Solution of the present value price determination equation yields a price per ton charged to the community:

	P_t
for i = 5%	$11.33 per ton
for i = 8%	$12.39 per ton

(B) Three-day Cycle System

All refuse received in a three-day period is initially stored at a transfer station, then loaded on board and transported to the burning and dumping site. The capacity of this system is approximately 780 tons per day, (two 500 tons per day furnaces used), operating 6 days per week.

Initial Capital Costs:

(1) Vessel

Division	Material	Labor
Hull Steel	1,032,000	2,300,000
Hull Outfit	145,000	170,000
Incinerator and Housing	4,000,000	Included
Hull Engineering Items	525,000	465,000
Machinery and Propulsion	1,235,000	445,000
Total	$6,937,000	$3,380,000

Total Material	=	6,937,000
Total Labor	=	3,380,000
Total Cost (1) Ship	=	$10,317,000

(2) Transfer Station

Building Structure	300,000
Ramp	200,000
Compactors—7 @ $40,000	280,000
Crane	100,000
Storage Racks, Winches, etc.	350,000
Scale	20,000
Total Cost (1) Transfer Station	$1,250,000

(3) Containers

Containers—190 @ $8,000 $1,520,000

Total Initial Cost:

Vessel (1)	=	10,317,000
Transfer Station (1)	=	1,250,000
Containers		1,520,000
		$13,087,000

Operating Costs (Annual):

(1) Vessel

Although slightly larger than a two-day cycle vessel, manpower requirements and other operation costs are estimated at approximately the same as the two-day cycle vessel—$1,039,800.

(2) Transfer Station

Total Personnel Cost	188,000
Heat, Power, Water, etc.	9,400
Maintenance	55,400
	252,800

Estimated Unit Cost:

Total Annual Tonnage

780 tons per day by 6 days per week by 50 weeks per year equals 234,000 tons per year.

Present Value Determination:

Capital Costs:

Item	Life	Total Cost
Vessel	20 yr	$10,317,000
Transfer Station	20 yr	1,250,000
Total		$11,567,000
Containers	10 yr	$ 1,520,000

Operating Costs (Annual):

Vessel	$ 1,039,800
Transfer Station	252,800
Total	$ 1,292,600

Solution of the present value price determination equation yields a price per ton charged to the community:

	P_t
for i = 5%	$10.10 per ton
for i = 8%	$11.08 per ton

(C) Liberty Ship Conversion System

Due to ship size constraints, the liberty ship incinerator will operate on a two-day cycle. The capacity of this system is approximately 433 tons per day (two tons per day furnaces used) operating 6 days per week.

Initial Capital Costs:

It is assumed that a municipality can obtain liberty ships from the Reserve Fleet at no cost by Federal action.

(1) Vessel

Division	Material	Labor
Drydocking, Sandblasting and Repairing Hull	Included	150,000
Removal of Machinery and Interior Steel	Included	
New Hull Steel	80,000	220,000

Division	Material	Labor
New Hull Outfit	22,000	33,000
Incinerators and Housing	3,200,000	Included
Hull Engineering Items	525,000	465,000
Machinery and Propulsion	1,235,000	445,000
Total	$5,062,000	$1,313,000

Total Material	=	5,062,000
Total Labor	=	1,313,000
Total Conversion	=	$6,375,000

(2) Transfer Station

Building Structure	250,000
Ramp	200,000
Compactors—5 @ $40,000	200,000
Crane	100,000
Storage Racks	250,000
Scale	20,000
	$1,020,000

(3) Containers

Containers—106 @ $8,000 $848,000

Total Initial Cost:

Vessel (1)	=	6,375,000
Transfer Station (1)	=	1,020,000
Containers	=	848,000
		$8,243,000

Operating Costs (Annual):

Although slightly smaller than a two-day cycle vessel, manpower and other operating costs are estimated at approximately the same as the two-day cycle vessel.

$1,039,800

(2) Transfer Station

Total Personnel Cost	188,000
Heat, Power, Water, etc.	9,400
Maintenance	37,400
	$234,800

Estimated Unit Cost:

Total Annual Tonnage

433 tons per day by 6 days per week by 50 weeks per year equals 129,000 tons per year.

Present Value Determination:

Capital Costs:

Item	Life	Total Cost
Vessel	20 yr	6,375,000
Transfer Station	20 yr	1,020,000
		$7,395,000
Containers	10 yr	$848,000

Operating Costs (Annual):

Vessel	1,039,800
Transfer Station	234,800
Total	$1,274,600

Solution of the present value price determination equation yields the following prices per ton charged to the community.

	Cost in Dollars per Tons for Interest Rate of	
	5%	8%
Two-day Cycle System	11.33	12.39
Three-day Cycle System	10.10	11.08
Liberty-ship Conversion System	14.97	16.08

The 3 day cycle appears to be the most economic of the systems proposed.

The figure of $10.10 for an incineration-at-sea cost is undoubtedly very optimistic in that it does not take into account the effect of days in which operation is not possible due to weather conditions. A sea disposal system should expect to be inoperative due to weather conditions 1 to 2 days per month in the summer and 3 to 6 days per month in the winter. Moreover, these days should be anticipated to occur in blocks.

Since the solid waste flow is essentially continuous, this condition places major backup requirements on any sea-based system, and especially on the waterborne incinerator system due to the specialized nature of the equipment. While the dumping-at-sea system can hire additional barges on a short term basis from other dumping operations to dispose of a backlog of solid waste, the waterborne incinerator system must bear the full costs of additional capacity.

A reasonable design criterion might be the requirement of disposal of material backed up by a three-day storm in two weeks. (It should be noted that any actual implementation of an incineration-at-sea system would require a careful study of local weather patterns to determine an appropriate back-up capability level.) This design criterion would probably require additional incineration capacity as well as storage space, since, if a seven-day week is worked after storms, it would take three weeks for the three-day cycle to adjust. Disruption of the normal operating schedule during the three weeks may entail further operation problems.

A conservative estimate of the necessary back-up requirements implies a 25% increase in incineration capability and suitable buffer storage. The rough costs of such back-up capability would entail a 25% increase in incinerator and housing costs, a 50% increase in container cost, and a 50% increase in storage costs. When these estimates are placed back into the present-value cost determination, the final cost for the most efficient waterborne incinerator is shown to be: \$10.89, (i = 5%); \$12.00, (i = 8%).

Weather is not the only problem faced by the waterborne incinerator. Any solid waste disposal system must have the ability to function during, or at least quickly recover from, any sort of system disruption.

Ocean dumping of the residue of landbased incinerators is clearly superior to a waterborne incinerator due to the former's system configuration of a large number of low-complexity barges. A damaged barge could be fairly easily repaired or replaced, and the total system could operate effectively with one or two barges down.

Only failure of your main propulsion tug(s) would represent complete system failure and these too are quickly replaceable. However, due to the system configuration associated with sea-borne incineration, namely one ship of high complexity, container manipulating components, etc., failure of any major component would probably result in the abortion of the trip and a complete slippage of the two- or three-day cycle time. Differences in system reliability have not been allowed for in the cost per ton figures.

Conclusions

As previously stated, there are two rather distinct philosophies which can be followed for dumping refuse at sea. One approach is to accept the ecological destruction of a designated area and attempt to confine all the effects of the refuse to that area. One, then, chooses areas which are judged to be ecologically unimportant. This has been the prevailing philosophy among marine biologists who argue for dumping off the continental slope at depths greater than 1,000 fathoms with the waste suitably contained to keep it from spreading.

The other philosophy is to view the ocean as a link in the natural process of returning the wastes to the life cycle. Holders of this point of view point out that the ocean has considerable regenerative powers, that in proper concentrations much waste material can be regarded as useful nutrients for the lower levels of marine life. They point to even such unattractive materials as the acid mine wastes dumped on outer New York Harbor, noting that in the center of this dump the water is biologically dead while around the edges marine life flourished at levels higher than those in the undisturbed waters.

Obviously, there are very substantive differences in these two philosophies. The first calls for dumping in very concentrated areas, the second calls for a much more even distribution of the wastes. Holders of the first view point out that almost no bacterial action takes place at such depths. An apple on a sunken boat after being submerged at 4,500 ft for close to a year was perfectly edible when the boat was raised, as was the meat in a bologna sandwich.

Thus, holders of the second view are led to recommending near-surface dumping in shallow water or at least in the photic zone. These are just the areas most valuable to man, and this places a much heavier burden on holders of the second philosophy since they must be much more concerned with the sociological effects of the garbage than those who call for confinement in deep waters. Their information requirements are much higher and presently this information does not exist. There are a variety of ways of dumping solid wastes at sea:

(1) Surface dump of the raw garbage;
(2) Surface dump of ground garbage;
(3) Compaction and dump of compacted bales with and without packaging, and
(4) Pump raw garbage to depths where natural compression renders all the garbage denser than the sea water.

The first alternative was at one time practiced by New York City among others. However, New Jersey was able to obtain an injunction on any dumping which might result in the garbage returning to shore and New York found it uneconomical to take the garbage far enough to prevent its return. The economics have changed in favor of longer trips. However, this alternative is dominated by compaction because the shorter trip which compaction allows more than compensates for the costs of compaction. Thus, dumping raw garbage appears neither feasible nor economic.

The second alternative is the one most consistent with the philosophy of using the ocean's biological powers to return wastes to the life cycle. It will run about $1.00 more per ton than the first alternative and probably require as long a trip. Therefore, at present it also appears to be dominated by alternative (3) with respect to costs of transportation. However, if the validity of the dispersal philosophy can be substantiated, then it will deserve serious attention.

Compaction of garbage to densities higher than that of sea water is at the time of this report at the edge of the state-of-the-art. Bales have been produced from municipal garbage with densities in the 70 to 75 pcf range. However, no one has demonstrated the ability to produce consistently heavier than water bales over a range of garbages. It appears, however, that this ability can be achieved at little more than baling costs at the time of this report as soon as the need is demonstrated.

Baling is the technological advance that has made dumping at sea feasible. No experimentation on the ecological effects of the dumped bales other than very short-term observations of a few bales on the bottom in shallow water have been undertaken. Nor has the differential in these effects associated with different packaging materials been studied. Covering the bales with polyethylene would cost about $1.00 per ton. Given a deep water dump, this would have little effect other than to prevent surface spalling during descent.

In summary, dumping of compacted bales at sea appears to be the most attractive of the marine alternatives with respect to solid waste disposal from the point of view of market costs. However, before it or any other sea based disposal system is placed in operation, much more should be known than is presently known about the ecological effects associated with this activity.

MUNICIPAL WASTE DISPOSAL BY SHIPBORNE INCINERATION

It seems clear that for the foreseeable future all but a small fraction of the total quantity of solid wastes will be returned to the environment as a means of final disposal. The current tendency to regard each element of the environment as a resource makes it very clear that each becomes most valuable when exploited to the maximum degree compatible with prudent safeguards for human health, well-being and preservation of property. A distinction must therefore be made between contamination of the environment—that is, addition of foreign substances, and pollution—that is, accumulation of contaminants to the point where undesirable effects begin to occur.

It is naive to maintain that nothing whatsoever can be added to the environment, for even if it became technically possible to achieve such an ideal, the human and material resources that would have to be committed to this single purpose would be so great that the quality of daily life would be substantially impovished.

Instead, the objective should be to utilize to the fullest the self-purifying properties of the environment, which are enormous when properly understood and managed. In Los Angeles County, for example, all solid-waste collections are used for landfill to transform steep, dry hills and canyons into public parks and golf courses. Most citizens appear well pleased with this decision, although a few express a preference for preservation of the waste lands.

Similar considerations and differences of opinion apply to the management of tidal flats. Even staunch conservationists may concede, in good conscience, that not every single acre of tide land is best utilized by being left alone. As poorly located and badly polluted tide lands can be drained, filled and utilized for many good purposes, so, in like manner, can new tidal lands be created in ideal locations to fill needs not now satisfied by natural geologic processes.

It is never possible to predict with certainty the exact effects of large-scale disposal efforts from small-scale pilot and theoretical studies. Therefore, when indications are uniformly favorable, large-scale demonstrations, for extended periods must follow if proper evaluations are to be made. This does not mean that the areas to be used for these large-scale experiments should be allowed to degrade to the point of serious damage. Rather, they should be monitored on a continuous and adequate basis so that incipient damage to the environment can be detected and protective measures employed in an orderly and timely way.

The Solid Waste Problem

Solid wastes, often called refuse, include a wide variety of discarded, and generally useless, materials resulting from normal community activities. Garbage and rubbish from households, institutions and commercial establishments comprise the largest fraction of municipal waste collections, but bulky objects, such as stoves and refrigerators and, especially, abandoned automobiles, represent more difficult municipal collection and disposal problems. Industrial refuse from factories, power plants and food-processing plants may contain waste materials not usually found in municipal collections, but more often the difference is quantitative rather than qualitative since industrial processes, by their very nature, generate large amounts of one, or a few, kinds of waste products. By contrast, ordinary municipal collections tend to be extremely diverse.

Prosperity has increased per capita waste production enormously. Modern packaging practices, a spate of periodical literature, almost instant obsolescence of an endless variety of material acquisitions and the enormous growth of industry have combined to increase the daily per capita production of solid wastes in the United States from 2.7 lb in 1920 to 5.0 lb in 1970.

Similar increases are now occurring in Western Europe. The present trend is equivalent to an annual increase of 25 lb of waste per capita. Paper consumption alone increased 16.6% during the same interval. Even more impressive is a figure of 86,000,000,000 lb for the total paper and paperboard products consumed during 1963. At a national conference on the use and disposal of single use items in health care facilities, it was pointed out that some hospitals already generate up to 20 lb of solid waste per patient per day and that when the concept of single use disposal items is extended to include all bed linens, food service, uniforms, and everything else to which it may be applied, hospitals can be expected to generate 50 lb of solid waste per patient per day.

The concept of disposing of solid wastes by transporting them to dockside in pickup and transfer vehicles, loading the material onto specially designed incinerator ships, burning at sea, and disposing of the residue (plus noncombustible bulky items) in designated areas of the ocean bottom was considered sufficiently novel to make it advisable to utilize conventional equipment throughout except when no suitable item was available commercially.

This has the disadvantage of eliminating innovative ideas as well as efficiencies inherent in specially designed vessels, burning grates, etc. The advantages are economy, the opportunity to place the new technology into prompt practice, and ability to make better cost estimates. The latter reasoning has prevailed. Consequently, preliminary engineering planning has been concerned with adapting existing vessels, incinerators, and mechanical handling equipment to the special requirements of an incinerator ship. For example, the capabilities of Liberty ships were evaluated with the aid of plans supplied by the U.S. Maritime Administration.

It was found that the decks and cargo spaces provide adequate room for raw waste storage bins and three rotary type incinerators, each having a daily capacity of 600 tons, for a ship's total of 1,800 tons per 24 hour day. The removal of most of the oil tanks (required only for overseas crossings) yielded additional cargo space for waste storage to give a total well over 2,000 tons.

The sea-going incinerator approach seems to have application for many coastal cities and towns as well as for Great Lakes communities. It provides a means whereby several communities can have an incinerator not located within their confines which has the capability of carrying waste to a remote location, burning it, and discharging the residue in a manner which will not return it to the community or produce shore or beach pollution. Further, a ship provides a vehicle for carrying large objects, i.e., bulky wastes such as abandoned automobiles, for sea disposal at the same locations.

A preliminary evaluation showed that a single ship could handle one half of Boston's waste (the other half is being incinerated at a conventional incinerator) plus that of several communities to the north, e.g., Beverly and Lynn, and to the south, e.g., Quincy and Braintree. Suitable harbor and dockside facilities are available and these are capable of providing terminal locations for waste collection trucks as well.

One operating plan provides for waste to be loaded on board during the daily eight hour waste collection period (and perhaps even incinerating at low burning rates at dockside) after which the ship would proceed to sea. Burning would begin as soon as the ship was three to four miles from the dock and continue until the ship had reached a U.S. Coast Guard-designated official burning and disposal ground ten to twenty miles offshore where the ship would discharge the residue as it was burned after slurrying it with sea water.

It was anticipated that the slurried ash would settle readily into the depths after floating material had been skimmed from the surface of the slurry tank. A typical one mile square depression in the ocean floor was calculated to be able to handle the incinerated waste from a 1,000 ton per day incinerator for the next 250 years without creating problems.

An alternative operating plan calls for storing waste collections in silos at dockside during the daily collection period and transferring the entire load to a ship in a short period of time. This would permit the ship to spend more hours at sea and would facilitate rapid loading at several ports in Massachusetts Bay.

Ship Incinerator Costs

In 1963, preliminary estimates were made of ship incinerator costs and these were compared with then-current costs of: a modern, land-based incinerator and incineration in a sanitary landfill located at four graduated distances from the center of the metropolitan core area in which a major fraction of the wastes were generated. This is believed to be a reasonable approach to cost comparisons because, as land available for filling becomes more and more remote from the core city, the total cost of sanitary landfill disposal is closely related to the haul distance between pickup and disposal sites.

Estimation of landfill costs reflect Los Angeles County's experience and include the economies achievable by the use of local transfer stations and 20-ton tractors to transport consolidated wastes to remote landfill sites. Land-based incinerator costs reflect current records of Boston's modern central incinerator which handles approximately one-half of Boston's total waste collections.

All of the cost figures are summarized in Table 5.6 where it may be seen that the estimated per-ton disposal cost for a ship incinerator is less than the cost of disposal in a new land-based incinerator of equal capacity located near the center of the collection district. The relatively low cost of ship incineration reflects two special conditions:

Boston already owns functional dock sites for waste transfer from shore to ship that were formerly used for barging raw garbage to sea.

At the time these cost estimates were prepared, there were surplus Liberty ships "in mothballs" that could be transferred by an act of Congress to a public agency, such as a city or state, for a nominal payment.

TABLE 5.6: COMPARATIVE SYSTEM COSTS FOR TRANSPORTATION AND DISPOSAL*

(Dollars per Ton of Refuse)

Cost Item	Sanitary Landfill in the Metro. Area		Sanitary Landfill Outside Metro. Area				Second Boston Incinerator	Mobile Ship Incinerator
Distance to Site	8 miles	13 miles	20 miles		30 miles			
Truck Size	4 ton	4 ton	4 ton	20 ton	4 ton	20 ton		
Transportation	2.55	4.05	6.15	2.20	9.15	3.30	1.05	1.05
Transfer Station	- --	-	-	0.35	-	0.35	-	0.50
Disposal Cost	1.00	1.00	1.00	1.00	1.00	1.00	4.00	2.90
Total	3.55	5.05	7.15	3.55	10.15	4.65	5.05	4.45

* Pick-up costs, the same for all systems, are omitted.

It is obvious that substantial ship modifications are required to convert a Liberty ship into an incinerator ship, but acquisition of a sea worthy vessel of this size would substantially reduce the final cost relative to building an entirely new ship because of the salvage value of hull, personnel quarters, cargo handling equipment, propulsion equipment, plus other important items.

It is interesting to note in Table 5.6 the cost dependence of sanitary landfill disposal on haulage distance and the use or nonuse of transfer stations for long hauls. It is evident from the table that the cost of landfill disposal is significantly less than incineration only when landfill sites are available close to the pickup areas.

An engineering feasibility and cost study of ship incineration was conducted for the City of New York Planning Commission in 1968. It drew heavily on the data accumulated in this study and concluded that "waterborne incineration may provide answers to some of the refuse disposal problems of the City." As New York City also owns several dockside facilities suitable for waste transfer activities from shore to ship, cost comparisons between Boston and New York City for ship incineration are based on closely similar requirements.

In addition, inquiry to the New York Office of the U.S. Corps of Engineers in 1968 revealed that "the Corps would have no objections to dumping of noncombustible incinerator residue in the waters outside of New York harbor provided that no flotables resulted from the operation, that the dumping would take place in at least 60 feet of water and clear of shipping lanes, and that the material dumped would not be considered objectionable by public health agencies generally and the Federal Water Pollution Control Administration particularly."

To determine the distance from shore which would be adequate to insure against polluting the air over land, a 1965 report by the New York City Department of Sanitation indicated that 5 miles will suffice during favorable weather conditions (offshore winds) and 10 miles would be satisfactory during inclement weather as a suitable distance for barge burning of oversized waste. For purposes of the New York City study, it was assumed that burning occurred 10 miles offshore.

An examination of the U.S. Department of Commerce Coast and Geodetic Survey Chart #1215 revealed that dumping spaces which met the U.S. Corps of Engineers physical limitations and the 10 mile offshore air pollution limits exist between the Barnegat to Ambrose sea lane and the Ambrose to Nantucket sea lane. These spaces start at approximately 18 miles from the Verrazano Narrows Bridge at depths of approximately 80 feet and continue to drop off to depths of 100 feet and more within less than 10 miles.

It is apparent that incineration and residue disposal could take place between the shipping lanes referred to without causing nuisance or pollution problems, and the locations suitable for this operation are approximately 18 miles from the Verrazano Narrows Bridge. For the New York City study, several methods of ship incinerator operation were examined for comparative costs:

(1) For the 24-hour cycle, all refuse would be loaded on board during the normal 8 hour daily collection period and burned at sea during another 8 hour period. This schedule would allow 8 hours for

loading, 8 hours for sailing, and 8 hours for incineration each day. The furnace capacity would have to be capable of burning a full day's collections in 8 hours. Based upon present incineration practice in the United States, the largest furnace which should be considered for this service is 500 to 600 tons in 24 hours. If the furnace capacity is 500 tons per day, a collection system load of 1,000 tons in 24 hours will require six 500 tons per day furnaces when burning takes place only 8 hours out of 24.

(2) For a 48-hour cycle, all refuse received during the two day period would be stored temporarily on land and then loaded into a vessel in 8 hours. This would permit 32 hours of burning out of each two days and only three 500 ton per day furnaces would be required for a 1,000 ton per day collection system.

(3) For a 72-hr cycle, all refuse received during the three day period would be stored temporarily on land and then loaded into a vessel in 8 hours. This schedule would permit 56 hours of burning out of each three days and only two and a half 500 ton per day furnaces would be required for a 1,000 ton per day collection system.

Considering the cost of incinerator furnace construction and operation, a 48- or 72-hr cycle was considered more economical than the 24-hr cycle. Thus, preliminary designs were based upon 48- and 72-hr cycles and costs were estimated. Another New York City cost study was based upon the use of a converted Liberty ship, as was the Boston study. Three methods were examined:

(1) Using the hold of the ship as a storage bin for refuse as received and the use of a clam shell traveling crane for feeding the furnace.

(2) Grinding and blowing refuse into the hold of the ship and then using a pneumatic suction system for charging refuse into the furnace.

(3) Storing refuse in containers using compactors and placing the containers into a conveyance system in the hold of the ship which would permit them to be moved to the furnaces for unloading.

Of the three methods examined, the one that appeared most suitable was the container system. Ships designed to handle and stow containers are in common usage as are compaction container refuse transfer stations. This design permitted considerable flexibility in land-based storage and handling as station capacities can be varied by adding or subtracting containers, and refuse transfer stations, using truck tractor trailers, can be integrated into the system.

Waste disposal in the ocean can be regarded as a highly selective erosion of the land in which man controls and directs the process. The crucial question appears to be one of the relative importance of human activities of this sort and the natural processes involving wind and water action which continuously introduces substances into the sea. In this regard, both rate of disposal and total waste load are variables which will have significant, immediate and long-term effects upon the receiving ocean environment.

From the point of view of the biotic component of the sea, natural erosion generally occurs at a rate compatible with the ability of resident plants and animals to adapt to the changing environment, whereas, waste disposal may place unusual stresses on living systems over relatively brief spans of time. However, it is important to recognize that total quantities of substances entering the sea from natural and man-generated sources may, in fact, be of comparable magnitude. For example, it was found that half of all the iron released each day into the New York Bight had its origin in the drainage of the major river in that area, the Raritan, with the other half coming from acid-iron waste disposal activities.

Similarly, it was estimated that the Sacramento (California) River drainage system contributes about the same amount of phosphorus and nitrogen to the ocean as does the Hyperion sewage treatment plant of the City of Los Angeles. Even though most present-day river effluents are grossly affected by human activities, having received their dissolved and particulate loads in a variety of ways, nevertheless, such comparisons are indicative of the relative importance of "natural" and "artificial" erosion.

The implications of the disposal of incinerator residue at sea are numerous and, at best, only partially recognizable and understood. Nevertheless, a few general conclusions may be drawn from existing information.

(1) Incinerator residue will dramatically modify the immediate environment. Beneficial effects which may be anticipated include improvement of existing habitats or creation of new ones with respect to particular species while harmful effects will include damage or obliteration of existing habitats of other species.

(2) Assimilation of the components of incinerator residue by the marine ecosystems will involve sedimentation, dissolution, biological uptake and cycling, and dispersal. The importance of each of these processes will vary with the particular component of the waste. The role of the biosphere in assimilation cannot be predicted accurately from available information. (The ability of marine organisms to concentrate elements has been studied in very few species and is known to show wide variation among species.)

(3) Response of organisms to the presence of incinerator wastes will not be uniform. Species difference may be expected as may differences related to stages in life histories and physiological state of the organisms. Therefore, extrapolation from one species to another or from larva to adult may be very misleading.

(4) Toxic effects from residue components (especially heavy metals) will depend upon many factors other than the simple presence of the toxin and a sensitive species. Uptake rates are influenced by associated physiological condition of the organism (i.e., age, reproductive stage) and chemical factors such as concentration and presence of other elements in the environment.

Selecting an Offshore Dumpsite for Incinerator Residue

Offshore areas of similar depths have individual characteristics. For this reason selected areas should be surveyed prior to any dumping operations. The survey should include the following.

Sediments and Bottom Contours: These area characteristics indicate the strength of natural forces which are acting on the bottom. An area of deposition would be indicated by a fairly level silt-clay bottom. An area of activity would be indicated by rapidly changing depths with rocks or clean sand and gravel formed into ripples.

The Biology: A preliminary survey of the flora and fauna would indicate the presence or absence of economically important species. Moreover, surveys before and during dumping operations would indicate changes in abundance. It is not altogether unlikely that incinerator residue would attract sport fishes in the same manner as artificial reefs. Fishes and crustacea were apparently attracted to one of the experimental dumps used in the study.

Depth, Bottom Currents and Wave Activity: These parameters will determine the degree of movement activity at a dumping site, a fact demonstrated at both the shallow water and the deep water sites studied. The shallow water site tidal currents were strong and while the wave activity was much less than offshore wave activity, the shallow depth allowed much of this activity to reach the bottom.

Consequently, residue was distributed widely. At the deeper water site bottom currents were less strong and even the strong wave activity was largely filtered out by the increased depth. Can movement was slight. Also, in the shallow water, objects were observed to be moved in the direction of wave propagation (toward the shore) whereas, in the deeper water, objects were observed to be distributed in the direction of predominating currents.

An incinerator residue dump site should probably be at least 61 m deep and the area at this depth should be at least 1 square mile to contain the residues after one year of movement. Bottom currents appear to be insufficient to move cans by themselves. However, extreme wave activity will reach the bottom at this depth and material will be shifted according to the direction of bottom currents. Therefore the site area must be large enough to contain this movement.

If the materials migrate into shallower waters, movement will probably be exponentially accelerated. During a one year period a percentage of the residue will decay and become less subject to movement. One other consideration is that repeated dumping episodes over the years will eventually shoal an area.

Residues should be thoroughly incinerated and crushed, otherwise, some unincinerated materials have a low density. Some are apt to float. Those unincinerated materials that reach the bottom will move faster and farther than cans. Crushing the cans and breaking bottles reduces their tendency to float or to move under extreme oceanic conditions.

NEW YORK AND MID-ATLANTIC BIGHT

The following sources were used for the information provided in this chapter.

PB 198 032
PB 224 793
PB 229 761
PB 240 917

EPA reports on Ocean Dumping in the New York Bight dated

July 1973
April 1974
July 1974
April 1975

The waste products of technology reach marine systems through a variety of pathways. These include: the ocean dumping of dredged sediments which may contain pollutants as well as natural materials; the direct discharge of municipal sludges and industrial waste products from barges or ships; and discharge of waste material from cities and industries directly to the oceans through ocean outfalls.

These types of ocean discharge are common in each of the major U.S. Coastal areas, the Pacific, the Gulf of Mexico, and the East Coast including Puerto Rico. The disposal of dredge spoil occurs in all areas. The discharge of municipal sewage sludge by barge dumping is prevalent on the East Coast in the Mid-Atlantic Bight. The disposal of industrial wastes by barge dumping is practiced mainly in the Gulf of Mexico and off the East Coast.

The discharges of dredge spoil and sewage sludge in the New York Bight have degraded the environmental quality in the area. This degradation is of local but not regional impact.

The disposal of dredge spoil will remain a problem as long as navigation of coastal and inland sea routes is maintained. The widening and deepening of existing channels will cause the volume of both virgin dredged materials and maintenance dredging spoil to increase. The concentration of contaminating substances in dredge spoil such as organic materials, heavy metals, and microorganisms should decrease as effluent control methods in the form of better sewage treatment and better industrial discharge control are realized.

The volume of sewage sludge to be disposed of can be expected to increase as higher levels of sewage treatment are reached. The impact of this higher volume of sludge will be substantial if disposed of at sea, unless sludge digestion reduces substantially the organic fraction and microorganism levels and care is taken to reduce to a minimum the heavy metal and chlorinated hydrocarbon content of the sludge. The reduction of heavy metal and chlorinated hydrocarbon content in the disposal sites will be achieved only as a result of elimination of the sources of these materials, since their removal from dredge spoil and sewage sludge is not expected to be feasible.

It can neither be demonstrated that the discharge of industrial waste chemicals to designated disposal areas using presently specified discharge methods is harmful or not harmful to the local ecosystems. The characteristics of the materials involved and the disposal sites indicate that the effects of disposal at the Puerto Rico, Gulf of Mexico chemical, and Atlantic deepwater sites may have regional as opposed to local effects. Inadequate information is available regarding the deepwater sites being used for industrial waste dumping, the history of discharge, and their nature.

Data on toxicity, bioaccumulation, biodegradability, and physical and chemical behavior of the materials and mixtures being discharged are needed. Proper evaluation of presently discharged materials will require substantially increased effort to evaluate social and economic factors of product need, ultimate treatment or disposal methods, and the effect of the waste components on the marine system.

BACKGROUND STUDY OF THE NEW YORK BIGHT

During the period 1960-1968, the mass of waste solids from New York City dumped in the ocean increased from an average of 6.8 million tons per year during the period 1960-1963 to 8.6 million tons per year for the period 1964-1968. As this material was dumped directly into coastal waters it is instructive to compare this mass of wastes dumped with that delivered to the Atlantic Ocean by rivers draining the Atlantic Coast of the United States. For many rivers, such as the Hudson or Connecticut, there are no modern data; the only estimate of sediment loads in the United States Atlantic Coast rivers was made early in the present century. Such data (Table 6.1) are useful, however, to indicate the order of magnitude of natural river sediment loads.

The available data indicate that no U.S. Atlantic river has a natural sediment load approaching the mass of solids dumped annually by the New York metropolitan region into the ocean. The Potomac River and its tributaries, has the largest sediment loads of rivers near the New York area, approximately 2.5 million tons, but its sediment load probably does not reach the ocean and is likely deposited either in the Potomac River estuary or in nearby Chesapeake Bay.

TABLE 6.1: SEDIMENT DISCHARGE OF SOME U.S. ATLANTIC COAST RIVERS

North Atlantic Region	Annual Discharge of (10^6 tons per year)	Suspended Sediment (tons per year/km^2 of drainage basin)
NORTH ATLANTIC REGION	6.1	15
Connecticut River	0.089	3.1
Hudson River	0.40	12
Raritan River	0.071	29
Delaware River	0.72	21
Susquehanna River	0.96	14
Potomac River	1.4	37
James River	1.0	36
SOUTH ATLANTIC REGION	21.8	68
Roanoke River	2.5	99
Peedee River	1.6	59
Santee River	3.4	90
Savannah River	2.6	90
Ogeechec River	1.2	87
Altamaha River	3.0	83

Furthermore, waste solids from the New York area exceed the sediment discharge of all rivers emptying into the Atlantic Ocean between the U.S.-Canadian border and Chesapeake Bay. If the 1909 river data are still valid, the New York metropolitan region alone discharged an amount of sediment equal to about 31% of the total river-sediment transported into coastal waters along the Atlantic Coast of the United States.

The amount of waste solids discharged by the New York metropolitan region is comparable to that of the world's major rivers as a source of sediment entering the coastal ocean, although it is by no means the largest source. Excluding the Amazon River which flows into the equatorial Atlantic Ocean and the Mississippi River which discharges into the Gulf of Mexico, the New York metropolitan region may well be the largest source of sediment entering the North Atlantic Ocean. No data are immediately available for comparison on the amount of waste solids discharged from European metropolitan regions.

To put the solid discharge in perspective, it is instructive to consider the waste-solids discharge on a per-capita basis and on an areal basis. For this purpose, two different assumptions are made for the calculations. Assuming that all the solids were discharged from the core of the metropolitan region (Manhattan, Bronx, Queens, and Brooklyn, New York; Essex and Hudson counties, New Jersey) the discharge for the 1960-1963 period was 0.75 tons per person per year (based on the 1960 census population of about 9 million). This is 2.1 kg (4.5 pounds) per person per day. On an areal basis, the waste solid discharge is about 6,100 tons per km^2 of core metropolitan area per year. Alternatively, it can be assumed that the waste solids were discharged by the entire metropolitan region lying within 50 km of the Battery (the southern tip of Manhattan Island). On this basis, per capita solid discharge (1960 census population about 16 million) for the region was 0.42 tons per person per year or 1.2 kg (2.5 pounds) per person per day. The solid discharge resulting from navigation dredging and for other activities seemed to be about equal for the period covered by this study. On an areal basis, the entire metropolitan region produced about 1,100 tons per km^2 of metropolitan area per year.

This sediment discharge per unit area greatly exceeds the sediment yield per unit of drainage area of major North American river basins and is exceeded only by rivers draining India and China. However, the yield of solids per km^2 drained falls well within the range of sediment yields observed for housing construction areas.

The impact of these solids on the coastal ocean can be appreciated only when it is realized that little riverborne sediment enters the Atlantic Ocean at present (1968). Nearly all the sediment transported by U.S. Atlantic Coast rivers is trapped and normally deposited in the estuaries. Consequently, the New York region is not only supplying a large quantity of solids but is dumping on the Continental Shelf where little other sediment is beng deposited to dilute or to bury the waste deposits.

Dumping of waste-solids appears to be significantly different from natural processes by which riverborne sediment is deposited on the ocean bottom. Waste dumping operations are highly localized in both space and time and usually involve the release of several thousand cubic meters of solids whenever a barge or dredge is emptied. Thus, the ocean bottom beneath the dump area may receive a relatively thick layer of waste almost instantaneously. The designated disposal areas are small (a few square kilometers) although it is probable that dumping actually occurs over much larger areas owing to navigational errors, and changes in operations necessitated by adverse weather conditions, and illegal dumping operations.

Effects of waste-solid discharge on the continental shelf are difficult to evaluate. Little is known about the distribution or characteristics of the deposits formed by these waste discharges, nor is the ocean bottom off the entrance to New York Harbor well known. Furthermore, there are no data to permit a comparison with conditions existing before waste disposal operations began in this area.

OCEAN DISPOSAL STUDY OF THE NEW YORK BIGHT

The Ocean Dumping Act (Public Law 92-532, Marine Protection Research, and Sanctuaries Act of 1972) specifically charges the National Oceanic and Atmospheric Administration (NOAA) of the U.S. Department of Commerce with responsibility for monitoring of dumping areas and for comprehensive research on effects of ocean dumping. The Middle Atlantic Coastal Fisheries Center is one of a series of seven centers established recently by the National Marine Fisheries Service (NMFS), an organization of NOAA. The Center is a consolidation and integration of the Sandy Hook Marine Laboratory, the Oxford Biological

Laboratory, the Milford Biological Laboratory, and the former Ann Arbor Technological Laboratory (now based at Milford).

The mission of the Center is to develop and establish a cooperative multidisciplinary research program on the biology and ecology of the living marine coastal organisms of the North Atlantic Ocean, especially in the zoo-geographic area known as the Middle Atlantic Bight (MAB).

The MAB includes the coastal and shelf areas between Nantucket Shoals, off the Massachusetts coast, to Cape Hatteras, North Carolina, and thus, falls outside the study area of this report. The New York Bight constitutes one of the most intensively used coastal environments in the world and this area is the major immediate responsibility of the Ecosystems Investigations section of the Sandy Hook Marine Laboratory.

Field and laboratory studies concerned with the effects of ocean disposal of sewage sludges, dredging spoils, industrial wastes, and thermal additions have been carried on at the Sandy Hook Laboratory. Cooperative cruises with personnel from other NMFS or NOAA facilities, or academic institutions or organizations, have been part of the recent and ongoing research programs.

Comprehensive biological reports/data have been prepared by the NMFS at Sandy Hook; U.S. Army Corps of Engineers at the Coastal Engineering Research Center (CERC); EPA, Edison, N.J.; the FDA Region II; the New Jersey DEP; and the New York State Dept. of Environmental Conservation. Studies of typical biological parameters have considered population trends of phytoplankton, zooplankton, nekton, benthos, and tests of coliform bacteria and other pathogenic organisms.

Additional tests included bioassay and toxicity, biomass, primary productivity, chlorophyll, BOD and nature-type of detritus material. Surveys also include statistical data on commercial and sport fisheries, indicator organisms, as well as radiological monitoring of the biota. A number of the larger crustaceans, such as crabs and lobsters, collected from the disposal area have been found to be diseased. Diseased (Finrot) finfish have been retrieved from inside the disposal areas. The large number of coliform bacteria found in the New York Bight indicates the presence of pathogenic bacteria.

Coliform bacteria were present in high concentrations throughout the areas receiving dredging spoils and sewage sludges. High concentrations have even been found outside the actual dumping areas. Additional studes are continuing in order to determine the effects of known disposal amounts of heavy metals on the physiology of larval and adult crustaceans.

Species diversity and total number of organisms was markedly reduced in those areas directly affected by sewage sludge and dredge spoil disposal. Dumping characterized a reduction in the number of species present, as well as reduced numbers of individuals of particular species.

Prolonged detrimental effect on the zooplankton and benthic organisms by ocean disposal of industrial acid wastes was not substantiated. Existing sewage sludge and dredge spoil practices in the New York Bight have degraded the marine benthic communities, produced large amounts of floatable materials, and resulted in deteriorated waters and marine sediments.

A complete assessment of the environmental studies conducted in the New York Bight was prepared by Coastal Engineering Research Center. Interdisciplinary, short-term investigations related to the effects of ocean dumping in the New York Bight were contracted by CERC as directed by the Office of the Chief of Engineers. Studies made by the Sandy Hook Laboratory of the NMFS, the State University of New York at Stonybrook, the Woods Hole Oceanographic Institution, and the Sperry Rand Corporation were reviewed by the Smithsonian Institute and CERC.

The studies included hydrographic, geological, chemical, biological investigations, and a feasibility study for a remote-controlled electronic sensing system that could assist regulating agencies in detecting the location and dump status of waste disposal vessels operating in the Bight.

The New York Harbor complex and the nearby offshore disposal sites rank as one of the largest grossly polluted areas in the United States. The problem has not been ignored, as demonstrated by the extensive bibliography collected on the physical, chemical, and biological studies conducted in the New York Bight (NYB).

Federal, state, and local agencies, along with educational institutions, have for years conducted water quality monitoring and sampling studies in the harbor and the offshore dumping grounds. The basic obstruction to a solution has been lack of communication. Failure to integrate these efforts into a viable program for interagency coordination and the exchange of water quality data has contributed to the belief that not enough is known about the effects of waste disposal in the NYB.

The EPA Water Quality Protection Branch, Division of Water Quality and Non-Point Source Control, through a contract with IEC, developed an Initial Network to provide assistance, coordination and indoctrination of local users in the philosophy of the EPA National Computer and Data Processing System. Under this proposed plan, all monitoring in the NYB would be coordinated to stimulate establishment of Information Centers at local, state and regional levels, in support of improved information exchange and use by all agencies involved.

The liaison established between the key contacts of the various agencies in formulating the NYB Initial Network established communications exchange which provided the main body of information contained in this report.

The U.S. Army District Engineer, New York, was designated Supervisor of New York Harbor under the River and Harbor Act approved by Congress in 1888. Pursuant to the provisions of that Act, the Supervisor designated certain areas off the entrance to the New York Harbor as waste disposal grounds, and conducted a program of issuing permits to towing firms that transported the waste materials. During the period from July 1, 1972 to February 28, 1973, 349 dumping permits were issued which permitted 12,160,464 cubic yards of material to be dumped in the designated areas.

Effective April 23, 1973, the Marine Protection Research and Sanctuaries Act of 1972 authorized the Administrator of the EPA to issue permits for ocean dumping and to establish and apply criteria for reviewing and evaluating permit applications. The U.S. Army Corps of Engineers will continue to issue permits or regulations for federal projects for ocean dumping of dredge materials upon concurrence by EPA to ensure that the criteria have been complied with.

Under this Act, the United States Coast Guard is authorized to conduct surveillance and enforcement activities to prevent unlawful dumping. EPA is also authorized to designate recommended sites and times for dumping, protect critical areas, and designate sites and times within which certain materials may not be dumped.

Under interim regulations, permits for dumping will be issued for the sites currently in use. Final regulations will be issued within one year, based upon comments made about the interim regulations and the information collected while they are in effect. The information collected from New York will be correlated with other regional inputs by the EPA Headquarters staff in an extensive review and evaluation of the existing problems on a national level, which will assist in establishing a plan for the implementation of final regulations to control ocean dumping.

Disposal Areas

Disposal areas have been established by the Supervisor of New York Harbor in three major localities: Hudson River, Long Island Sound, and the Atlantic Ocean off the entrance to the New York Harbor. Seven areas in the Hudson River and nineteen areas (seven presently active) in the Long Island Sound are designated primarily for the disposal of materials dredged from local harbors and waterways. An area off Eatons Neck in Long Island Sound has been used for the disposal of clean cellar dirt and wrecks, particularly when inclement weather and rough seas make trips to the ocean disposal sites too hazardous.

The scope of this report concerns the six separate dumping grounds in the Atlantic Ocean, which provide for the disposal of mud and one-man stone, cellar dirt, sewer sludge, wrecks, waste acid, and chemical (toxic) wastes.

Disposal Site Geography and Uses

The disposal sites are located in a part of an area called the New York Bight (NYB). The NYB is the shallow ocean area shoreward off the limits of the continental shelf, along an indentation of the Atlantic Coast extending about 200 miles from Cape May, New Jersey, to Montauk Point (the eastern end of Long Island), New York. The five dumping areas nearest to shore vary from about 10 to 22 miles south of the Long Island shore, and from about 5 to 14 miles east of the New Jersey shore. The chemical dumping ground is located 106 miles offshore on the edge of the continental shelf. The descriptions that follow are referenced to the Ambrose Channel Light.

Mud Dumping: A mud dumping ground is located at a point not less than 4 nautical miles, bearing 198°00' true from Ambrose Light in not less than 60 feet of water. Substances to be dumped in this area consist of material dredged from vessel berths, anchorage grounds, and channels; clean earth; and steam ashes from fossil-fueled electric power generating stations. Most of the materials deposited result from improvement and maintenance of channels and anchorage areas by the Corps of Engineers under projects authorized by Congress. The material is transported in bottom dump scows owned and operated by dredging and marine construction contractors, and seagoing hopper dredges owned and operated by the Corps of Engineers.

The original Mud Dumping Ground was established in 1888, shortly after enactment of the Supervisor Act. The site was selected to avoid creation of a hazard to navigation. As the designated area decreased noticeably in depth, its location was changed a number of times, finally to its present site more than 33 years ago.

Cellar Dirt Dumping: A cellar dirt dumping ground is located at a point not less than 4.7 nautical miles bearing 170°00' true from Ambrose Light, in not less than 90 feet of water. The material disposed of in this area consists primarily of earth and rock from cellar excavations and broken concrete, rubble, and other nonfloatable debris from building demolition and highway construction work.

Most of this material originates on the island of Manhattan where, because of its built-up condition, there are no upland disposal sites available. Drilled and blasted rock from channel improvement work is also disposed of in this area under contract with the Corps of Engineers. The material is transported to this area in dump scows owned by marine contractors and towing companies.

The original Cellar Dirt Dumping Ground was selected in 1908 so as not to endanger navigation, but has been changed several times as the depths decreased. The present area has been in use for more than 33 years.

Sewer Sludge Dumping: A sewer sludge dumping ground is located 4.5 nautical miles, bearing 124°30' true from Ambrose Light, in about 72 feet of water. The sewage wastes are either in raw or treated state or are in a digested form, and are disposed of at this dumping ground by cities in New York and New Jersey.

The Sewage Sludge Dumping Ground was selected in 1924 pursuant to a stipulation reached by the Supreme Court of the United States, in an action brought by the City of New York, to prohibit the deposit of sewage by the Passaic Valley Sewage Commission into the waters of Upper Bay, New York Harbor. The site was chosen to avoid offensive discoloration and prevent solids from washing up onto Long Island and New Jersey beaches, as well as to avoid endangering navigation.

Wreck Dumping: A wreck dumping ground is located at a point 14.3 nautical miles bearing 168°30' true from Ambrose Light, in not less than 200 feet of water. This area is utilized for the disposal of obsolete vessels, wrecks, and other submerged obstructions to navigation. The Corps of Engineers carries out its obligation under the law to remove and dispose of sunken vessels and other obstructions to navigation and contracts for their disposal in this area.

Waste Acid Dumping: During the winter season, a waste acid dumping ground is located with its northwesterly corner at a point not less than 9.2 nautical miles, bearing 145°00' true from Ambrose Light. The area extends south of latitude 40°20'N and east of longitude 73°43'W. During the summer season, the area is located with its northwesterly corner at a point not less than 10.7 nautical miles bearing 135°00' true from Ambrose Light, and extends south of latitude 40°20'N and east of longitude 73°40'W. Depths in both dumping areas are about 90 feet.

The Waste Acid Dumping Ground was established in 1948 and is used for the disposal of dilute acid wastes containing various dissolved solids, including iron compounds. These wastes originate in a number of industries, principally in New Jersey, and are transported in specially constructed, rubber-lined tank barges. The wastes are released under water while the vessel is under way to attain maximum dilution and dispersion. The vessels, after reaching the dumping ground, head on a southeasterly course while discharging half of their cargo and, after a wide U-turn, proceed on a northwesterly course discharging the balance of their cargo in the dumping ground.

Waste Chemical (Toxic) Dumping: A waste chemical (toxic) dumping ground is located at the edge of the Continental Shelf with its northwesterly corner approximately 106 nautical miles, bearing 145°00' true from Ambrose Light. It is defined as the area lying south of latitude 39°00'N; west of longitude 72°00'W; north of latitude 38°30'N; and east of longitude 72°30'W. Depths are greater than 7,000 feet.

The Waste Chemical Dumping Ground was established in 1965 following the receipt of requests from industries to dispose of chemical wastes which State health authorities refused to allow to be disposed of in sanitary landfills or into streams because of possible contamination of the potable groundwater supplies. The actual limits of the area were recommended by the U.S. Fish and Wildlife Service, which was one of several Federal agencies consulted in determining where disposal of such wastes should be permitted in open waters. Because of its distance offshore, the cost of disposal is high, which limits the use of this area.

Radioactive Waste Dumping: A radioactive waste dumping ground is located at a point not less than 141 nautical miles, bearing 145° true from Ambrose Light, in not less than 200 fathoms of water.

High Explosives and Chemical Dumping: A high explosive and chemical dumping ground is located at a point not less than 110 nautical miles, bearing 130° true from Ambrose Light. Small quantities of toxic wastes and high explosives have been disposed of intermittently in past years; however, data on quantities of the wastes and their sources are not readily available.

Alternative Sewage Sludge Dumping: A proposed alternative sewer sludge dumping site is tentatively located at latitude 40°25.7'N, longitude 73°11.5'W, which is a point 29.2 nautical miles bearing 094° true from Ambrose Light, in about 100 feet of water. This area is 3 nautical miles square, centered at the lighted whistle buoy BW "NB" which is 12.3 nautical miles, bearing 174° true from Fire Island Light (12.1 nautical miles, from Great South Beach at Fire Island) and 25.3 nautical miles, bearing 089° true from the center of the present sewer sludge dumping ground. This site was tentatively selected as an experimental location where a selected amount and type of digested domestic sewage sludge will be discharged under varying controlled conditions. The overall direction of this research project is provided by the staff of the National Coastal Pollution Research Program of EPA, who also are the principal scientific participants in the field and laboratory work. The NOAA Sandy Hook Marine Laboratory is providing assistance as a base of operations for field studies and some vessel time. Additional vessel time, sampling assistance, analytical service, and liaison with the Corps of Engineers and the City of New York are being provided by the Surveillance and Analysis Division of EPA at Edison, New Jersey.

It is estimated that 2,230,000 cubic yards of sludge are added annually to the New York Harbor complex because of the discharge of 480 MGD of raw sewage from the east and west side of Manhattan, Red Hook, Brooklyn, and Staten Island. These sludge accumulations are dredged along with the other bottom materials and deposited in the mud dumping ground. For the past 40 years, it has been the common practice of 15 New Jersey coastal communities to store accumulated sludge during the summer season, and discharge this sewage sludge into the Atlantic Ocean via effluent outfall pipes approximately 1,000 feet from the shoreline (less than 5 miles from the mud dumping ground).

In February 1972, the Federal Court issued a permanent injunction discontinuing this practice. Consequently, 5,764,000 gallons of sludge is barged to sea until an adequate technical solution for an alternative method of disposal can be achieved.

Regional Economy

Population: The population of the 31-county New York Region is approximately 20 million. It is expected that by 1980, the population will be 23 million and by 1995, approximately 29 million. The distribution of population shown in Table 6.2 represents the 5 counties that border the dumping areas described above. A reasonable estimate by the Tri-State Regional Planning Commission in New York City indicates a projected increase of more than 668 thousand by 1985.

TABLE 6.2: STATISTICAL DATA – COUNTY POPULATION

	Population	
County	1970 Census	1985 (Projected)
Nassau	1,428,080	1,700,000
Queens	1,986,473	2,090,000
Kings	2,602,012	2,470,000
Richmond	295,443	480,000
Monmouth	459,379	700,000
Totals	6,771,387	7,440,000

Estuarine Economics: Beach Recreation – Approximately 5 nautical miles west of the mud dump ground and 10 nautical miles north of the sewer sludge dumping ground is the shoreline of the New York-Northern New Jersey estuarine region which supports an annual $2 billion recreation industry. The shoreline is mostly fronted by low sandy beaches and the shore development is primarily recreational and residential with some commerce and industry. Shore ownership is Federal, public, and private. The shoreline provides 47.8 miles of public beaches where more than 65 million visits were recorded during the 1970 beach season which begins in the last week of May and ends the second week in September (approximately 113 days). Statistics are shown in Table 6.3.

TABLE 6.3: STATISTICAL DATA–BEACH RECREATION

	- - -Shore Ownership, (miles) - - -			Total Shore Length,	1970 Beach
County	**Federal**	**Private**	**Public**	**(miles)**	**Attendance**
Monmouth	6.1	9.4	11.3	26.8	6,940,000
Richmond	0.3	3.7	9.0	13.0	698,000
Kings	0.02	1.6	3.5	5.12	21,818,100
Queens	1.0	2.0	7.0	10.0	22,372,000
Nassau*			17.0	17.0	13,900,000
Totals	7.42	16.7	47.8	71.92	65,728,100

*Includes Jones Beach, and approximately 10 miles of beaches in Suffolk County including Captree State Park at Fire Island Inlet.

The National Park Service has proposed setting aside five areas totaling 20,000 acres of land and water for the Gateway National Recreation Area. When completely developed, this area would be capable of serving more than 50 million visitors annually.

Commercial Fishing and Shellfishing – The continental shelf extends from the New York-New Jersey region, offshore to the 100-fathom (600-foot) contour. Off New Jersey, the 100-fathom contour ranges between 60 and 105 miles offshore. Commercial fishing and shellfishing for much of the northeast coast of the United States relies heavily on the continental shelf. Surf clams, lobsters, and 40 species of fish are commercially important to New Jersey.

Table 6.4 represents the New Jersey dockside weights. They do not include foreign or out-of-state landings of fish and shellfish caught off the New Jersey coast. Values, likewise, are representative of New Jersey only, and are dockside prices as opposed to generated values. Generated values often reach three to five times the dockside values.

TABLE 6.4: STATISTICAL DATA–NEW JERSEY COMMERCIAL FISHING

Year	Total Weight	Total Value	Species of Greatest Value
1956	513,807,546 lb	$15,238,931	Menhaden & Surf Clams
1957	464,924,418 lb	12,224,923	Menhaden & Surf Clams
1968	126,369,000 lb	10,609,000	Surf clams & lobsters
1969	92,529,380 lb	10,893,371	Surf clams & lobsters

Sport Fishing – The New York Bight is an important hatchery and nursery ground for numerous fish (33 species) of recreational importance. Many of these fish do not spawn in the Bight, but the eggs and larvae are transported there by currents. Some of the former dumping grounds for dredged materials, cellar dirt, garbage, and other wastes are now favorite fishing spots, locally known as "The Mud Hole," "The Tin Can Grounds," "The Subway Rocks," and "The Acid

Grounds." Thousands of private and party charter boats fish for migratory species that move through these areas at different times of the year. The most important sport fish (food fish) are bluefish, weakfish, codfish, Atlantic mackeral, and scup (porgy). Winter flounder, striped bass, whiting, summer flounder (fluke) and blackfish (tautog) are found inshore.

The State of New Jersey in 1954 estimated that in the months of April through September, 44.28% of the total catch was by sportsmen, or 13,302,154 pounds (sport) versus 16,735,033 pounds (commercial). Sport fishing in the deeper waters has been limited to the catching of sailfish, tuna, marlin, and dolphin.

Permit System

To assure that waste materials are disposed of in the approved dumping grounds, permits are issued on a routine quarterly basis to towing firms that transport the waste materials to sea. This permit system was one of the functions of the New York District Corps of Engineers under provisions of the River and Harbor Act of 1888. The Corps of Engineers Deputy Supervisor of New York Harbor, during January 1973, advised the current permittees that under the new Marine Protection Research and Sanctuaries Act of 1972 (Ocean Dumping Act), requests for dumping permits to cover the period after April 23, 1973 should be addressed to the EPA Region II Administrator in New York City, who became the authorized official to issue permits for dumping or transporting for dumping of all materials, except dredged material, into the NYB.

Applications for deposit of dredge material will continue to be processed by the Corps of Engineers. The following is a description of the program which was conducted by the Corps of Engineers as related to dumping of waste materials in the Atlantic Ocean.

Supervisor of New York Harbor: The permit program required by the Act of 1888, as amended July 12, 1952, is an ongoing activity of the Supervisor of the Harbor, administered by the Harbor Supervision and Compliance Section. During the three-month period ending June 30, 1971, 127 individual permits were issued for the disposal of material in the designated dumping areas.

During the period July 1, 1972 to February 28, 1973, 349 dumping permits were issued. The Compliance Section maintains the permit records and forms. Data are directly extracted from the permit application and entered into a ledger. The permittee mails a supplemental sheet which certifies that the scows have delivered or discharged materials at the location and time specified on the permit. Supplemental sheets are usually returned after the expiration date of the permit (issued quarterly) and, at that time, the amount (cubic yards) is entered into the ledger.

Surveillance of the dumping operations is undertaken by a 65-foot patrol boat with inspectors aboard who note the time a vessel leaves and the time of its return in order to determine whether the intervening elapsed time was sufficient to go to the approved site. The patrol boat checks the actions of the vessels at the dump site on a spot-check basis depending on weather conditions. The patrol boat is used primarily for inspections of waterways in lower New York Bay and patrols the entrance channels to keep them clear of interference by fishing craft or other boats in order to ensure safe navigation of deep-draft vessels. Other patrol boats operate in upper New York Bay and Long Island Sound. Inspections of shorefront facilities, such as industrial plants, oil refineries and shipyards, are conducted by Inspectors utilizing government vehicles equipped with two-way radios to ensure that industrial waste or refuse is not being discharged or deposited into the navigable waters.

It is estimated that 2,230,000 cubic yards of sludge are added annually to the New York Harbor complex because of the discharge of 480 MGD of raw sewage from the east and west side of Manhattan, Red Hook, Brooklyn, and Staten Island. These sludge accumulations are dredged along with the other bottom materials and deposited in the mud dumping ground.

For the past 40 years, it has been the common practice of 15 New Jersey coastal communities to store accumulated sludge during the summer season, and discharge this sewage sludge into the Atlantic Ocean via effluent outfall pipes approximately 1,000 feet from the shoreline (less than 5 miles from the mud dumping ground). In February 1972, the Federal Court issued a permanent injunction discontinuing this practice. Consequently, 5,764,000 gallons of sludge are barged to sea until an adequate technical solution for an alternative method of disposal can be achieved.

Analysis of Dumping Operations

Problems of Dumping: The six-mile radius sludge dump closure area in the NYB, and the six-mile radius dump closure off Cape May, are the two areas in the Atlantic Ocean off the New York-New Jersey coastlines that are officially closed (since 1970) to shellfishing by the Food & Drug Administration (FDA), under the National Shellfish Sanitation Program (NSSP). This program requires that all shellfish growing areas not remote from pollution sources be classified for sanitary quality. The classification must be made on the basis of a comprehensive sanitary survey and laboratory analysis in accordance with the NSSP Manual of Operations provisions.

During 1971-1972, such a study was planned and initiated by the FDA Region II, and was conducted jointly with the FDA Northeast Technical Services Unit, Davisville, Rhode Island; the Sandy Hook Marine Laboratory; the New Jersey Department of Environmental Protection; and the New York State Department of Environmental Conservation. Based on the survey's bacteriological data, the offshore areas between land and the six-mile radius sludge dump were closed to shellfishing. The pollution sources that have made this interim closure necessary are as follows:

(a) Thirty-three sewage treatment plants discharging through ocean outfalls between Sandy Hook and Beach Haven Inlet.

(b) One large chemical firm discharging industrial wastes 3,500 feet offshore.

(c) The combined storm-sanitary wastes and untreated sanitary wastes from the New York City metropolitan area flowing along the coastlines (400 MGD untreated and 1,100 MGD treated but not chlorinated during the nonsummer months).

(d) The sewage and dredge spoil dump sites which have an undetermined impact on the water quality outside the six-mile closure. Exceptions were noted during the last survey to several bottom water samples which exceeded the surface water sample results. Other than the possibility of short dumping and errors in navigation by sewage sludge barges, a ready explanation of this data is not available.

Mud Dumping: It is estimated that 45% of the dredge spoil deposited is polluted from industries, municipalities, and other sources near the harbors and channels being dredged. Pollution factors include biochemical and chemical oxygen demand, volatile solids, oil and grease, phosphorous, nitrogen, iron, silica, color, and odor. In dredge spoil deposited at the mud dump, average concentrations are estimated as follows: Copper, 200 parts per million (ppm); silver, 143 ppm; tin, 570 ppm; and chromium, 400 ppm.

Sewer Sludge Dumping: About 90% of the national total of sewage sludge dumped in the ocean is disposed of at this locality. The material contains significant quantities of heavy metals and oxygen-demanding materials. Preliminary analyses of sludge samples indicate heavy metals, chromium, copper, lead, tin, and zinc. Samples of clams taken up to three miles from the center of this dump contained coliform counts that exceeded permissible levels, and the area six miles in radius is closed to the harvesting of shellfish for human consumption. Slightly less than 4 million cubic yards were dumped in 1972.

Upgrading the present treatment facilities to secondary treatment, plus treatment of the present 480 MGD of raw sewage will significantly increase the volume of sludge to be disposed of. It is estimated that the total sludge volume will increase to approximately 15 million cubic yards. Unless alternative sludge disposal methods are developed, the additional sludge will be dumped at this site.

Waste Acid Dumping: The material dumped at this site (3,050,414 cubic yards in 1972) is difficult to identify, considering the extreme variation in physical and chemical properties of these liquid wastes. Not enough data are available to characterize and identify the various types of waste liquids.

(a) It was concluded as a result of a 1972 report on the effects of waste disposal in the NYB that disposal of dredge spoils and sewage sludges has had a significant, and often deleterious, effect on the living resources of the NYB.

(b) The wastes from the New York metropolitan area are now the largest source of sediments discharged directly into the North Atlantic Ocean from the North American Continent.

(c) The potential danger of highly polluted and toxic wastes disposed of less than five miles from the bathing beaches could cripple the estuarine tourist industry. Because of the wide publicity given to dumping, it is estimated that if only 10% of the potential visitors believe the waters polluted and avoid the shore areas, the cost to the estuarine economy would exceed $20 million per year.

(d) New York fish and shellfish landings amounted to 40,800,000 pounds, valued at $14 million and the New Jersey surf clams and lobster landings exceeded $10 million in 1969. Data indicate that there are higher concentrations of fecal coliform in sediments and shellfish adjacent to the dump areas. Finfish feed at the periphery of the waste disposal areas and are exposed to the toxic and pathogenic contents of these wastes.

A potentially valuable resource has been affected by present dumping practices, as evidenced by the FDA six-mile closure and the more recent interim three-mile closure to shellfish harvesting.

(e) It would be imprudent to shift dumping locations because evidence is not given to indicate that it would be less harmful to dump the sewage sludge and dredge spoil elsewhere than where these wastes are presently dumped.

(f) Harbor dredgings dumped at the mud disposal site are finding their way to the New Jersey coastline and the invasion of red tide (a proliferation of toxic microorganisms) at the beaches may have its genesis in the nutrient materials at the dump site.

During 1970, a labor strike of tugboat operators forced the Governor of New Jersey to proclaim a state of emergency. The state was obliged to commandeer three ocean-going barges and their crews to effect the disposal of sludge from six of the state's largest sewage treatment plants to prevent the release of 500 MGD of untreated sewage and industrial wastes into the rivers and bays.

During 1972, the New Jersey State Department of Environmental Protection held a public hearing on a proposed New Jersey Ocean Disposal Control Regulation. The Governor has proposed that dumping of waste products on the continental shelf be prohibited and should require a minimum distance of 100 miles offshore for dumping. The ocean disposal control regulation was not adopted and the original transcription of the hearing and recommendations were turned over to the EPA Region II.

Water Quality Monitoring and Sampling

Water quality monitoring is defined as having three major components: (1) The acquisition of data at approximately the same location at some repeat time frequency (arbitrarily established as at least once per year); (2) The processing of data into a usable format; and (3) The use of that data/information for a purpose. The agencies that maintain a monitoring program in the NYB conduct water quality surveillance programs in the adjacent waters of the New York Metropolitan region. The ocean disposal areas were excluded from the routine monitoring programs because of territorial jurisdictions and the lack of funds for personnel and ocean-going vessels.

Sampling is considered to be a one-time occurrence of the collection of information, and storage of that information in the form of reports. Comprehensive studies and extensive water quality sampling in the dumping areas have been conducted by many federal agencies and research institutions. The major studies, conclusions and recommendations of these studies, and the ongoing and proposed programs related to the dumping areas are summarized in this section. Most of these studies were restricted because of limited funds, and additional follow-up surveys to obtain synoptic data over a comparatively long period were not performed for the same reason.

Reevaluation of Ocean Dumping by the NY District Corps of Engineers

The economics, design problems, and the time needed to implement alternatives to dumping at sea have been submitted by the chemical companies to the New York District Corps of Engineers, under an evaluation program conducted during 1971, on the effects of disposal activities on water quality and water chemistry in the NYB. As part of this analysis, it was requested that the various chemical companies applying for dump permits provide the following information.

(a) Hypothetical analysis of behavior of waste materials subsequent to dumping in proposed locations, including specifically:
   - (1) Fractions of load which would float, would sink immediately, or would dissolve immediately, and the composition of each fraction.
   - (2) Rate of hydrolysis.
   - (3) Rate and pattern of dispersal from time of release until no longer identifiable.
   - (4) Particle size of insoluble fraction.
   - (5) Kinds and amounts of substances that would leach out of insoluble fraction, and rate of leaching.

(b) Operational data, including:
   - (1) Volume and weight loaded per ship.
   - (2) Volume and weight dumped per ship.
   - (3) Number of trips per year and frequency.
   - (4) Total amount of material to be disposed of annually.
   - (5) Description of dumping mechanism and procedures to be followed during dumping operation (i.e., movement of ship, one release, or a series, etc).

Dumping permits were held in abeyance by the Corps of Engineers pending submittal of the requested information. The companies responding during 1971 emphasized that alternative procedures will require time as well as large expenditures, and are working diligently on alternative means of disposal. In the meantime, the companies will continue the practice of disposal at sea. (More than 3 million cubic yards of chemical wastes were dumped at the acid grounds in 1972.)

Alternative to Dumping of Spent Caustic at Sea: The alternative methods studied for disposal of spent caustic at sea were:

(a) Build a sulfide oxidizer to convert spent caustic into waste products harmless to the environment. The sulfide oxidizer process converts spent caustic with high oxygen demand sulfides to low oxygen demand wastewater. Thirty-four hundred barrels per day of odorless wastewater, having a 1 ppm sulfur concentration and a 7.0 pH, would be produced. The sulfide oxidizer converts sulfides to thiosulfates and mercaptans to disulfides. In nature, oxidation of thiosulfates to sulfates proceeds very slowly; hence, process conversion of sulfides to thiosulfates is sufficient to meet oxygen demand requirements for a wastewater stream. An initial investment of $1 million, and an operating cost of $250,000 per year has been estimated.

(b) Build a sulfide saturation plant to convert spent caustic to an unfinished product for sale. Spent caustic disposal in any form would be eliminated entirely because all spent caustic would be converted to a useful product for use in other industry. Initial investment would involve $500,000 and an operational cost of $100,000 per year.

(c) Contract with an outside company with facilities to dispose of the spent caustic. Operation costs per year would be approximately $825,000.

Alternative to Dumping of Acid-Iron Industrial Waste at Sea: The principal wastes disposed of at the waste acid dump ground are gangue solids, iron (Fe), and sulfate (SO_4); these wastes represent substantial quantities of the elemental materials, iron and sulfur. Recovery of these elements for reuse presents attractive possibilities: iron for steelmaking or powder metallurgy, and sulfur for recycling in the manufacture of sulfuric acid in the captive facilities used to produce the acid required for extracting titanium.

Extraction and separation of titanium from the complex titanium-iron crystal, ilmenite, is accomplished inorganically by dissolving the ore in concentrated sulfuric acid to form a solution of titanium and iron sulfates with the insoluble gangue residue or "mud" to be filtered off for disposal. The original process (1934) included concentrating, dehydrating, and roasting facilities for recovery and recycling of waste sulfate materials.

However, there were many technological difficulties in the large-scale operation which proved to be technically and feasibly insurmountable. The low efficiency of the recovery process, and the inherent liquid and atmospheric emissions, necessitated finding alternate means of handling the wastes foom the manufacture of titanium dioxide. The plan with the least objectionable environmental impact was to dispose of the waste materials at sea. Ocean dispersal of the acid-iron wastes began in April 1948, and has continued on a daily basis with only minor interruption.

Studies have been conducted through the years on the waste dispersal operation and its effects. These studies concluded that repeated industrial acid-iron waste disposal off the New Jersey coast has not appreciably affected the marine environment in the acid dump ground area.

There are no known alternative methods for disposal of these wastes that would offer as ecologically acceptable a solution as the present method of ocean dispersal. The usual practice for small quantities of such materials would be neutralization, precipitation, and removal of all solids to a landfill operation. The tremendous volume of solids generated (48 acre feet per year) by such a treatment of these wastes would present a landfill problem that would result in a minor ecological disaster; therefore, efforts have been directed toward reducing the amount of waste generated, and to recovery of elemental values from the wastes. A great amount of effort has been expended through consultants and by support of research in various institutions.

In these efforts, principal developments have included: (1) beneficiation of ilmenite ore, (early 1950s), to remove a substantial portion of the iron before the sulfate extraction process, and (2) the chloride extraction process (late 1950s) which requires an initially high-grade ore (rutile), and permits recycling the chlorine used to extract titanium. Neither of these developments provide a total answer to the waste problem; there are unresolved technological problems in each, as well as long-term questions regarding their feasibility.

As a result, there are no immediate plans to eliminate the present method of ocean dispersal. Until a feasible method is developed, any requirement to change the present practice substantially will necessitate a major production curtailment with its resultant profound economic impact on the plant and community.

Alternative Methods of Disposal of Fermentation Residue: The end products from the manufacture of penicillin are two solids, mycelium and filteraid. Mycelium was trucked from the manufacturing plant to an open dump, filling in a swamp from June 1948, until 1952. Nutrients from the mycelium leached into the swamp and finally into a creek causing biological growth which became odorous and led to many complaints. An alternative method of disposal was sought at that time resulting in the present method, barging to the Long Island Sound.

By 1957, the company was dumping approximately 100,000 cubic yards of wet mycelium a year (36,000 cubic yards in 1972). Results of laboratory analysis indicated that the residue from the fermentation process consisted of a gray-brown, putty-like mass, with an oily texture and a decidedly disagreeable, sour-mash, nauseous odor. Chemical analysis indicated percentages of copper, chromium, and zinc. Spectrographic analysis also showed evidence of aluminum, calcium, iron, magnesium, manganese, and silica. Bioassay result of a 0.1% solution was not lethal to fish in a 48-hour observation. Results of the laboratory examination show that this material is probably safe for landfill disposal.

During the 19-year period (1952-1971), there has never been any evidence that the mycelium was harmful to fish life; on the contrary, the growth of bluefish and fishing in general in the Sound has been tremendous, a commonly known fact in this area.

Alternative methods of disposal that could be utilized in the Connecticut area are sanitary landfill or incineration. Landfill disposal would increase Pfizer's annual disposal costs by approximately $250,000; in addition, it would involve a number of serious problems. The high water content of the material makes conventional covering operations difficult, if not impossible. It would be necessary to study the use of specialized methods and equipment.

Incineration would involve a capital expenditure in the order of $1.5 million, in addition to approximately $500,000 annual operating expenses. This method also involves environmental problems.

Alternative Methods of Sludge Disposal: The various disposal areas in the NYB have had a measurable effect on the New York-Northern New Jersey estuarine region, but sludge disposal effects are possibly of little consequence when compared with the justification of disposing sludge at sea, still the most dependable and economical method. Because of the conclusions of many recent studies, it is evident that alternatives to sludge disposal must be studied and proposed methods must be carefully examined for their environmental impacts and costs. The following information presents three major alternatives studied (since 1970) by the New York City Environmental Protection Administration (NYCEPA).

Sludge Disposal 100 Nautical Miles Offshore — The purpose of an EPA study during 1970 was to examine the problems and ramifications associated with disposing of sludge 100 miles offshore in self-propelled sludge vessels, and to determine the costs of such operations. It was estimated that two Owls Head class vessels with a capacity of 60,000 cubic feet, and four Newtown Creek class vessels with a capacity of 95,000 cubic feet would be needed to transport approximately 7 million cubic feet of sludge 100 miles offshore each month.

The estimated annual operating costs would be more than $5 million, which represents a 456% increase above present operating costs. In addition to this increase in annual operating costs, it would require a redesign and construction time of 3½ years for three additional Newtown Creek vessels at a cost of $18 million. No attempt has been made to estimate the cost of modifying the existing fleet of vessels for 100-mile-offshore operations.

Sludge Disposal 25 Nautical Miles Offshore — Sludge vessels currently off-load their cargos at not less than 11 nautical miles from the nearest point of land. An extension of the dump area to a point 25 nautical miles from the nearest point of land would require the vessels to steam 3.5 to 5 hours longer (depending on speed of vessel and sea/weather conditions) for each trip to sea. Round-trip transit time will be increased to an average of 9½ hours. The present complement of 58 marine personnel would be increased to 94. Based on 1968-1969 price criteria, the increase in annual operating costs is an estimated $704,761, utilizing present equipment.

Sludge Incineration — A minimum lead time of 5 years is envisioned for the budgeting, planning, design, and construction of sludge incineration to serve all New York City facilities. On the assumption that existing means of sludge disposal at sea are abandoned, incinerators will be designed to adjoin every existing pollution control plant utilizng a fluidized-solids methodology for on-site sludge incineration.

At the Red Hook plant now under design in New York City, consideration is being given to installation of equipment which would prepare the sludge for incineration in a very large adjacent municipal refuse incinerator, also under design. If such an installation is decided upon, it will be the first New York City plant not dependent on ocean disposal, and may be used as a process evaluation center, aided by the availability of huge furnaces almost within the same structure.

Estimates of total costs range from $5 to $11 million and make no provision for solution of such problems as disposal of incinerated residue, which would present scrubber liquor problems. Other operating problems include odor production and the necessity for difficult sludge dewatering techniques, such as vacuum filtration.

Conclusion – The total sludge disposal costs would increase by a factor of 1.5 to 1 for 25-nautical-mile disposal at sea, 4.8 to 1 for 100-nautical-mile, and 3.4 to 1 for sludge incineration, within the near future if such plans are implemented. By the year 2015, the relative cost for 25-nautical-mile disposal would increase to 2.4 to 1, but other ratios would remain constant.

The absolute costs, however, would increase in the year 2015 by an increment of $14.5 million for 100-nautical-mile disposal, $5.5 million for 25-nautical-mile disposal, and $9.3 million for sludge incinerations, compared with an increment of $2.6 million if present methods are continued.

Recommendations – Studies into alternative methods for ocean disposal will require many years, and most of the reports offered by studies carried out contain recommendations for long-term changes to solve the complex problem of ocean disposal in the New York Bight. Utilizing the information gathered during the field study, this section will review the major problems associated with each of the dump grounds and recommend the actions that may be implemented in a realistic and reasonable time scale.

NEW YORK BIGHT IN DEPTH—EPA FACTS AND FIGURES

Summary of Waste Discharges in the New York-New Jersey Metropolitan Area

In the New York Metropolitan Area, there are 31 major drainage areas discharging approximately 2 billion gallons of wastewater per day. These discharges are to the waters of New York Harbor, Jamaica, Newark and the west end of Raritan Bays, and the Arthur Kill-Kill Van Kull.

Three of these drainage areas (all in New York City) discharge 480 MGD of untreated waste, 16 areas provide primary treatment for 540 MGD and 12 areas (all in New York City) provide intermediate treatment for 920 MGD.

All of the 31 major sewered drainage areas have combined sewers with overflow points to the waters of the Metropolitan Area. During periods of rainfall, the volume of flow in the combined sewers significantly exceeds the dry weather flow, resulting in the discharge of untreated waste. For a storm with an intensity of 1" per hour, the sewer flow can exceed the dry weather flow by a factor of 15 to 60, depending on the characteristics of the particular drainage area. This excess flow results in the direct discharge of greater than 90% of the organic load which would normally pass through the treatment plant.

Treatment plants in most of these major drainage areas dispose of their sludge to the ocean at the New York Bight dumping grounds (approximately 12 miles beyond New York Harbor entrance). Close to 5,000,000 cubic yards were dumped in 1971. Upgrading present treatment facilities to secondary (90% reduction of BOD and Suspended Solids), plus treatment of the present raw sewage discharges, will significantly increase the volume of sludge to be handled. It is estimated that the sludge volume will increase to approximately 15,000,000 cubic yards/year. Unless environmentally acceptable alternative sludge disposal methods are developed, this additional sludge must be dumped in the ocean.

It is estimated that 2,230,000 cubic yards of "sludge" are added annually to the New York Harbor complex due to the discharge of 480 MGD of raw sewage. These sludge accumulations are dredged along with other bottom materials and deposited in the "mud" spoils area, located only 5 miles from New Jersey beaches.

Along the New Jersey shore, from Sandy Hook to Long Beach Island there are 32 municipal waste treatment facilities discharging to the Atlantic Ocean. These installations, 2 secondary and 30 primary treatment plants, discharge 41 MGD to the Atlantic Ocean via outfall lines 1000' or less in length from shore.

Year-round disinfection of effluents is provided. Sludge handling practices, however, are somewhat unique in that most municipalities permit the sludge to accumulate during the year in their settling and/or storage tanks. During the winter months, November through March, this one-year buildup is discharged through the effluent outfall lines. In February 1972, the Federal Court issued a permanent injunction discontinuing this practice, which had been going on for the past 40 years. Alternative sludge disposal methods, including landfill and dumping in the approved site 12 miles off the coast, are now being practiced.

Sludge Handling Problem—New Jersey Coastal Communities

Volume of sludge on hand, March 1972: 5,764,000 gallons

Volume of sludge to be handled: Since sludge in storage at the various communities varies in concentration from less than 1% to greater than 40%, water, in volumes equal to twice that of the sludge, must be added to make the sludge pumpable; therefore, total volume of sludge equals 17,300,000 gallons.

Disposal Methods: Adequate incineration and/or landfill areas are not available; therefore, it appears that the soundest technical solution would be to barge the sludge to sea. Logistics involve the following:

Tank trucks (3,000 to 9,000 gallons capacity) - 1,900 to 5,400 truck loads

Sludge barge (500,000 gallons to 1.9 MG) - 9 to 35 barge trips

Approximate cost, based on 4¢/gallon - $692,000

Position Statement—1974

Ocean disposal has been, and will continue to be for at least the next five to ten years, the soundest available environmental alternative for sludge disposal in the New York Metropolitan Area. In the United States, approximately 80% of all ocean disposal of municipal sludge, acid and industrial wastes takes place off the coasts of New York and New Jersey. Disposal of sewage sludge began approximately 45 years ago, with volumes of industrial wastes increasing since the late 1950s. In 1973, approximately 5.8 million cubic yards (1,176.8 MG) of sewage sludge and 3.73 million cubic yards (754 MG) of industrial wastes were dumped in the ocean waters of the New York Bight. Six discrete disposal areas exist: dredge spoil, 6 miles from shore; construction rubble, 8 miles; derelict vessels, 12 miles; sewage sludge, 12 miles; waste acid, 15 miles; and industrial waste, 106 miles.

Since EPA assumed responsibility for the issuance of ocean disposal permits in April 1973, the following positive actions have been taken, or are in the process of being initiated:

> A total of 112 permit applications have been received and processed. Seventy-six of these applications, 27 municipal and 49 industrial, have been received in conformance with the Final Regulations which were promulgated October 15, 1973.
>
> Four industrial applications have been denied, and after final review, it is anticipated that others will also fall into this category. An additional 12 industries have, after discussions with EPA's technical staff, chosen alternate waste disposal methods.
>
> Implementation schedules for phasing out ocean disposal, where feasible, will be part of the special conditions of all industrial permits.
>
> Industrial wastes, previously dumped at the 12-mile sewage sludge site, are now being disposed of off the continental shelf at the 106-mile industrial waste site.
>
> Digester clean-out, which contains a high percentage of floatables, must now be disposed of at the industrial waste site.
>
> No new industrial or municipal dumpers, other than those using this method of ultimate disposal prior to the passage of the Ocean Dumping Bill, have been approved by this Region for using any of the sites in the New York Bight. (See Figure 6.1).
>
> A biweekly monitoring program of the bathing beaches along New York and New Jersey, as well as the waters contiguous to the 12-mile sewage sludge dumping grounds, has been initiated.
>
> Dumpers are now required to provide EPA with a detailed chemical and biological analysis of the waste materials being discharged into the ocean. No such requirement existed prior to April 1973, the effective date of the Ocean Dumping Bill.
>
> With the cooperation of the U.S. Coast Guard, which is responsible for police-type monitoring of vessels using the dumping sites, EPA has initiated a vigorous enforcement program. Violations of permit conditions, including failure to notify, premature dumping, and non-segregation of municipal and industrial wastes, have been discovered and appropriate legal action initiated.
>
> In cooperation with NOAA, areas which could be used as alternate sites for sewage sludge disposal are being investigated.
>
> Municipalities in the metropolitan area, have been notified of EPA's intention of moving the present sewage sludge dumping grounds if our monitoring programs indicate an environmental threat.

The EPA plan for development and implementation of the most environmentally

FIGURE 6.1: PRESENT OCEAN DISPOSAL SITES IN THE NEW YORK BIGHT

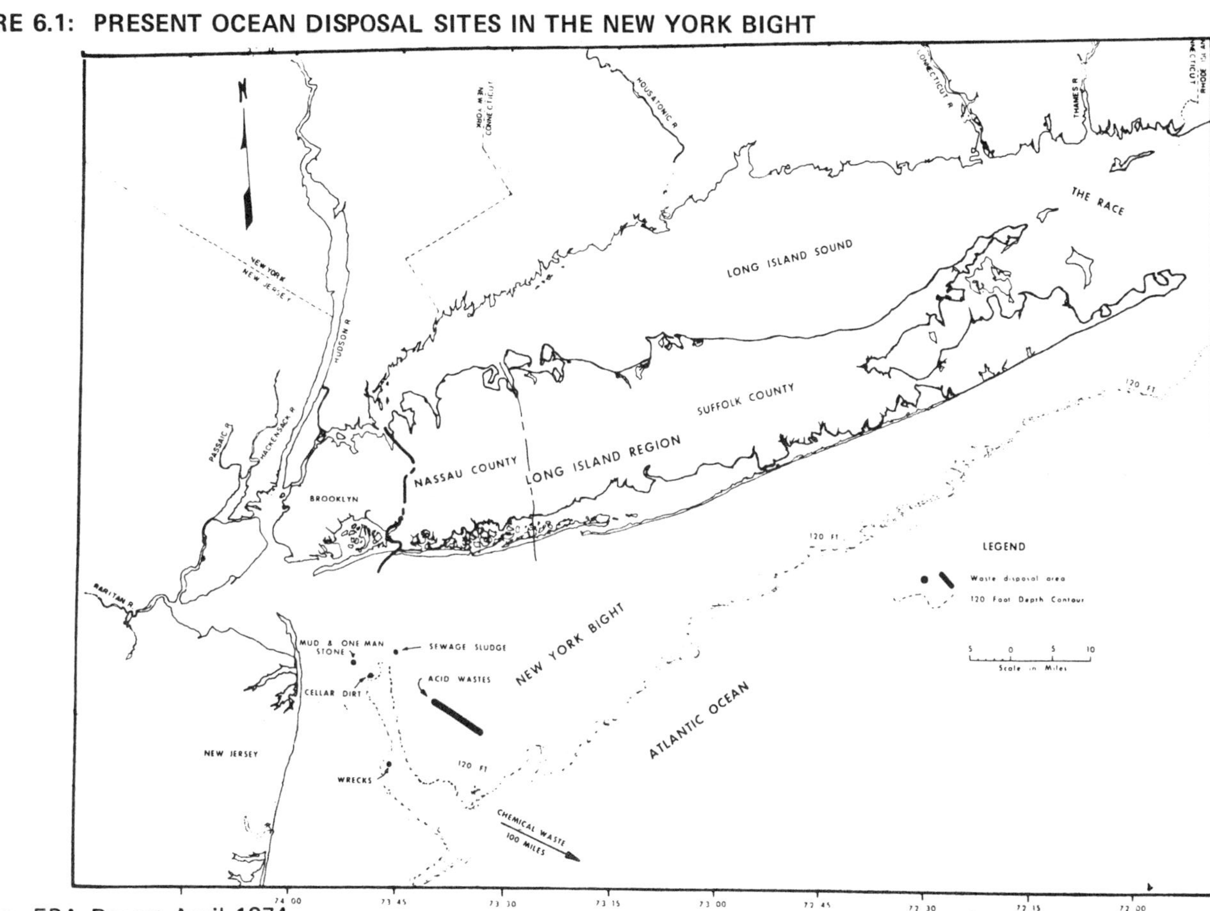

Source: EPA Report April 1974

acceptable method of ultimate disposal of domestic wastes in the New York Metropolitan Area which has already been recommended and accepted by the New York State Department of Environmental Conservation and the New Jersey Department of Environmental Protection, advocates the formation of a regional sludge management authority(s).

Under EPA and state guidance, an agency capable of environmental planning would develop the most acceptable long-term alternative, or alternatives, for the management of this environmental problem. That agency, or one with operating authority, would implement the most effective alternative(s) to permit the disposal of sludge with minimal environmental impact. A sludge management authority would have options not available to individual sewerage authorities, i.e., pipelines, remote disposal sites, cost effective energy recovery, etc.

EPA will act as the catalyst for this project because of its responsibility and policies relative to ocean dumping and the potential availability of Federal funding for these activities.

On the basis of technical information, the following potential alternatives to current ultimate disposal practice appear to be the most attractive on the basis of a regional authority approach:

(1) Disposal in a regional incinerator or incinerators, located at a site where air quality impact would be minimal. Offshore sites merit serious consideration. Power generation and the burning of other solid wastes should be considered to enhance the economic feasibility of this type project.

(2) Controlled disposal to the marine environment. Other sites and disposal techniques are presently being investigated by EPA and NOAA. About $7,000,000 of Federal funds are being expended to study the acute environmental stresses associated with this practice as well as the long-term effects, including the impact on marine organisms and the food web.

(3) Disposal at remote landfill sites. When large quantities of sludge are disposed on landfills, pretreatment techniques such as sludge dewatering, and leachate treatment will be required.

(4) Disposal as a soil conditioner and fertilizer. EPA presently has a demonstration project with Ocean County, New Jersey relative to sludge disposal on sandy soils in the Pine Barrens.

It must be recognized that for sewage sludge to be disposed of by any of the abovementioned techniques, it must be properly pretreated to meet EPA requirements for heavy metals and toxic components, as mandated in the 1972 FWPCA Amendments. All land disposal alternatives must consider the potential pollution problems of groundwaters in the disposal areas. In addition to requiring sophisticated techniques to meet stringent air quality standards, sludge incineration will still produce an ash, 0.35 million cubic yards by 1977, that must be handled. Thus, without further study of these alternatives, it's impossible to predict whether any of the approaches proposed are less damaging to the environment than the present practice of ocean disposal.

Present estimates indicate that a planning agency and a planning/operating authority could be formed, studies made and the program ready for implementation by mid-1976. Design and construction of facilities are estimated to require an additional four years; thus, the project would have a target completion date of 1980.

The Federal Water Pollution Control Act Amendments of 1972 require that all sewage treatment plants in operation on July 1, 1977, whether or not they are built with the aid of a Federal grant, and no matter when they were built, must provide a minimum of secondary treatment. This upgrading of treatment will substantially increase the volume of sludge that must be ultimately handled. If the practice of ocean dumping continues, it's estimated that the volume of sludge dumped at sea will triple.

Recognizing this problem, EPA, with the cooperation of State and Interstate Agencies, along with other Federal agencies, particularly NOAA, is looking for alternate disposal sites, since it is recognized that the present site would not be suitable for handling these increased volumes of sludge. Unless alternative means of disposal are found to handle these increased volumes of sludge, the present sludge dumping site, located 12 miles off the coast, should eventually be moved.

Reports that the present sewage sludge dumping site is causing environmental harm to the beaches of Long Island, are unfounded and lack technical substantiation. Joint studies, by EPA, NOAA, FDA, New York State Department of Environmental Conservation and Nassau County, all of which were conducted independently, concluded the following:

(a) There is no massive movement of sludge from the present dumping grounds to the shore of Long Island. In fact, at 19 stations in which EPA had scuba divers search the bottom, no evidence of sludge was found. Chemical and bacteriological analyses of sediment and water column samples, collected by the divers, confirmed these visual observations that sewage sludge was absent.

(b) The waters of the beach area along Long Island and the New Jersey coast are still of excellent quality, and there is no indication that there will be a deterioration of this high quality because of dumping at the 12-mile site. Residents and bathers may be assured that these waters meet the stringent bathing bacteriological requirements, and thus are safe for recreation.

(c) Environmental problems appear due to inshore occurences, rather than with the movement of sewage sludge from the present dumping ground.

A completely separate study, conducted by the Department of Conservation and Waterways, Town of Hempstead, Long Island, New York, reported that "...sewage sludge as deposited at the designated dump site, loses its identity 7.5 miles south of Rockaway Inlet. Sediments beyond a 3-mile radius, north and east of the dump site, contain background levels of heavy metals and organic matter; however, in the vicinity of Atlantic Beach and East Rockaway Inlet, level of metals, organic matter, and bacteria are slightly in excess of background. This results from material being flushed from the waste region of Hempstead Bay by diurnal tidal transport."

Based upon present and past studies, as well as increased enforcement activities and knowledge of the types and volumes of wastes being disposed of in the ocean, the decision to issue interim 12-month permits to municipalities for the utilization of the present sewage sludge disposal site was made recognizing the fact that the practice over the past 45 years has created a "dead sea" in the general area of this site. It is also recognized that this site does not adversely affect the recreational waters of New York and New Jersey; therefore, a planned program for reducing the volume of wastes and/or finding alternate sites or methods of handling these wastes has been implemented.

Municipal/domestic type waste permits were issued for the transport of approximately 6,000,000 yd^3/yr of sludge. New York City accounts for 60% of the total volume, Nassau and Westchester 7%, and New Jersey communities the remaining 33%. Industrial permits were issued for the 8-mile rubble site, 15-mile acid site, and the 106-mile chemical site. Only two industries qualify for special permits to use the rubble site. All other permits are "Interim" and issued for only 12 months. Only two companies are disposing of wastes at the acid site 15 miles off the New Jersey Coast. All other industrial permit holders must discharge their waste off the continental shelf at the 106-mile site.

Permit for Use of 12-Mile Sewage Sludge Dumping Ground

UNITED STATES ENVIRONMENTAL PROTECTION AGENCY
REGION II
26 FEDERAL PLAZA
NEW YORK, NEW YORK 10007

MARINE PROTECTION, RESEARCH AND
SANCTUARIES ACT (OCEAN DISPOSAL) PERMIT

Permit No.__________

Name of Permittee__________

Effective Date__________

Expiration Date__________

Reapplication Date__________

In reference to the following application:

Application Number:__________

Name and Address of Applicant__________

____

____

hereinafter referred to as the applicant, for a permit authorizing the transportation and disposal of any material in compliance with the provisions of the Act of Congress enacted October 23, 1972, entitled The Marine Protection, Research and Sanctuaries Act of 1972, 33 U.S.C. 1401 et seq., hereinafter referred to as the Act,

__________,

hereinafter referred to as the permittee, is authorized to transport material for disposal from the facilities of

____

____

____

and to discharge to ocean waters, all in accordance with the following general and special conditions:

Permit Number:

General Conditions:

1. All transportation and disposal authorized herein shall at all times be undertaken in a manner consistent with the terms and conditions of this permit.

2. a. Transportation to, and disposal at any location other than that authorized by this permit shall constitute a violation of the Act and of the terms and conditions of this permit.

b. Transportation and disposal of any material more frequently than or in excess of that identified and authorized by this permit, or disposal of material not authorized by this permit, shall constitute a violation of the Act and of the terms and conditions of this permit.

c. The permittee shall comply with each and every condition, provision and limitation in this permit and compliance with one or more but less than all conditions, provisions and limitations shall not constitute a ground or grounds of defense in any proceeding against permittee for violation of one or more of such conditions, provisions or limitations.

3. The applicant may not apply for, nor the permittee simultaneously hold, a permit from another EPA Regional Office for any of the material to which this permit is applicable, nor may the applicant or permittee transfer material from one EPA Region to another if a permit for the transportation or disposal of such material has been denied by one EPA Region.

4. Nothing contained herein shall be deemed to authorize, in any way, the transportation from the United States for the purpose of disposal into the ocean waters, into the territorial sea, or into the contiguous zone, of the following material:

a. High-level radioactive wastes.

b. Materials in whatever form produced for radiological, chemical or biological warfare.

c. Persistent synthetic or natural materials which may float or remain in suspension in the ocean.

5. After notice and opportunity for a hearing, this permit may be modified or revoked, in whole or in part, during its term for cause including, but not limited to, the following:

a. Violation of any term or condition of the permit;

b. Misrepresentation, inaccuracy, or failure by the applicant to disclose all relevant facts in the permit application;

c. A change in any condition or material fact upon which this permit is based that requires either a temporary or permanent reduction or elimination of the authorized transportation or disposal including, but not limited to, changes in conditions at the designated disposal site, and newly discovered scientific data relative to the granting of this permit;

d. Failure to keep records, to engage in monitoring activities, or to notify appropriate officials in a timely manner of transportation and disposal activities as specified in any condition of this permit.

6. This permit shall be subject to suspension by the Regional Administrator if he determines that the permitted disposal has resulted, or is resulting, in imminent and substantial harm to human health or welfare or the marine environment. Such suspension shall be effective subject only to the provisions of section 223.2(c) of the Final Regulations promulgated pursuant to the Act.

7. Any person who violates any provision of the Act, the Final Regulations issued thereunder, or any term or condition of this permit shall be liable to a civil penalty of not more than $50,000 for each violation. Additionally, any knowing violation of the Act, Final Regulations, or permit may result in a criminal action being brought with penalties of not more than a $50,000 fine or a year in prison, or both.

Permit Number:

8. Any toxic material effluent standard promulgated under section 307(a) of the Federal Water Pollution Control Act Amendments of 1972 (Pub. L. 92-500 (1972)) is incorporated in the terms and conditions of this permit. Material authorized to be transported and discharged under this permit shall at all times comply with each and every such effluent standard.

Similarly, the discharge of hazardous substances regulated under section 311(b)(2)(A) of the Federal Water Pollution Control Act (Pub. L. 92-500 (1972)) is subject, under the terms of this permit, to the same limitations as imposed under section 311 of Public Law 92-500.

9. This permit, or a true copy thereof, shall be placed in a conspicuous place in the vessel which will be used for the transportation or disposal authorized by this permit. If the disposal vessel is an unmanned barge the permit or true copy of the permit, shall be transferred to the towing conveyance, or an additional true copy shall be available onboard the towing conveyance.

10. The permittee shall at all times maintain in good working order and operate as efficiently as possible all facilities, including vessels, used by the permittee in achieving compliance with the terms and conditions of this permit.

11. Unless otherwise provided for herein, all terms used in this permit shall have the meanings assigned to them by the Act or the Final Regulations issued thereunder.

12. The issuance of this permit does not convey any property rights in either real or personal property, or any exclusive privileges, nor does it authorize any injury to private property or any invasion of rights, nor any infringement of Federal, State or local laws or regulations, nor does it obviate the necessity of obtaining State or local assent required by applicable law for the activity authorized.

13. This permit does not authorize or approve the construction of any onshore or offshore physical structures or facilities or, except as authorized by this permit, the undertaking of any work in any navigable water.

14. The permittee named herein, if a person, firm or corporation other than the applicant, must at all times during the term of this permit have the legal right, independent of this permit and enforceable at law or in equity, to transport for the purpose of disposal in the ocean waters the materials described herein. If such legal right should cease to exist during this permit's term for any reason, including but not limited to, the expiration of a contractual relationship between the applicant and the named permittee, this permit shall revert, for the remainder of its term, to the applicant.

15. This permit may, at the discretion of the Regional Administrator, be transferred to a person, firm or corporation other than the permittee named herein, provided that a request for such a transfer be made, in writing, by the applicant at least 30 days prior to the requested transfer date.

16. If material, which is regulated by this permit is discharged due to an emergency to safeguard life at sea in locations or in a manner not in accordance with the terms of this permit, the permittee shall make a full report, in accordance with the provisions of 18 USC 1001, within 10 days to the Regional Administrator detailing the conditions of this emergency and the actions taken.

17. Telephone notification of sailing two (2) hours prior to vessel departure for approved site, will be provided by the permittee to the U. S. Coast Guard, Captain-of-the-Port, New York. Calls should be made to 212-264-8753 during working hours (8:00 AM to 4:30 PM Monday through Friday) and to 212-264-8770 during non-working hours, weekends, and holidays. The following information shall be provided in the notification of sailing:

a. Name of towing vessel and barge or tank vessel

b. Name of transporter

Permit Number:

c. Description of the vessel's contents, including volume

d. Place of departure

e. Location of disposal site

f. The time of departure

g. Estimated time of arrival at the disposal site

h. Estimated time of return to port.

18. In accordance with 33 USCA 445, every scow or boat engaged in the transportation of material for the purpose of ocean disposal shall have its name or number and owner's name painted in letters and numbers at least fourteen inches high on both sides of the scow or boat. These names and numbers shall be kept distinctly legible at all times, and no scow or boat not so marked shall be used to transport or discharge any such material.

19. Permittees shall maintain and submit Coast Guard Form CCGD 3-278, Monthly Transportation and Dumping Log, to Captain-of-the-Port, USCG, c/o New York Station, Governors Island, New York, N. Y. 10004. Permittees shall enter on this form under the column titled "Dump Site", the latitude and longitude at which the actual dumping occurred. These forms are to be mailed to the Coast Guard during the first week of the succeeding month for which they were prepared. If additional forms are required they may be obtained by forwarding a written request to Commander (mep), Third Coast Guard District, Governors Island, New York, N.Y. 10004. Copies of these logs will be forwarded, on a quarterly basis to: U.S. Environmental Protection Agency, Surveillance and Analysis Division, Edison, N.J. 08817.

Special Conditions:

1. This permit shall expire at midnight on ___________ This permit is nonrenewable. Application for a new permit must be submitted to EPA at least 150 days prior to expiration of this permit.

2. During the term of this permit, the type and quantity of material permitted for transportation for the purpose of ocean disposal shall be in accordance with the following:

   ____

   ____

   ____

3. Transportation for the purpose of ocean disposal shall terminate at, and waste disposal shall be confined to, the area described below:

   Latitude:

   Longitude:

4. Method of Disposal. The permittee will use only the following vessel(s)/barge(s) for transportation and disposal of wastes authorized under this permit.

   ____

   ____

   Waste is to be discharged at a uniform rate over a distance of at least _______ nautical miles within the disposal site designated in Special Condition No. 3. Vessel/barge traverses shall be at least 0.5 nautical mile apart. If two or more vessels/barges are discharging simultaneously, or if any two vessel/barge trips are to occur within one hour of each other, the latter discharge is to be at least 0.5 nautical mile from the previous discharge.

Permit Number:

If the waste cannot be uniformly discharged as required above, the permittee shall, within 30 days of issuance of this permit, provide to EPA in writing, detailed technical information, certified by a naval architect or marine engineer, as to why this condition cannot be met. A time period of not more than one year from the date of issuance of this permit will be allowed for the installation of equipment or systems necessary to meet the uniform discharge requirement.

5. Analyses shall be conducted by the permittee on a representative sample of ________ vessel/barge load.

Parameters to be analyzed:

Bioassay, using the organism Artemia salina or substitute organisms designated to be more appropriate by EPA, Region II.

Mercury, liquid and solid phase

Cadmium, liquid and solid phase

Density

pH

Oil and grease, using liquid-liquid extraction with trichlorotrifluoroethane

Petroleum hydrocarbon (oil), using tentative IR procedure

Arsenic	Lead
Copper	Zinc
Vanadium	Selenium
Beryllium	Chromium
Nickel	Ammonia nitrogen
B.O.D.	T.O.C.
C.O.D.	Total Solids
Dissolved Solids	Suspended Solids
TKN	Nitrate nitrogen
Total acidity ($CaCO_3$)	Total alkalinity ($CaCO_3$)
Total Phosphorous	Phenols

Analytical data shall be submitted to EPA on a monthly basis, with the first report due no later than 30 days following the initial discharge.

All analyses shall be conducted according to approved test procedures contained in the Federal Register, October 16, 1973, Vol. 38, Number 199, Part II, "Guidelines Establishing Test Procedures for Analysis of Pollutants" or according to specific analytical procedures distributed by EPA, Region II.

6. The applicant, permittee, or their contract laboratory shall maintain a viable analytical quality control program and routinely submit — on at least a quarterly basis — these data with the results of sample analysis to the Environmental Protection Agency. Upon request, the laboratory shall participate in EPA sponsored quality control programs relative to analyses required under this permit.

7. The permittee and/or the applicant will be required, during the term of this interim permit, to conduct or participate in a monitoring program of the impact of the permitted waste disposal on the marine environment at the designated disposal site, pursuant to the Federal Register, October 15, 1973; Vol. 38, Number 198, Part II, "Ocean Dumping - Final Regulations and Criteria", Section 223.1(f).

Standard Permit Issued to Industrial Firms Requesting Use of the 106-Mile Chemical Site

UNITED STATES ENVIRONMENTAL PROTECTION AGENCY
REGION II
26 FEDERAL PLAZA
NEW YORK, NEW YORK 10007

MARINE PROTECTION, RESEARCH AND
SANCTUARIES ACT (OCEAN DISPOSAL) PERMIT

Permit No.____________________

Name of Permittee____________________

Effective Date____________________

Expiration Date____________________

Reapplication Date____________________

In reference to the following application:

Application Number:____________________

Name and Address of Applicant____________________

____

____

hereinafter referred to as the applicant, for a permit authorizing the transportation and disposal of any material in compliance with the provisions of the Act of Congress enacted October 23, 1972, entitled The Marine Protection, Research and Sanctuaries Act of 1972, 33 U.S.C. 1401 et seq., hereinafter referred to as the Act,

____,

hereinafter referred to as the permittee, is authorized to transport material for disposal from the facilities of

____

____

____

and to discharge to ocean waters, all in accordance with the following general and special conditions:

Permit Number:

General Conditions:

1. All transportation and disposal authorized herein shall at all times be undertaken in a manner consistent with the terms and conditions of this permit.

2. a. Transportation to, and disposal at any location other than that authorized by this permit shall constitute a violation of the Act and of the terms and conditions of this permit.

b. Transportation and disposal of any material more frequently than or in excess of that identified and authorized by this permit, or disposal of material not authorized by this permit, shall constitute a violation of the Act and of the terms and conditions of this permit.

c. The permittee shall comply with each and every condition, provision and limitation in this permit and compliance with one or more but less than all conditions, provisions and limitations shall not constitute a ground or grounds of defense in any proceeding against permittee for violation of one or more of such conditions, provisions or limitations.

Permit Number:

3. The applicant may not apply for, nor the permittee simultaneously hold, a permit from another EPA Regional Office for any of the material to which this permit is applicable, nor may the applicant or permittee transfer material from one EPA Region to another if a permit for the transportation or disposal of such material has been denied by one EPA Region.

4. Nothing contained herein shall be deemed to authorize, in any way, the transportation from the United States for the purpose of disposal into the ocean waters, into the territorial sea, or into the contiguous zone, of the following material:

a. High-level radioactive wastes.

b. Materials in whatever form produced for radiological, chemical or biological warfare.

c. Persistent synthetic or natural materials which may float or remain in suspension in the ocean.

5. After notice and opportunity for a hearing, this permit may be modified or revoked, in whole or in part, during its term for cause including, but not limited to, the following:

a. Violation of any term or condition of the permit;

b. Misrepresentation, inaccuracy, or failure by the applicant to disclose all relevant facts in the permit application;

c. A change in any condition or material fact upon which this permit is based that requires either a temporary or permanent reduction or elimination of the authorized transportation or disposal including, but not limited to, changes in conditions at the designated disposal site, and newly discovered scientific data relative to the granting of this permit;

d. Failure to keep records, to engage in monitoring activities, or to notify appropriate officials in a timely manner of transportation and disposal activities as specified in any condition of this permit.

6. This permit shall be subject to suspension by the Regional Administrator if he determines that the permitted disposal has resulted, or is resulting, in imminent and substantial harm to human health or welfare or the marine environment. Such suspension shall be effective subject only to the provisions of section 223.2(c) of the Final Regulations promulgated pursuant to the Act.

7. Any person who violates any provision of the Act, the Final Regulations issued thereunder, or any term or condition of this permit shall be liable to a civil penalty of not more than $50,000 for each violation. Additionally, any knowing violation of the Act, Final Regulations, or permit may result in a criminal action being brought with penalties of not more than a $50,000 fine or a year in prison, or both.

8. Any toxic material effluent standard promulgated under section 307(a) of the Federal Water Pollution Control Act Amendments of 1972 (Pub. L. 92-500 (1972)) is incorporated in the terms and conditions of this permit. Material authorized to be transported and discharged under this permit shall at all times comply with each and every such effluent standard.

Similarly, the discharge of hazardous substances regulated under section 311(b)(2)(A) of the Federal Water Pollution Control Act (Pub. L. 92-500 (1972)) is subject, under the terms of this permit, to the same limitations as imposed under section 311 of Public Law 92-500.

9. This permit, or a true copy thereof, shall be placed in a conspicuous place in the vessel which will be used for the transportation or disposal authorized by this permit. If the disposal vessel is an unmanned barge the permit or true copy of the permit, shall be transferred to the towing conveyance, or an additional true copy shall be available onboard the towing conveyance.

10. The permittee shall at all times maintain in good working order and operate as efficiently as possible all facilities, including vessels, used by the permittee in achieving compliance with the terms and conditions of this permit.

Permit Number:

11. Unless otherwise provided for herein, all terms used in this permit shall have the meanings assigned to them by the Act or the Final Regulations issued thereunder.

12. The issuance of this permit does not convey any property rights in either real or personal property, or any exclusive privileges, nor does it authorize any injury to private property or any invasion of rights, nor any infringement of Federal, State or local laws or regulations, nor does it obviate the necessity of obtaining State or local assent required by applicable law for the activity authorized.

13. This permit does not authorize or approve the construction of any onshore or offshore physical structures or facilities or, except as authorized by this permit, the undertaking of any work in any navigable water.

14. The permittee named herein, if a person, firm or corporation other than the applicant, must at all times during the term of this permit have the legal right, independent of this permit and enforceable at law or in equity, to transport for the purpose of disposal in the ocean waters the materials described herein. If such legal right should cease to exist during this permit's term for any reason, including but not limited to, the expiration of a contractual relationship between the applicant and the named permittee, this permit shall revert, for the remainder of its term, to the applicant.

15. This permit may, at the discretion of the Regional Administrator, be transferred to a person, firm or corporation other than the permittee named herein, provided that a request for such a transfer be made, in writing, by the applicant at least 30 days prior to the requested transfer date.

16. If material, which is regulated by this permit is discharged due to an emergency to safeguard life at sea in locations or in a manner not in accordance with the terms of this permit, the permittee shall make a full report, in accordance with the provisions of 18 USC 1001, within 10 days to the Regional Administrator detailing the conditions of this emergency and the actions taken.

17. The permittee shall provide telephone notification of sailing to Captain-of-the-Port, (COTP) New York at 212-264-8753 during working hours (8:00 AM to 4:30 PM Monday through Friday) and to 212-264-8770 during non-working hours, weekends, and holidays not later than twenty-four (24) hours prior to the estimated time of departure. The permittee shall confirm the exact time of departure within thirty (30) minutes of the actual departure time, and immediately notify the COTP upon any changes in the estimated time of departure greater than one hour. Within two (2) hours after receipt of the initial notification the transporter will be advised as to whether or not a Coast Guard shiprider will be assigned to the voyage.

18. Surveillance will generally be accomplished by a Coast Guard shiprider who will be on board the towing conveyance for the entire voyage. His quarters and subsistence while on board shall be provided by and shall be at the expense of the permittee. He shall be treated courteously and afforded free and immediate access to all navigational capabilities on the vessel which can provide information on position, course, speed, depth of water, bearings, etc. The notification procedures which will permit the timely assignment of a shiprider are specified in general condition 17. The following information shall be provided in the notification of sailing:

a. Name of the towing vessel and barge or tank vessel

b. Name of the transporter

c. Description of the vessel's contents including volume

d. Place of departure

e. Location of the disposal site

f. The time of departure

g. Estimate time of arrival at the disposal site

h. Estimate time of return to port.

Permit Number:

19. In accordance with 33 USCA 445, every scow or boat engaged in the transportation of municipal sludge or industrial wastes shall have its name or number and owner's name painted in letters and numbers at least fourteen inches high on both sides of the scow or boat. These names and numbers shall be kept distinctly legible at all times, and no scow or boat not so marked shall be used to transport or dispose any such material.

20. Permittees shall maintain and submit Coast Guard Form CCGD 3-278, Monthly Transportation and Dumping Log, to COTP, USCG, c/o New York Station, Governors Island, New York, N. Y. 10004. Permittees shall enter on this form under the column titled "Dump Site", the latitude and longitude at which the actual dumping occurred. These forms are to be mailed to the Coast Guard during the first week of the succeeding month for which they were prepared. If additional forms are required they may be obtained by forwarding a written request to Commander (mep), Third Coast Guard District, Governors Island, New York, N.Y. 10004. Copies of these logs will be forwarded, on a quarterly basis to: U. S. Environmental Protection Agency, Surveillance and Analysis Division, Edison, N. J. 08817.

Special Conditions:

1. This permit shall expire at midnight on ___________ This permit is nonrenewable. Application for a new permit must be submitted to EPA at least 150 days prior to expiration of this permit.

2. During the term of this permit, the type and quantity of material permitted for transportation for the purpose of ocean disposal shall be in accordance with the following:

   ____

   ____

   ____

3. Transportation for the purpose of ocean disposal shall terminate at, and waste disposal shall be confined to, the area described below:

   Latitude: 38° 40' N to 39° 0' N

   Longitude: 72° 0' W to 72° 30' W

4. Method of Disposal. The permittee will use only the following vessel(s)/barge(s) for transportation and disposal of wastes authorized under this permit.

   ____

   ____

   Waste is to be discharged at a uniform rate over a distance of at least _______ nautical miles within the disposal site designated in Special Condition No. 3. Vessel/barge traverses shall be at least 0.5 nautical mile apart. If two or more vessels/barges are discharging simultaneously, or if any two vessel/barge trips are to occur within one hour of each other, the latter discharge is to be at least 0.5 nautical mile from the previous discharge.

   If the waste cannot be uniformly discharged as required above, the permittee shall, within 30 days of issuance of this permit, provide to EPA in writing, detailed technical information, certified by a naval architect or marine engineer, as to why this condition cannot be met. A time period of not more than one year from the date of issuance of this permit will be allowed for the installation of equipment or systems necessary to meet the uniform discharge requirement.

Permit Number:

5. Analyses shall be conducted by the permittee on a representative sample of _______ vessel/barge load.

   Parameters to be analyzed:

   Bioassay, using the organism Artemia salina or substitute organisms designated to be more appropriate by EPA, Region II.

   Mercury, liquid and solid phase

   Cadmium, liquid and solid phase

   Density

   pH

   Oil and grease, using liquid-liquid extraction with trichlorotrifluoroethane

   Petroleum hydrocarbon (oil), using tentative IR procedure

Arsenic	Lead
Copper	Zinc
Vanadium	Selenium
Beryllium	Chromium
Nickel	Ammonia nitrogen
B.O.D.	T.O.C.
C.O.D.	Total Solids
Dissolved Solids	Suspended Solids
TKN	Nitrate nitrogen
Total acidity ($CaCO_3$)	Total alkalinity ($CaCO_3$)
Total Phosphorous	Phenols

   Analytical data shall be submitted to EPA on a monthly basis, with the first report due no later than 30 days following the initial discharge.

   All analyses shall be conducted according to approved test procedures contained in the Federal Register, October 16, 1973, Vol. 38, Number 199, Part II, "Guidelines Establishing Test Procedures for Analysis of Pollutants" or according to specific analytical procedures distributed by EPA, Region II.

6. The applicant, permittee, or their contract laboratory shall maintain a viable analytical quality control program and routinely submit — on at least a quarterly basis — these data with the results of sample analysis to the Environmental Protection Agency. Upon request, the laboratory shall participate in EPA sponsored quality control programs relative to analyses required under this permit.

7. The permittee and/or the applicant will be required, during the term of this interim permit, to conduct or participate in a monitoring program of the impact of the permitted waste disposal on the marine environment at the designated disposal site, pursuant to the Federal Register, October 15, 1973; Vol. 38, Number 198, Part II, "Ocean Dumping - Final Regulations and Criteria", Section 223.1(f).

Permits for the 15-mile acid site and the 8-mile rubble site are a combination of municipal and industrial "standard" permits. Again, special conditions are expanded to include specific requirements.

Environmental Findings—1974

EPA's position that the present sewage sludge disposal ground be moved by 1976 is still firm. In addition, it is recommended that the dredge spoil site, which presently influences and impacts the sludge site, be moved at the same time. From a technical standpoint, to move one without the other would be shortsighted, and to say the least, environmentally unwise.

An orderly and environmentally acceptable plan has been developed with the ultimate goal of phasing out ocean disposal of municipal sludges by 1981. A recommendation to immediately move the present disposal site, could prove environmentally disastrous.

That the sludge mass is moving toward the Long Island shore at an unprecedented rate has not been substantiated. In fact, the report presents contradictory data, not from the standpoint of disclaiming the findings of "black mayonnaise" one-half mile from shore at random locations, but in technically disagreeing completely with the conclusions drawn from these findings.

Surf zone studies along the beaches of Long Island and New Jersey clearly indicate that the water is safe for contact recreation. The absence of pathogens in the surf zone waters provides further verification of excellent water quality.

Nearshore cruises have also indicated excellent quality water, and the absence of "sewage sludge" in the sediments. The black mayonnaise found at random locations one-half mile from the beach has been identified as organic material of natural origin.

Results of transect cruises give further support to EPA's stated position that the organic material nearshore is related to inland occurrences and not associated with a massive movement of material from the disposal site. If, in fact, there was a massive movement toward shore, one would expect to find a gradual diminution of pollutant levels from the disposal site to the shore area. The data presented clearly indicate that this is not the case.

The leading edge of the sludge mass, associated with the sewage sludge disposal site, is still located approximately 5½ to 6 miles from the shore of Long Island, thus negating the urgency to move the present disposal site before the scheduled 1976 closure.

Volume of Waste Materials—1973: Twenty-one million cubic yards of waste materials were disposed of in the New York Bight during 1973. Municipal sludges amounted to 5.6 million cubic yards or roughly 70% of all municipal wastes disposed of via ocean dumping in the United States. Approximately 58% of these sludges were from New York City, 33% from sources in New Jersey, and the remainder from Nassau and Westchester Counties in New York. Industrial wastes amounted to 3.8 million cubic yards or about 60% of the total volume disposed of in the ocean within the United States. The remainder, 11.8 million cubic yards, results from the disposal of dredged spoils.

Long-Range Program for Phase-Out of Ocean Disposal in New York Bight by 1987:

By 1976 –

SOURCES

Start Implementation of Alternatives to ocean disposal as recommended by EPA-ISC study (site selection, EIS, construction)

SITES

EIS on new sites completed

Close existing sewage sludge site

Move to alternate site

By 1981 –

SOURCES

Goal - ocean disposal phased out

Complete implementation of alternatives

SITES

Decision to keep alternate site active or move off shelf, if controlled ocean disposal is considered most suitable environmental alternative as per PL 92-532

Technical Findings (April-July 1974): In April 1974, EPA, Region II, initiated a comprehensive three-phase monitoring program to investigate the quality of the water and bottom sediments in the New York Bight and along the Long Island and New Jersey beaches. A summary of the data and a brief discussion of the results of this monitoring program to date follows.

Monitoring Program of Sludge/Dredge Spoil Dumpjng Sites: A comprehensive sampling program was initiated by the EPA, Region II in April 1974 to monitor the quality of water and bottom sediments in the apex of the New York Bight and along the beaches of Long Island and New Jersey.

Type I	Surf Zone	Biweekly
Type II	Nearshore	Monthly
Type III	Transect	Monthly

Type I samples are collected in the surf zone at selected sites along the Long Island and New Jersey coastlines. Type II are collected by boat approximately 100 yards from shore and in water about 15 to 20 feet deep. Similarly, Type III samples are collected by boat in three transects, each running from the 12-mile sewage sludge dump site, one to the Long Island beaches, a second to the New Jersey beaches, and a third to the New York Harbor entrance.

Water-quality data, including sediment characterization, collected at these sampling sites from April to early July 1974 and selected parameters are illustrated in Figures 6.2 to 6.18.

Long Island Surf Zone and Nearshore — Data from the samples collected in the surf zone and nearshore (Figures 6.2 and 6.3) indicate low total and fecal coliform densities. The levels of fecal coliform at all sampling stations are significantly below the geometric mean density standards for primary and secondary contact recreation waters under New York's Class SB standard of 200 organisms per 100 ml. The effect of poorer quality runoff from New York Harbor and through East Rockaway inlet is evident from elevated values observed at LIC-1 and LIC-6, respectively. It is important to note that attempts to isolate Salmonella (enteric pathogens) at four sampling stations were unsuccessful.

Data on coliform densities, total organic carbon content, and heavy metals in bottom sediments are illustrated in Figures 6.4 and 6.5 for nearshore sampling stations. Highest values are again found at stations LIC-1 and LIC-6, reflecting the influence of New York Harbor and East Rockaway inlet. Comparison of data presented in these two graphs (Figures 6.4 and 6.5) with similar data collected in the vicinity of the sewage sludge dump site (Figures 6.7 and 6.8) indicate wide variations in quality.

For example, the highest geometric mean fecal coliform recorded at Long Island nearshore sampling stations was 400/100 g, while that near the dump site was 962/100 g. Similarly, along the Lond Island nearshore the highest mean total organics observed was 2.6% and that for metals was 134 mg/kg, while those near the dump site were 4.6% and 426 mg/kg, respectively. (Note about 80 mg/kg in the metals value reflect the use of absolute values when summing individual metals reported as less than analytical sensitivity.) Salmonella were not isolated in bottom sediment samples collected at LIC-6 even though high fecal coliform densities were observed.

New York Bight Transects – Water and bottom sediment samples were collected along three transects, each originating in the vicinity of the sewage sludge dump site and extending to the Long Island coast, the New Jersey coast, and the New York Harbor entrance.

Geometric mean fecal coliform densities observed in samples collected from the water column in the Long Island transect (Figure 6.6) and New York Harbor transect (Figure 6.9), indicate that the New York Class SB standard of 200/100 ml is not contravened, except at station NYB30 (436 fecal coliform/100 ml). This station is located closest to New York Harbor and thus, reflects a combination runoff and wastewater discharges from the Hudson River and Raritan Bay.

Even at sampling stations within the sewage dump site, observed fecal coliform densities in both surface and bottom water samples were low. A similar review of bacteriological quality in water column samples collected along the New Jersey transect (Figure 6.12) indicate that the more stringent New Jersey Class CW-1 standard for primary contact recreation of a geometric mean of 50 fecal coliform/100 ml is not contravened except by the 61/100 ml at the nearshore station (NYB20).

Review of the data illustrated for mean coliform densities at bottom sediment sampling stations in the three transects (Figures 6.7, 6.10 and 6.13) indicate extreme elevated counts of both total and fecal coliform in the vicinity of the sewage sludge dump site. Total coliform densities (geometric means) generally exceed 15,000/100 g and fecal coliform, 500/100 g in the vicinity of the dump site. A distance of 5½ to 6 miles of low coliform densities separates the high counts in the vicinity of the dump site and the slightly elevated values found at nearshore sampling stations. Salmonella were not found at six of the twenty-two transect sampling stations, even though elevated fecal coliform densities were observed at some of the six sites tested.

Results of total organic carbon and heavy metal analyses on bottom sediments at the transect stations (Figures 6.8, 6.11 and 6.14) also show wide variations in quality. Highest mean values are found in the vicinity of the dump site, with total organic carbon content generally exceeding 4% and metals exceeding 300 mg/kg. The data illustrated show a "clean water-sediment" zone of about 5½ to 6 miles separating the dump site from the New Jersey and Long Island bathing beaches. Total organic contents are in the main less than 1% and metals under 100 mg/kg in this clean water-sediment zone.

New Jersey Surf Zone and Nearshore — Results of sampling in the surf zone and nearshore (Figures 6.15 and 6.18) indicate low total and fecal coliform densities. The level of fecal coliform at all sampling stations generally are far below the geometric mean density standards for primary contact recreation under New Jersey's Class CW-1 standard of 50 organisms/100 ml. Elevated coliform values observed at JC14 (Figure 6.15) are related to an ocean outfall from a local municipal treatment plant.

Data on coliform densities, total organic carbon content, and heavy metals in bottom sediments are illustrated in Figures 6.16 and 6.17 for nearshore sampling stations. Comparison of data presented in these two graphs with similar data collected in the vicinity of the sewage sludge dump site (Figures 6.7 and 6.8) indicate wide variations in quality.

For example, the highest geometric mean fecal coliform recorded at New Jersey nearshore sampling stations was 382/100 g, while that near the dump site was 962/100 g. Likewise, along the New Jersey nearshore the highest mean total organics observed was 1.6% and that for metals was 123 mg/kg, while those near the dump site were 4.6% and 426 mg/kg, respectively. Note that while tested at two sites, Salmonella were not found in the bottom sediments.

Based upon sampling in the surf and nearshore waters along the Long Island and New Jersey beaches, it is evident that water quality remains excellent with respect to coliform density and is acceptable for contact recreation. More important, there is no evidence of a trend towards increased coliform density and thus, no indication of degradation. Occasional elevated coliform counts appear randomly distributed in time and location, and do not indicate a systematic change or degradation of water quality.

Review of data from sampling in the Bight of the water column and bottom sediments indicates the general location of the sludge mass associated with the sewage sludge and dredge spoil dump site. A clean water-sediment zone of about 5½ to 6 miles separates the leading edge of the sludge mass from the New Jersey and Lond Island coasts.

Slightly elevated organic carbon content and bacterial counts at selected nearshore sampling stations can be related to inland occurrences, such as runoff and wastewater discharges.

FIGURE 6.2

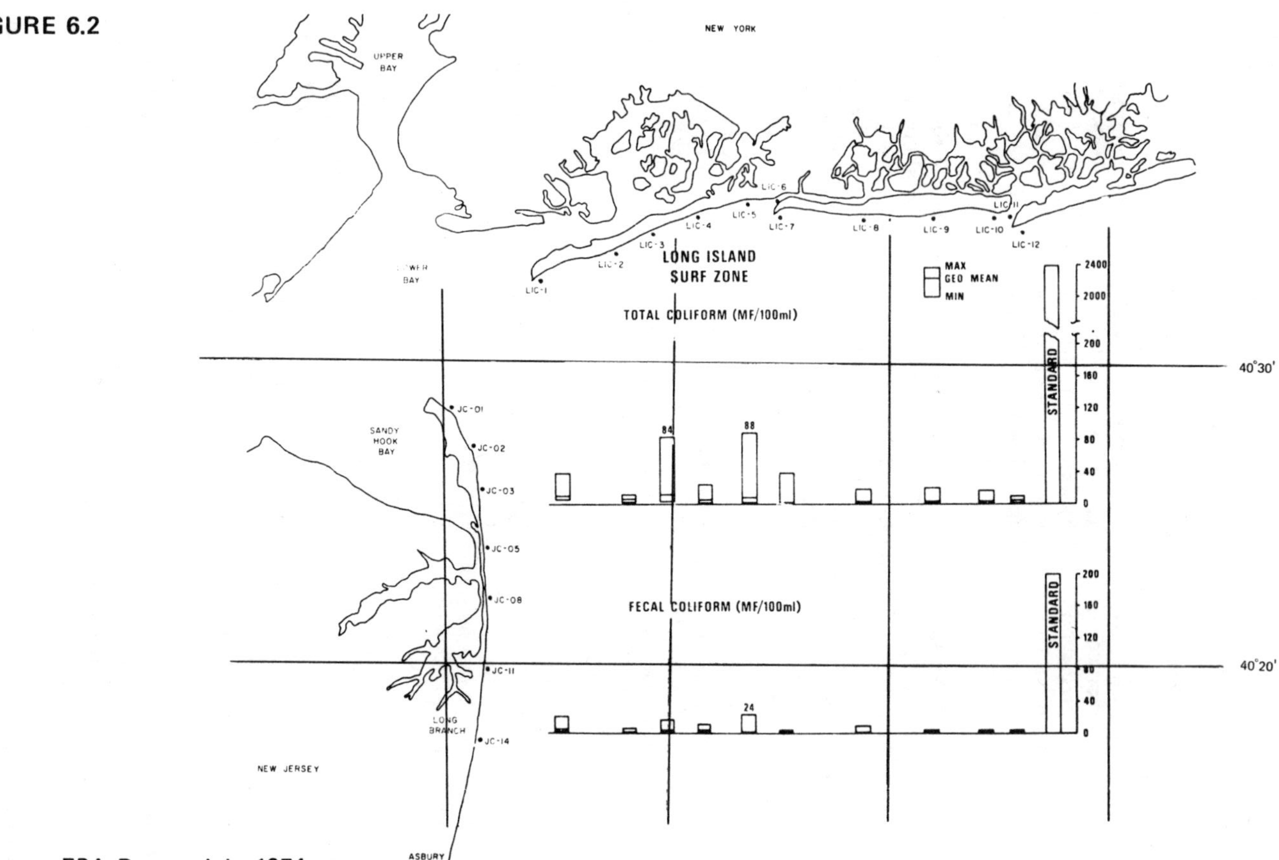

Source: EPA Report July 1974

FIGURE 6.3

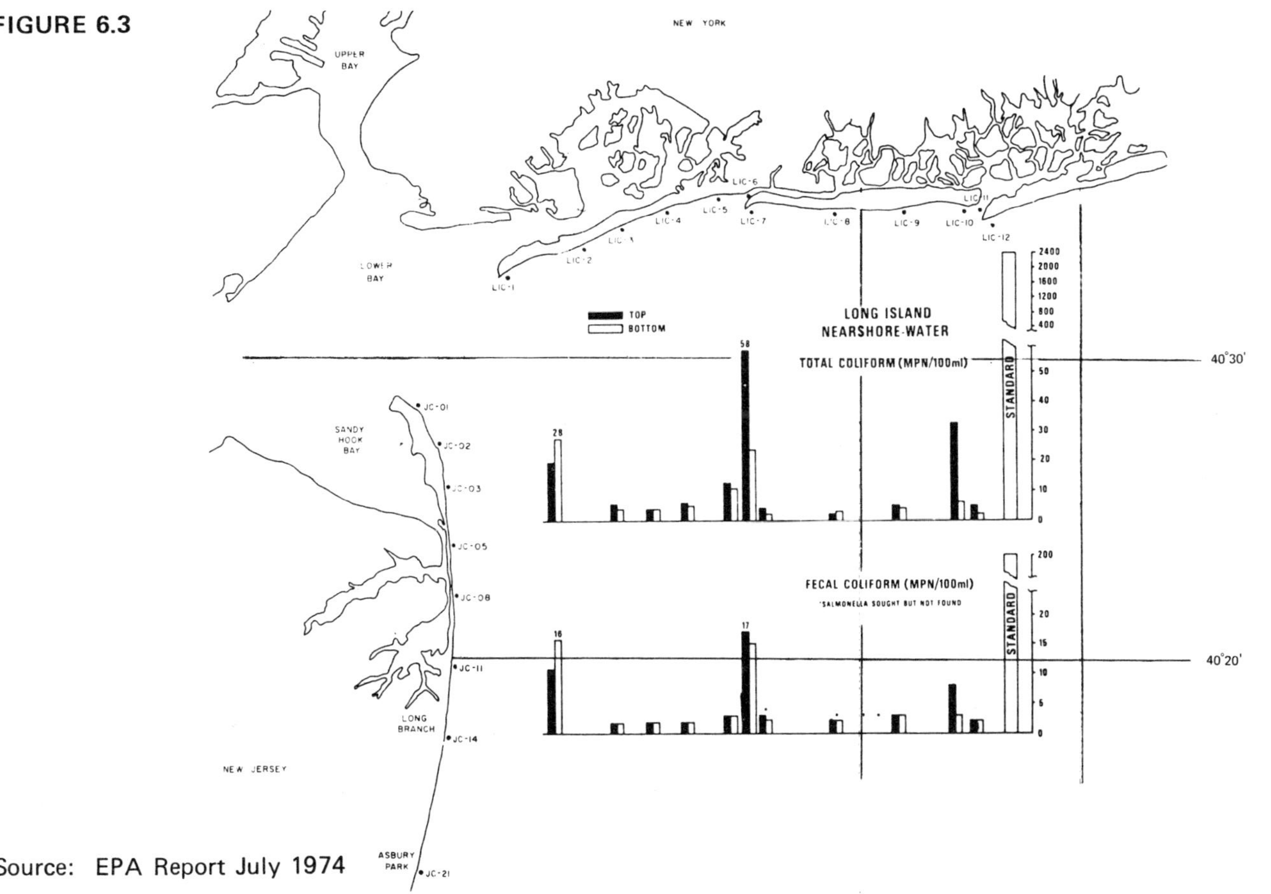

Source: EPA Report July 1974

FIGURE 6.4

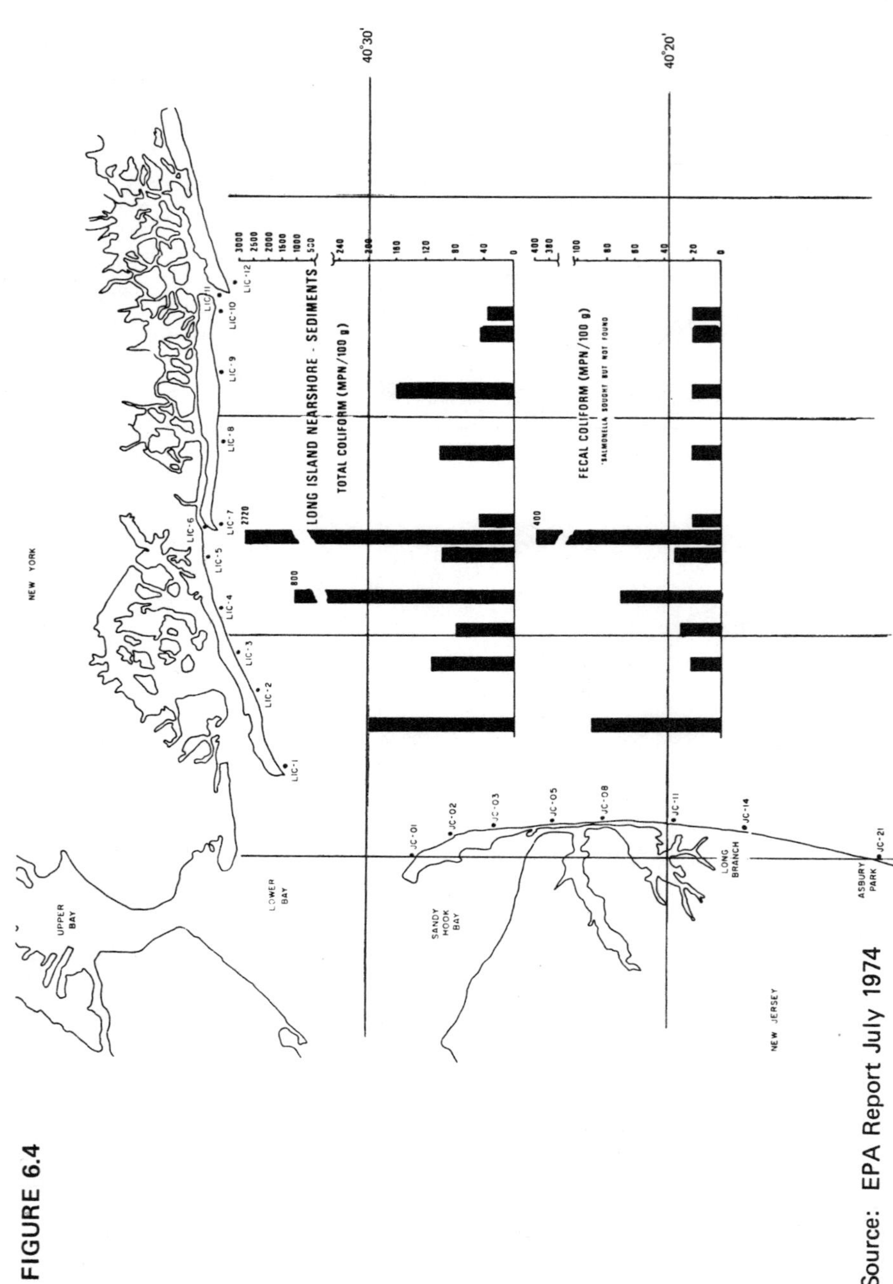

Source: EPA Report July 1974

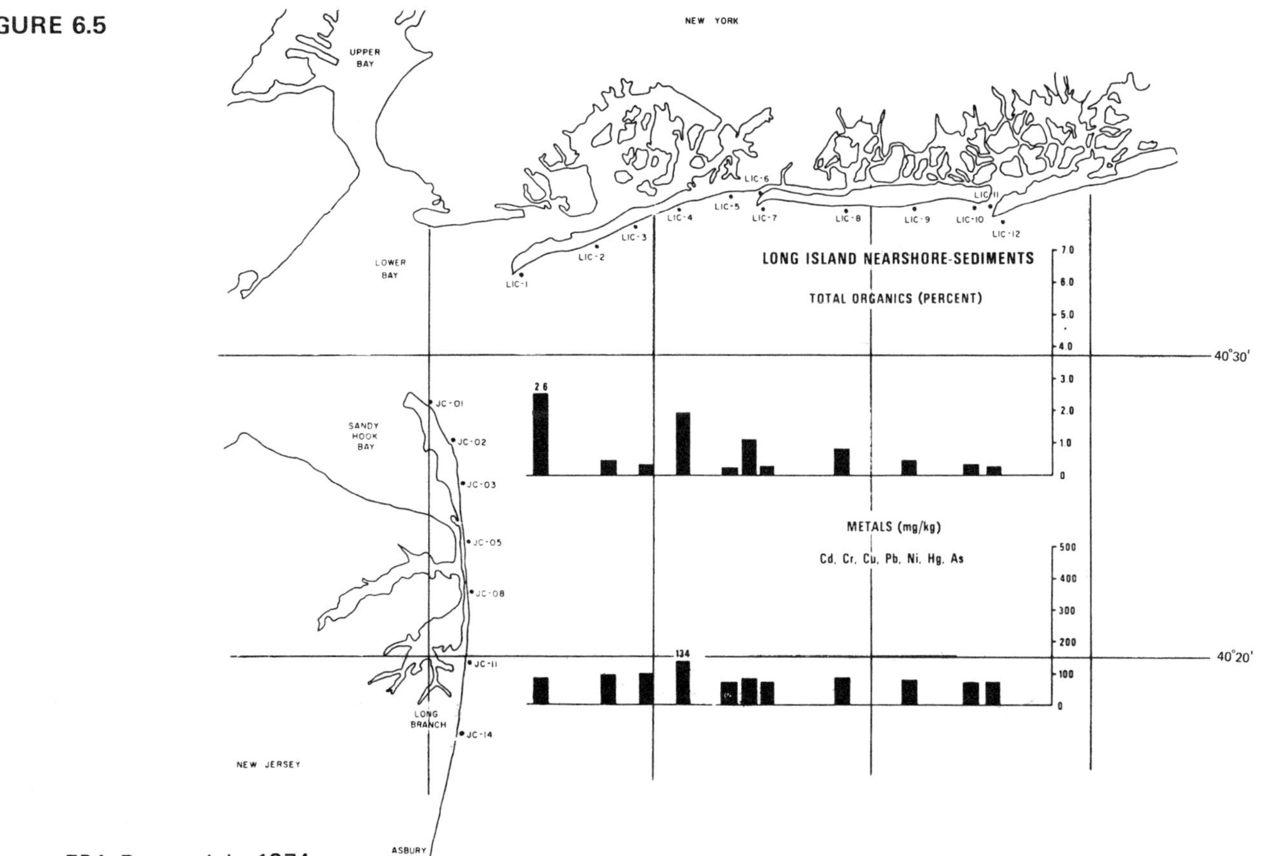

FIGURE 6.5

Source: EPA Report July 1974

FIGURE 6.6

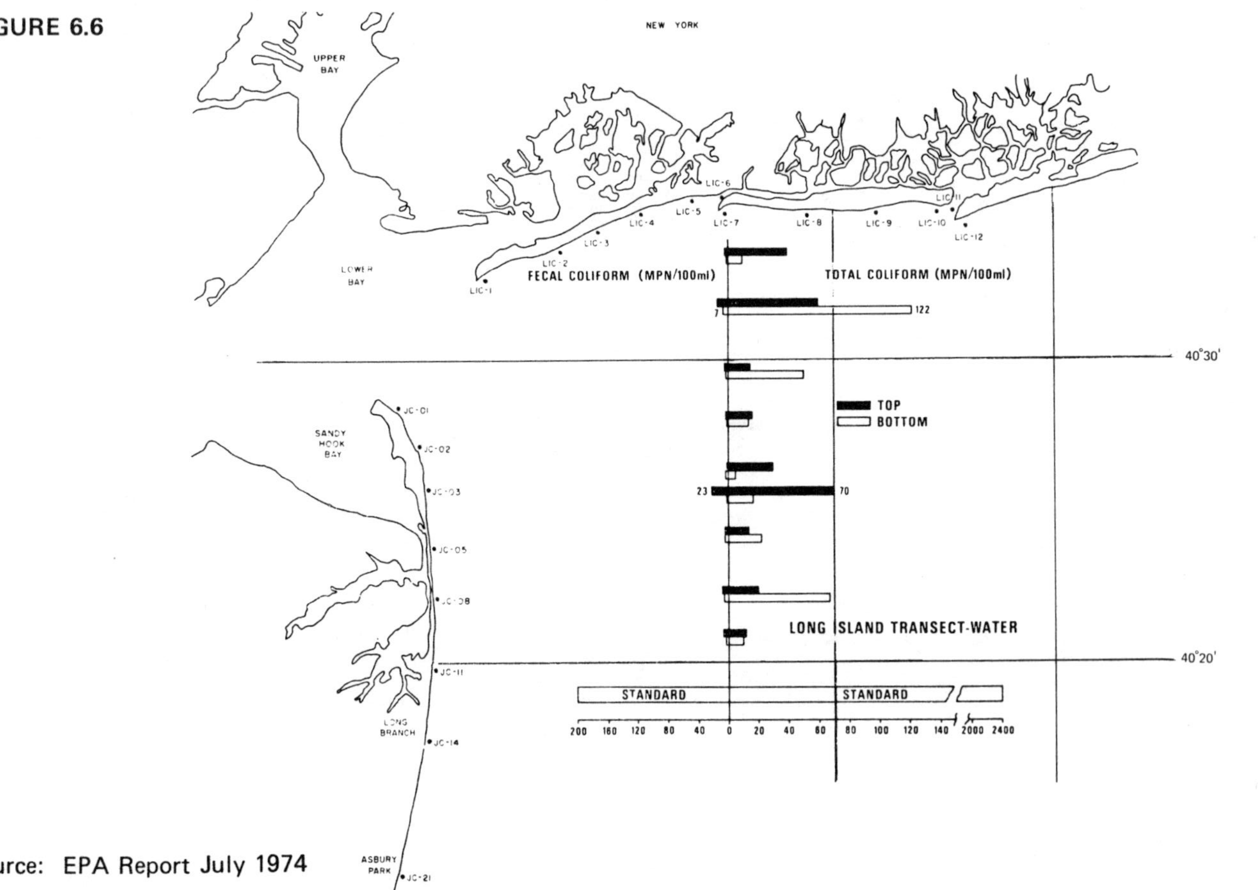

Source: EPA Report July 1974

FIGURE 6.7

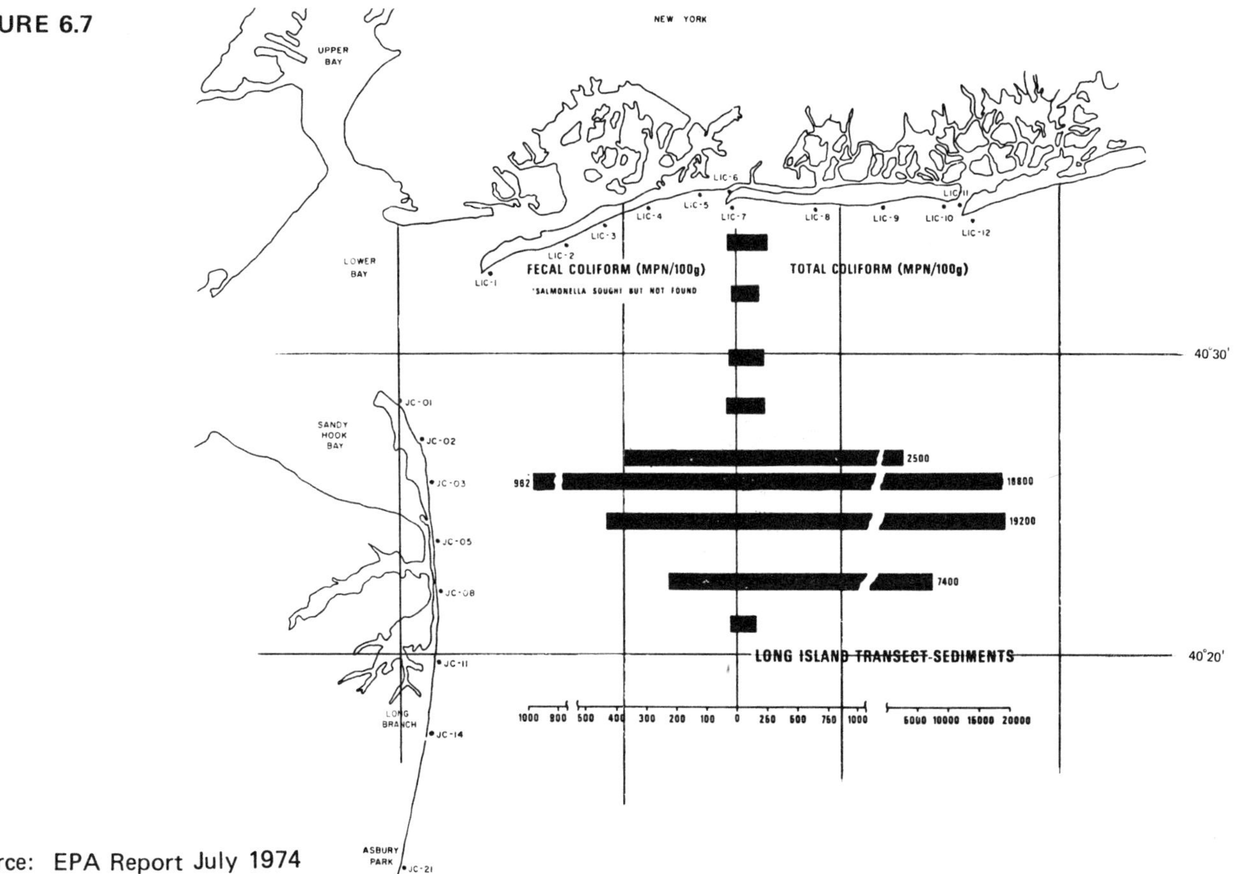

Source: EPA Report July 1974

FIGURE 6.8

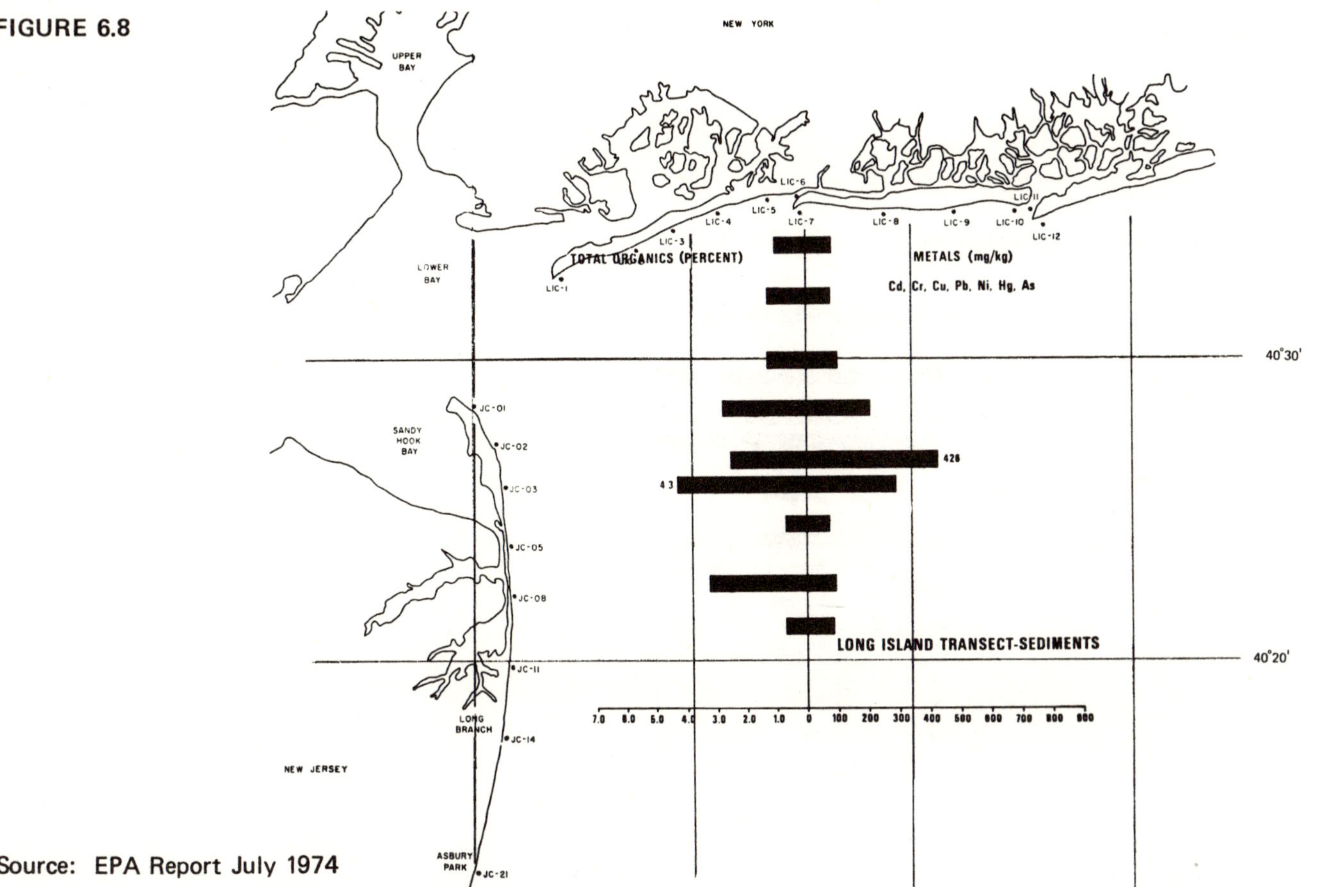

Source: EPA Report July 1974

FIGURE 6.9

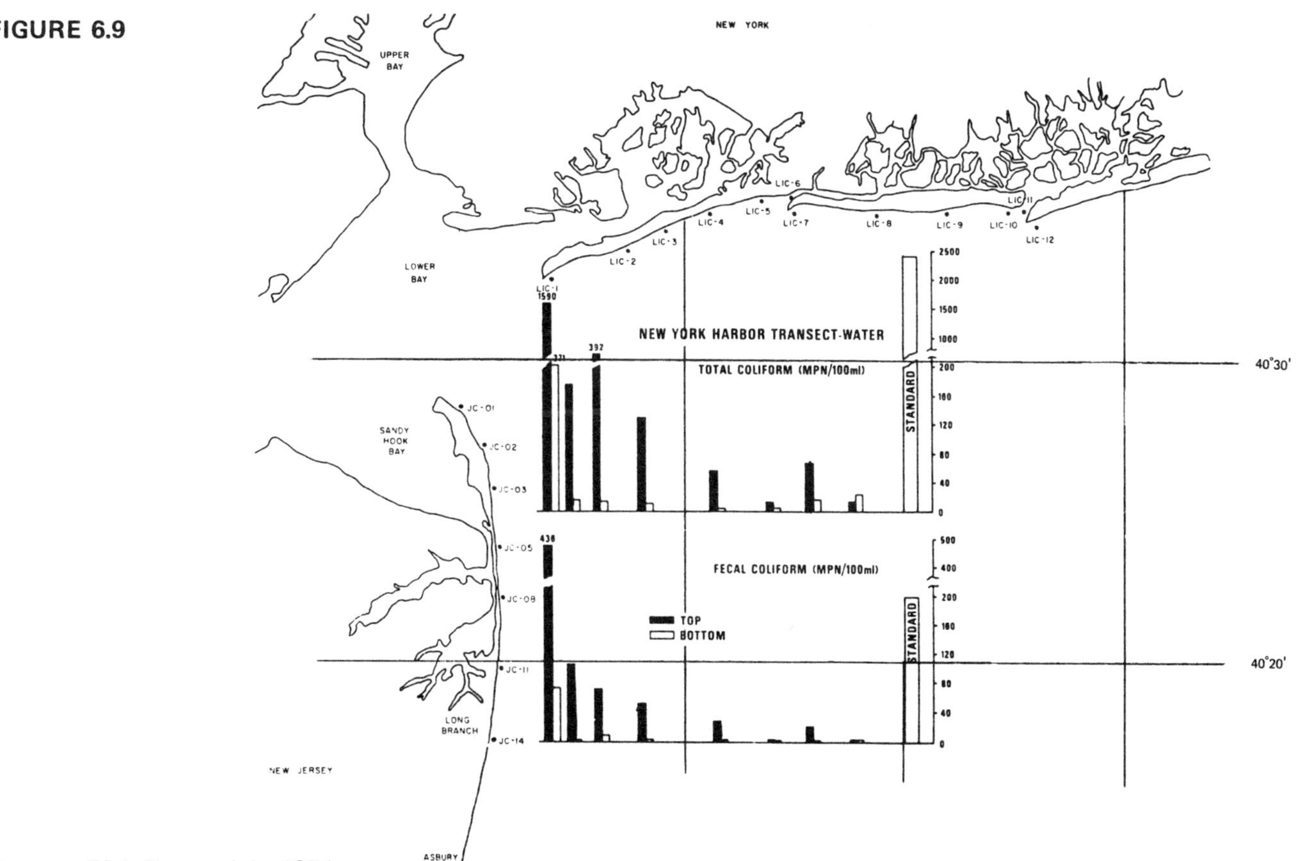

Source: EPA Report July 1974

FIGURE 6.10

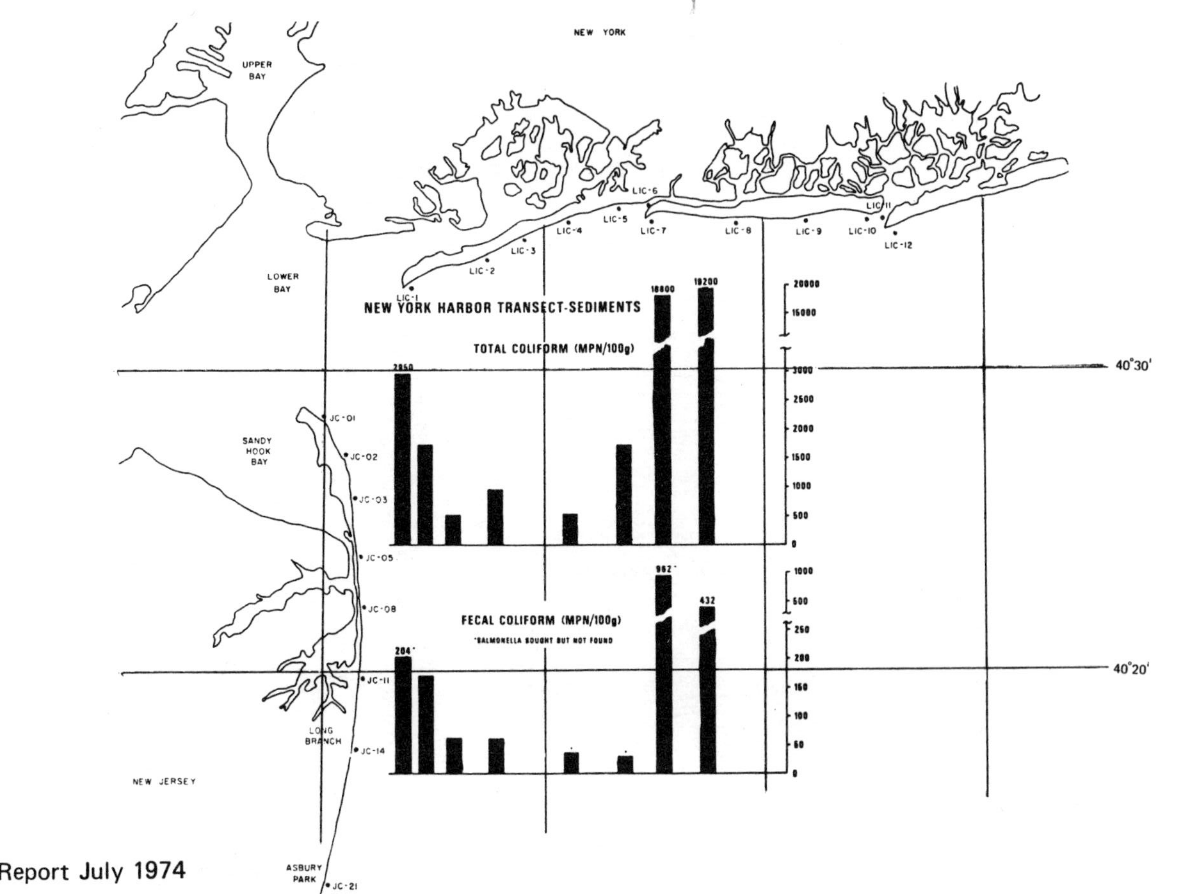

Source: EPA Report July 1974

FIGURE 6.11

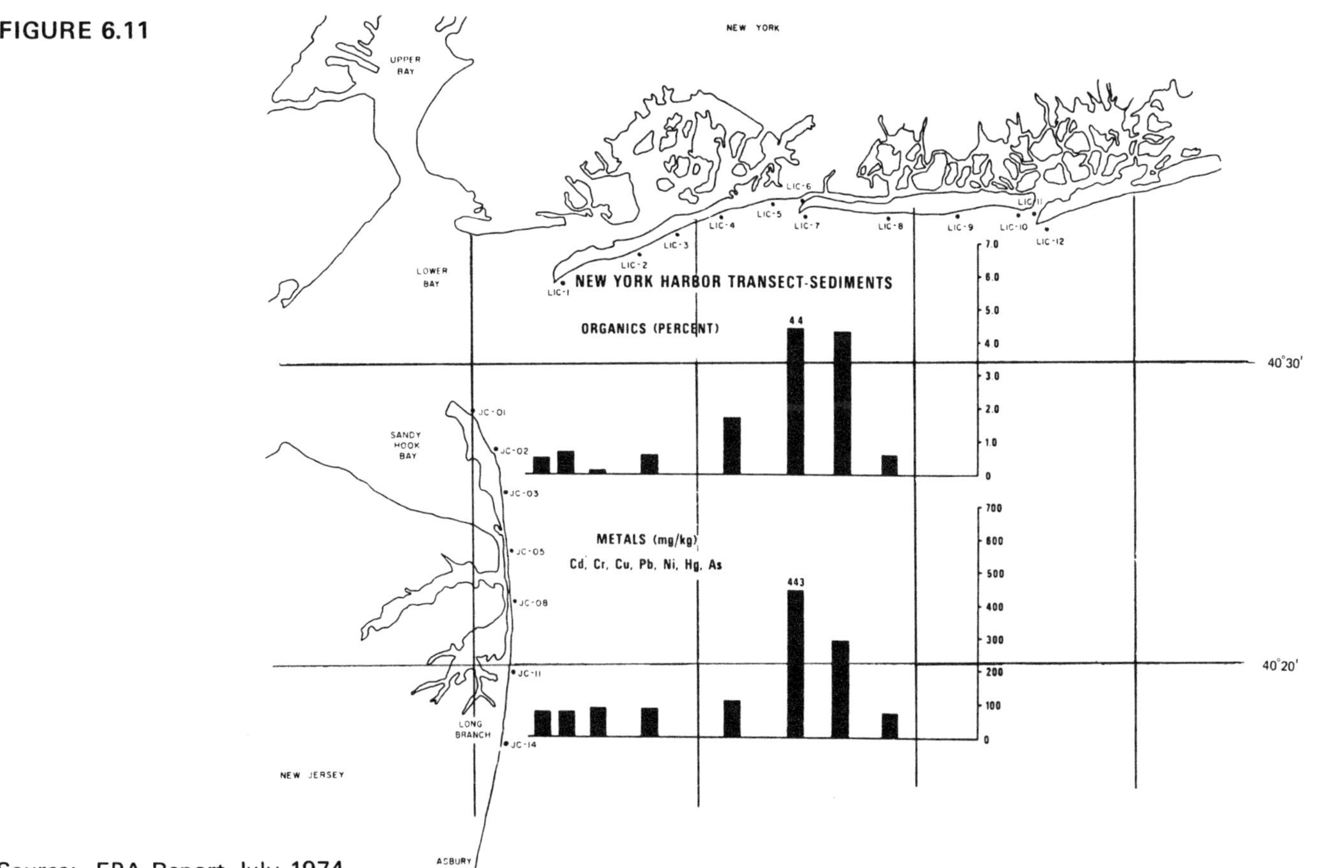

Source: EPA Report July 1974

FIGURE 6.12

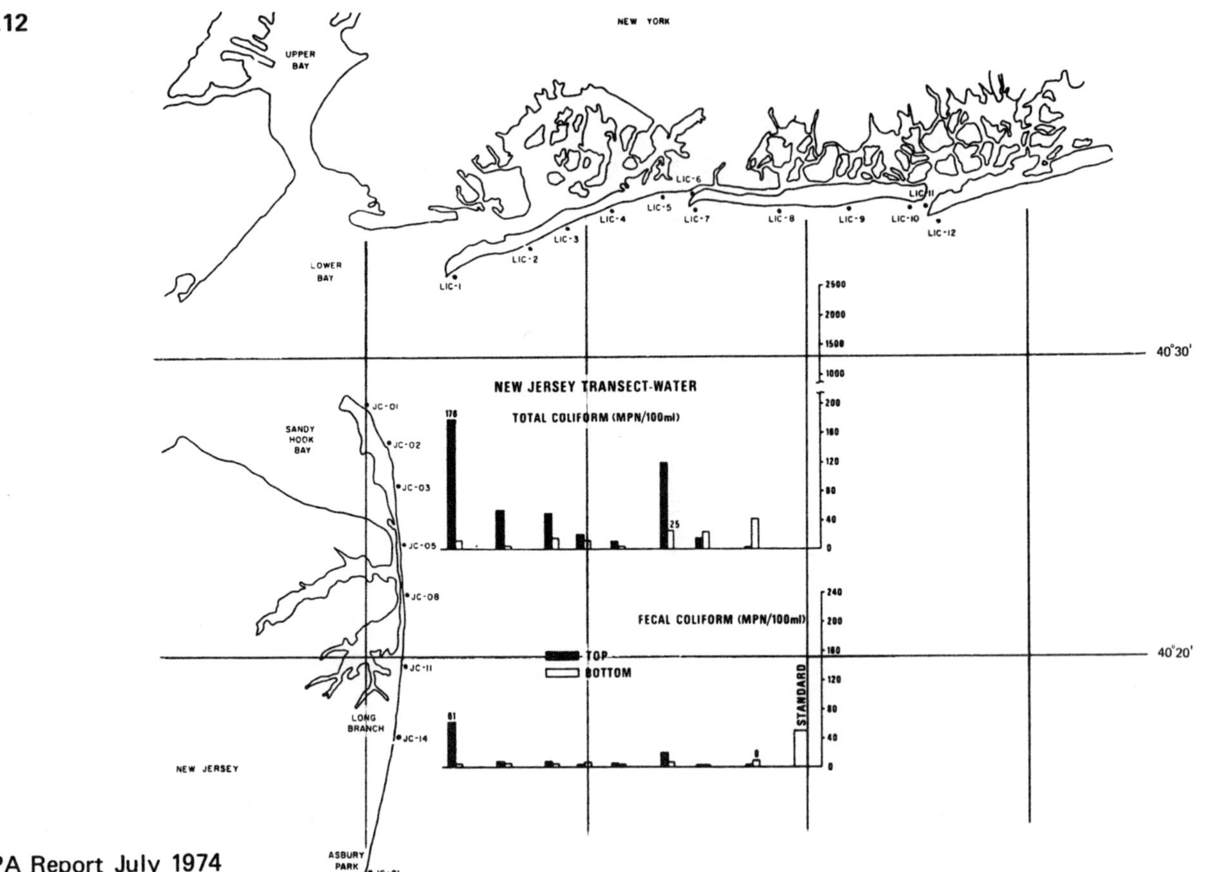

Source: EPA Report July 1974

FIGURE 6.13

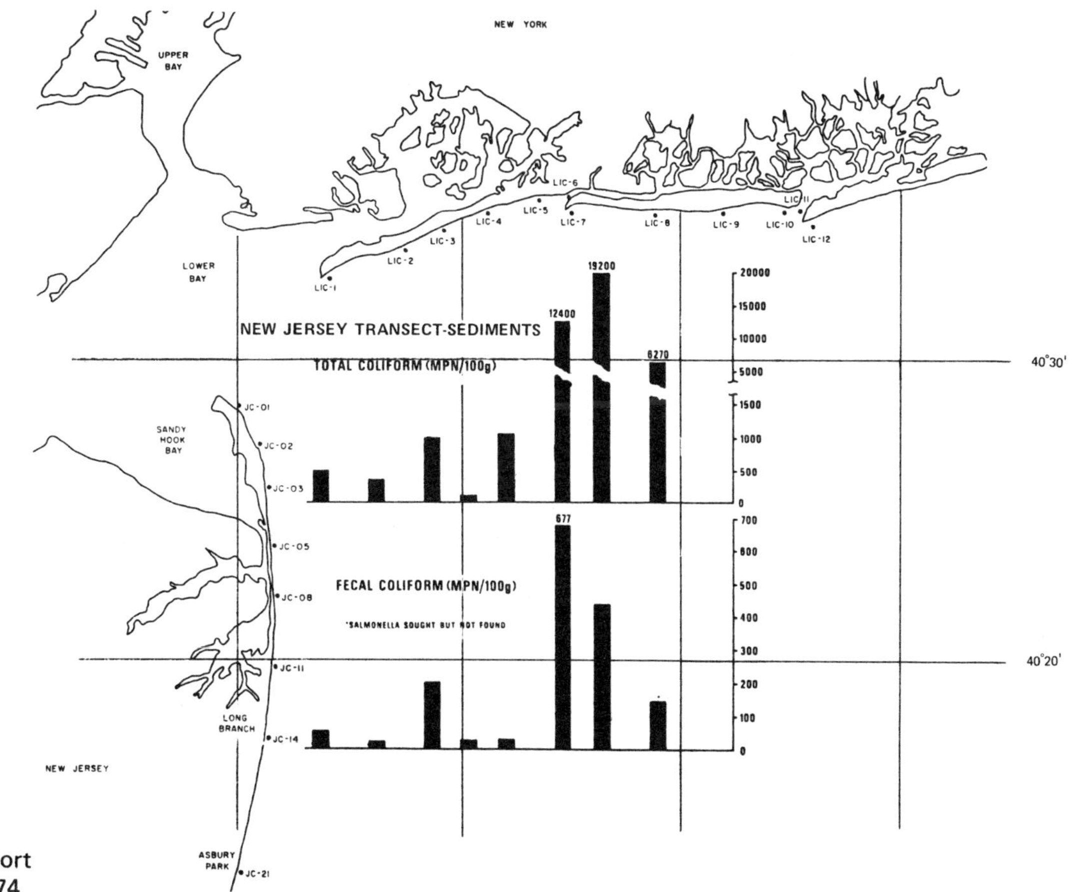

Source: EPA Report
July 1974

FIGURE 6.14

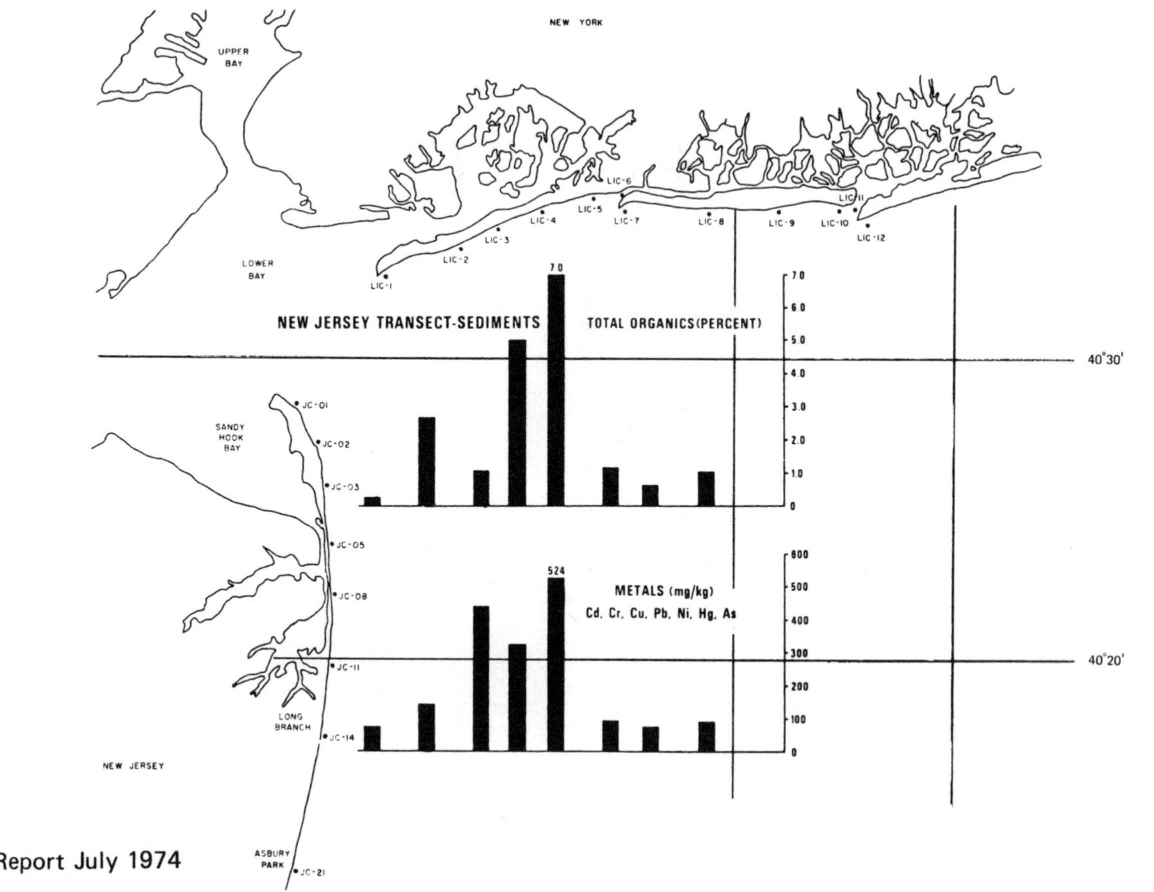

Source: EPA Report July 1974

FIGURE 6.15

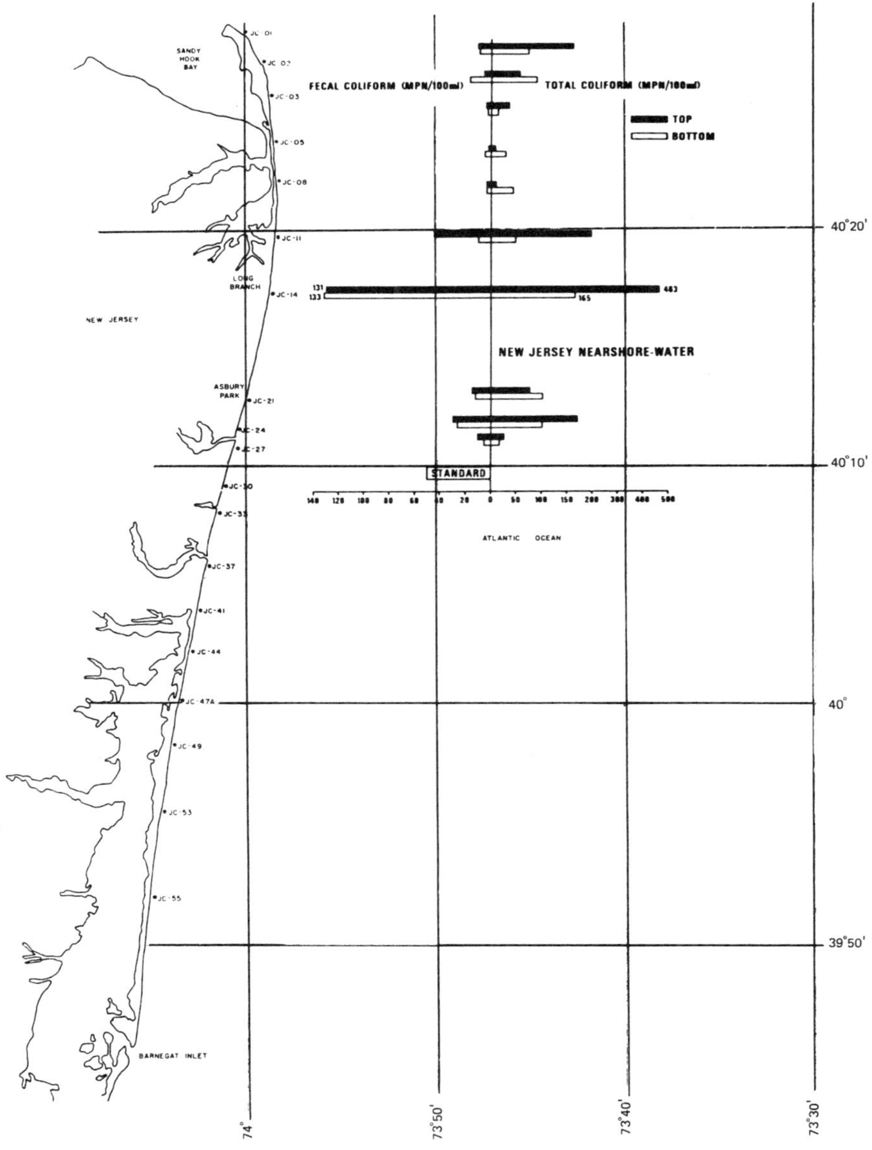

Source: EPA Report July 1974

FIGURE 6.16

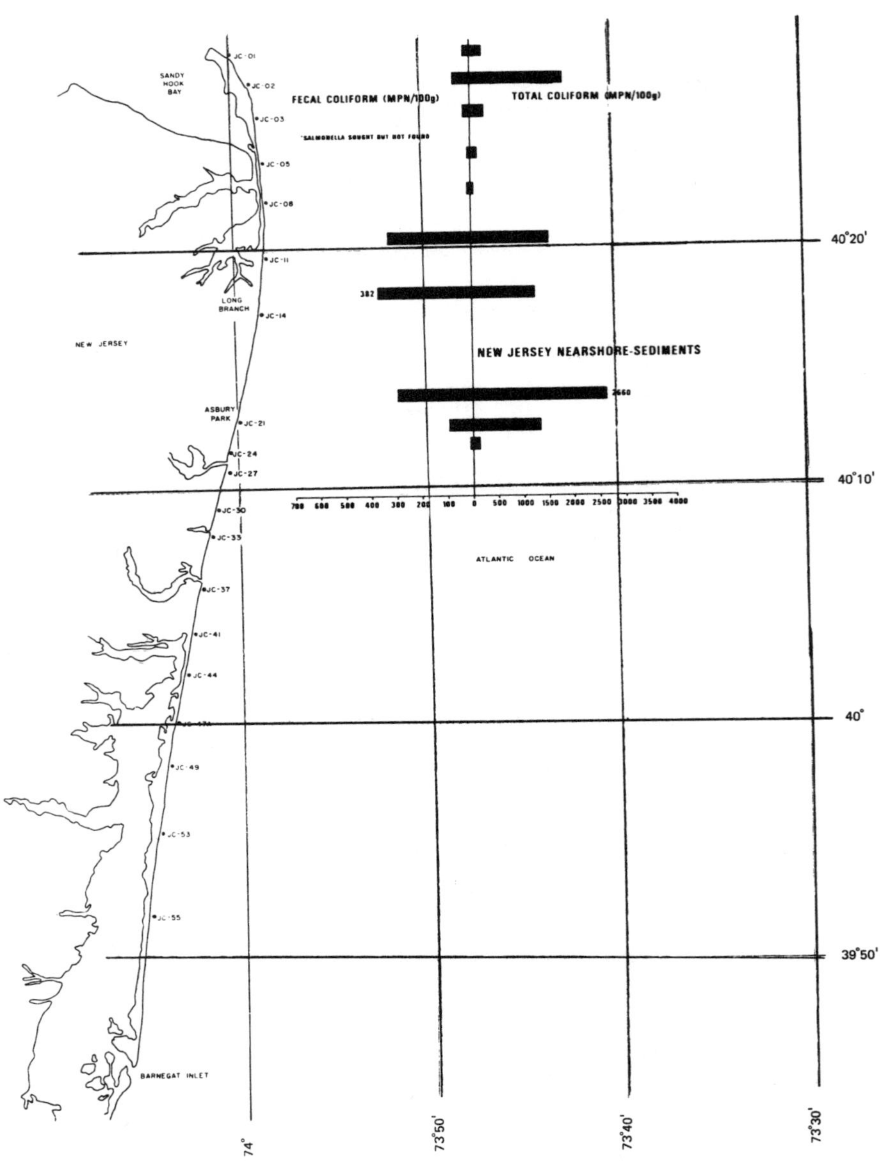

Source: EPA Report July 1974

FIGURE 6.17

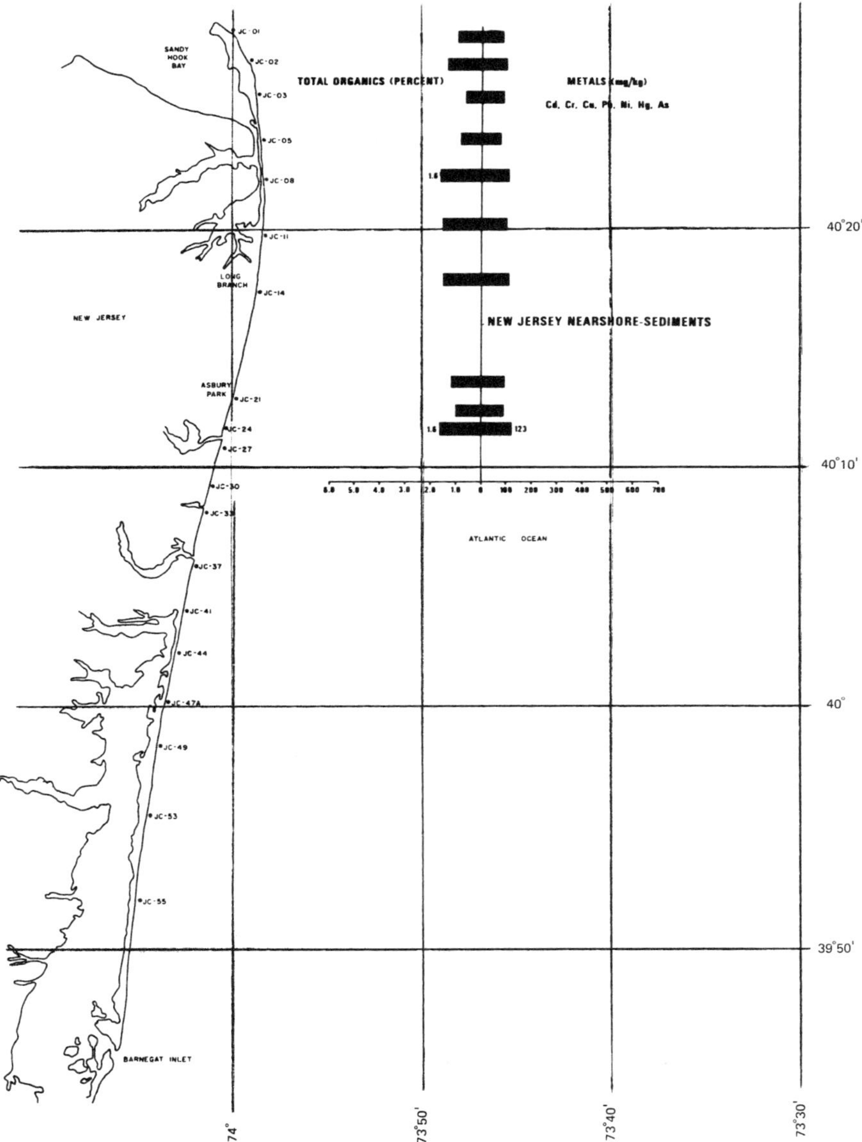

Source: EPA Report July 1974

FIGURE 6.18

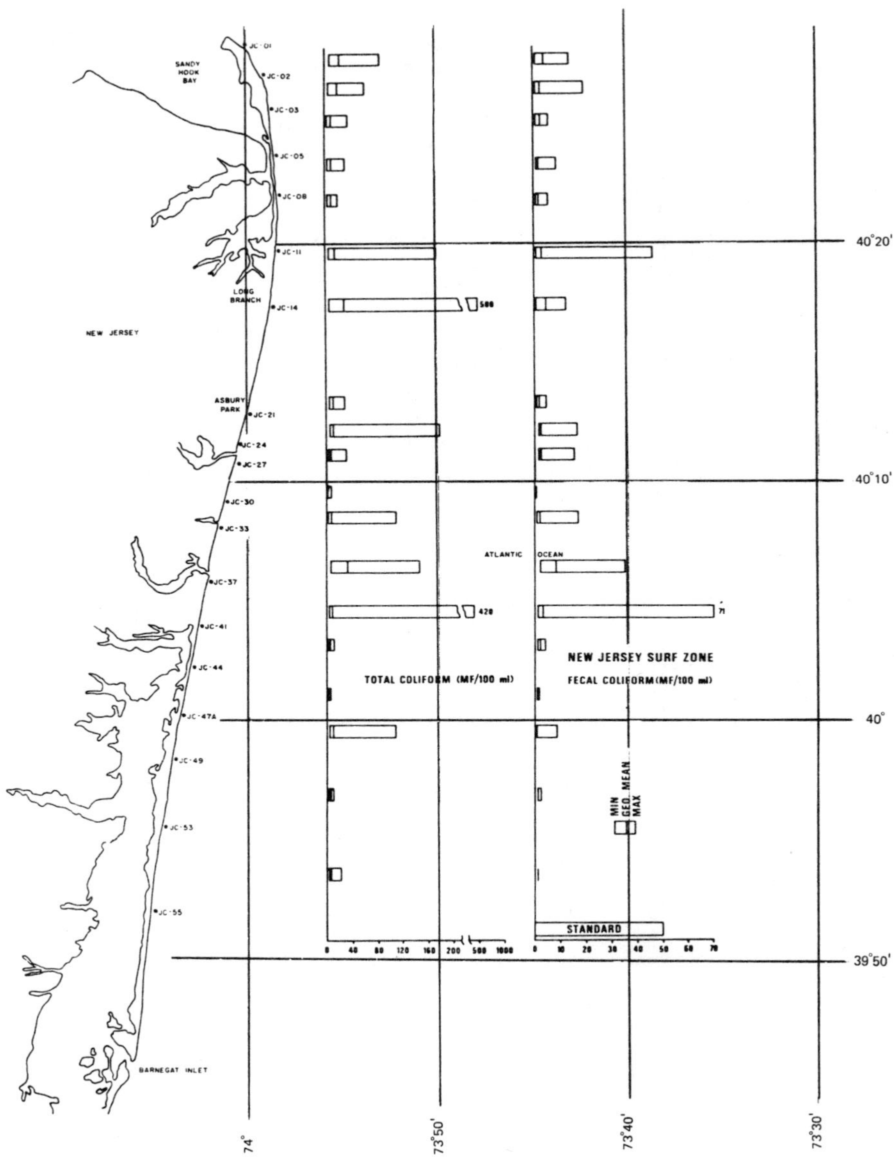

Source: EPA Report July 1974

Environmental Findings—1975

With the Environmental Protection Agency, Region II, expressed desire to halt indiscriminate dumping of waste materials in the marine environment, the Region has implemented a series of programs with the stated ultimate goal for the phase-out of ocean dumping of both industrial and municipal wastes in the New York Bight by 1981.

The initial phase of this comprehensive program was to establish a permit program which included the regulation of waste approved for dumping at designated sites, and the permit requirement for the submission of an implementation schedule or engineering studies leading to the phasing out of ocean dumping within a finite time period.

A listing of the existing ocean disposal permits, and detailed descriptions of waste volumes and characteristics are included in this report for public review. These data indicate a decrease in volumes of both municipal sludges, from 5.6 million cubic yards in 1973 to 4.8 million in 1974, and industrial wastes, from 3.8 million cubic yards in 1973 to 3.6 million in 1974. Most of the reduction in sewage sludge volumes is attributable to construction and repair operations at several New York City treatment plants. The reduction in industrial wastes was small; however, under implementation schedules an additional 12 industrial dumpers will be phased out during 1975 resulting in a three hundred thousand-cubic yard reduction.

An additional step in the Region's program to phase out ocean dumping was the initiation in 1974 of an EPA-funded investigation by the Interstate Sanitation Commission to identify environmentally acceptable, technically feasible, and viable alternatives to ocean disposal. The first quarterly report of this investigation titled, "*New York-New Jersey Metropolitan Area Sewage Sludge Disposal Management Program*," is included in this report. In this quarterly report is an outline of the technical investigation to be accomplished in this evaluation of alternatives.

A comprehensive sampling program was initiated in April 1974 to monitor the chemical and bacteriological quality of water and bottom sediments in the apex of the New York Bight and along the beaches of Long Island and New Jersey. Data collected in this sampling program are augmented by data collected by the National Oceanic and Atmospheric Administration (NOAA), State, and local agencies.

Results of this monitoring program indicate that water along the Long Island and New Jersey beaches is safe for contact recreation and that the leading edge of the sludge mass associated with the sewage sludge dump site is still located approximately 5½ to 6 miles from the Long Island shoreline. Plans are being formulated to expand this monitoring program to include the collection of samples for virological analyses.

The most recent phase was the decision to terminate use of the existing 12-mile sewage sludge dump site in 1976 and move to an alternate site(s). Although an alternate dump site(s) has not yet been selected, two areas designated by EPA and NOAA are being studied for interim use pending completion and implementation of an environmentally acceptable alternative. In accord with regional policy, a notice of intent to prepare an environmental impact statement (EIS) for the designation of an "interim" sewage sludge dump site or sites in the New York Bight was issued in January 1975.

Volumes and Characteristics of Waste Materials Dumped: A total of 14.6 million cubic yards of waste materials was disposed of in the New York Bight during 1974. Municipal sludges amounted to 4.9 million cubic yards, of which 49% was from New York City, 43% from sources in New Jersey, and the remainder from Nassau and Westchester Counties in New York State. Industrial wastes amounted to 3.6 million cubic yards. The remainder, 6.1 million cubic yards, resulted from the disposal of dredged spoils (84%) and construction debris (16%).

A graph summarizing the volumes of waste materials disposed of in the Bight during 1960-74 is presented in Figure 6.19. Summarized in Table 6.5 are the volumes of waste materials dumped into the New York Bight (by dump site) 1960-1974.

FIGURE 6.19: NEW YORK BIGHT-WASTE DUMPING

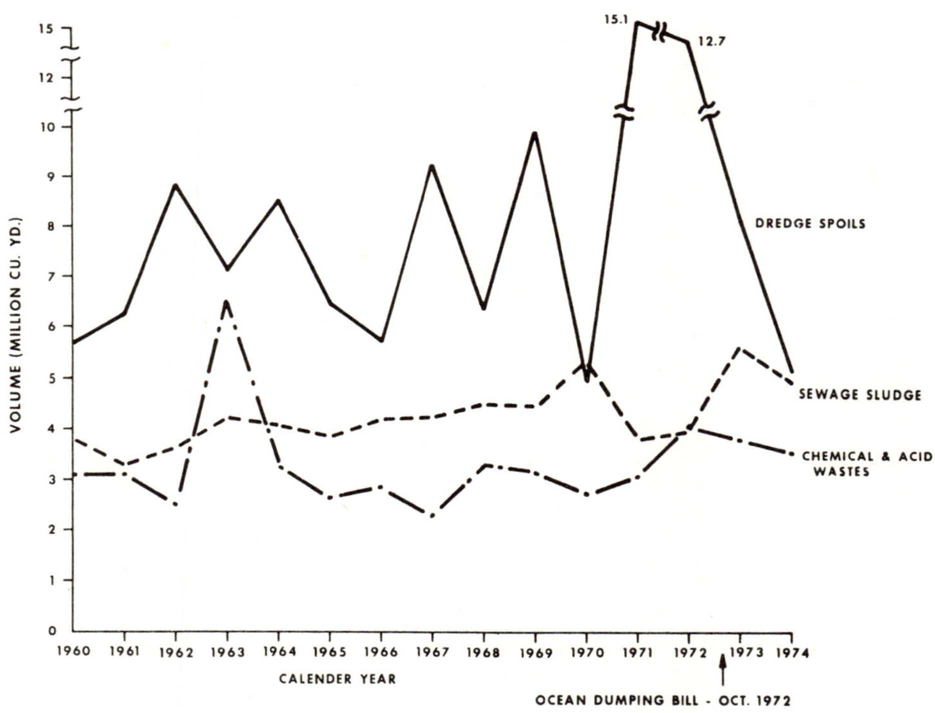

Source: EPA Report July 1974

TABLE 6.5: VOLUMES OF WASTE MATERIALS DUMPED INTO THE NEW YORK BIGHT (By Dump Site)

(Cubic Yards)

	1960	1961	1962	1963	1964	1965	1966	1967
Mud (Dredge Spoils)	5,611,174	6,238,000	8,816,000	7,186,000	8,540,000	6,480,000	5,687,000	9,214,000
Sewage Sludge	3,819,855	3,354,000	3,626,000	4,259,000	4,078,000	3,863,000	4,201,000	4,265,000
Chemical Wastes	-	7,000	13,000	690,000	-	-	-	150
Acid Wastes	3,071,000	3,100,000	2,551,000	5,797,000	3,310,000	2,634,000	2,857,000	2,297,000
Cellar Dirt	738,000	810,000	807,000	988,000	756,000	912,000	319,000	509,000

	1968	1969	1970	1971	1972	1973	1974
Mud (Dredge Spoils)	6,361,000	9,880,000	4,905,000	15,062,000	12,785,000	8,130,000	5,113,000
Sewage Sludge	4,481,000	4,455,000	5,287,000	3,830,000	3,955,000	5,600,000	4,860,000
Chemical Wastes	83,000	86,000	160,000	222,000	943,000	664,000	585,000
Acid Wastes	3,143,000	3,054,000	2,605,000	2,828,000	3,112,000	3,136,000	2,970,000
Cellar Dirt	400,000	632,000	796,000	676,000	969,000	827,000	1,048,000

Long-Range Goal for Phase-Out of Ocean Disposal in New York Bight by 1981: In accordance with its mandated responsibilities under the Marine Protection, Research, and Sanctuaries Act of 1972 "to prevent or strictly limit the dumping into ocean waters of any material which would adversely affect human health, welfare, or amenities, or the marine environment, ecological systems, or economic potentialities," EPA, Region II has committed itself to an orderly and environmentally acceptable plan with the ultimate goal of phasing out ocean disposal of both industrial and municipal wastes by 1981.

The following correspondence is a letter of modification to municipalities in the New York Metropolitan area stating that ocean dumping permits will not be renewed for the continued use of the existing sewage sludge dump site in the latter part of 1976.

> Earlier this year we notified you of our intention of moving the sewage sludge disposal site from its present location, 12 miles off the coast, to a new location approximately 65 miles from the apex of the New York Bight. Our position on the use of a site(s), within the designated areas is now firm; therefore, in 1976 your ocean disposal permit will not be renewed for the continued use of the present site. The specific location of the new "interim" site(s), will be designated in 1975, and will be used until such time as environmentally acceptable alternatives are implemented. Figure 6.20 provides specific details on the location of the new designated areas.
>
> As you know, EPA and NOAA's monitoring studies still continue to indicate that use of the present sewage sludge site does not pose any immediate threat to the waters of Long Island or New Jersey. It is our opinion, however, that the existing site cannot accommodate the anticipated three-fold increase in the volume of sludge, which will result from an upgrading of treatment facilities in this area. Thus, until satisfactory alternate means of disposal are developed, a new "interim" ocean disposal site(s) must be used.
>
> Studies to support an Environmental Impact Statement, which must be undertaken before any new site can be utilized, have already been initiated. Various sites within both designated areas are being investigated, thus, pending completion of these studies, the actual site(s) cannot be selected. If for some reason sites within these areas are found to be unsuitable for the disposal of sludge, sites outside of these areas, however within the same "steaming distance," will be selected and available by 1976.
>
> A detailed study aimed at evaluating the possible alternatives to ocean disposal in the greater New York-New Jersey Metropolitan area has already been initiated. Should this investigation demonstrate that other alternatives are viable, technically feasible and environmentally acceptable, our goal will be to phase out ocean disposal by 1981.

The actual choice of an interim site(s) within one, or both of the areas, will completely depend upon the results of detailed environmental studies now being conducted by EPA and the National Oceanic & Atmospheric Administration (NOAA). These investigations are scheduled to be completed by August 1975 and a decision regarding the specific interim site(s) location will be made during the latter part of 1975.

It is important to understand that the interim site(s), which will be used only until such time as acceptable environmental alternatives are developed and implemented, cannot be utilized for any dumping until an Environmental Impact Statement (EIS) is prepared and made public. Simply translocating the present problem to a new area is not what is intended; minimizing the environmental impact, while at the same time protecting the high-quality water along our beaches is what is hoped for.

A New York consulting firm has been retained by Region II to prepare the EIS. The report will include, but not necessarily be limited to, background information on the Bight, ocean dumping practices, federal control programs and international legal implications; a presentation of sewage sludge disposal methods other than ocean dumping; alternatives to the proposed interim disposal site(s) including that of "no action" on moving the existing site; a detailed description of the proposed interim site(s); an evaluation of the primary and secondary environmental impact in terms of the physical, chemical and biological characteristics; analyses of the adverse environmental effects which cannot be avoided should use of the proposed site(s) be implemented; the relationship between local short-term uses of man's environment and the maintenance and enhancement of long-term productivity; and the irreversible or irretrievable commitment of resources which would be involved in the proposed action, should it be implemented.

The final draft of this EIS should be completed by early 1976 and a public hearing held before April of that year.

With regard to the development of ocean disposal alternatives, and the need to meet the goal of phasing out the present practice by 1981, about 30% of "Alternative-Evaluation Study," under the auspices of the Interstate Sanitation Commission and EPA, has already been completed.

A final decision on the availability, acceptability and practicability of an environmentally suitable alternative(s), which depends upon the results of the EIS, as well as a thorough engineering and scientific evaluation of the many technical approaches for handling sludge, is not expected until July 1976. The five-year "gap" between 1976 and 1981, is the time needed to implement the alternatives recommended.

All of these studies and activities are progressing concurrently, and are presently on schedule; therefore, hopefully the 1976 deadline for decision making, i.e., to move to a new interim disposal site(s) and to recommend and implement environmentally acceptable alternatives to the present practice of ocean disposal will be met.

MESA Project Statement: The following is excerpted from the statement of Dr. R.L. Swanson, Manager, New York Bight Marine EcoSystems Analysis (MESA) Project, Environmental Research Laboratories, National Oceanic and Atmospheric Administration, United States Department of Commerce, before the New York State Select Committee on Environmental Conservation, February 20, 1975.

> NOAA's activities in the New York Bight are being conducted under the Marine EcoSystems Analysis, or MESA, program, which is managed by the Environmental Research Laboratories. The MESA New York Bight Project, which began in July 1973, is a seven-year study, designed to yield certain important information tools for managers of the coastal environment. These include: determining the fate and effect of pollutants on the New York Bight

FIGURE 6.20

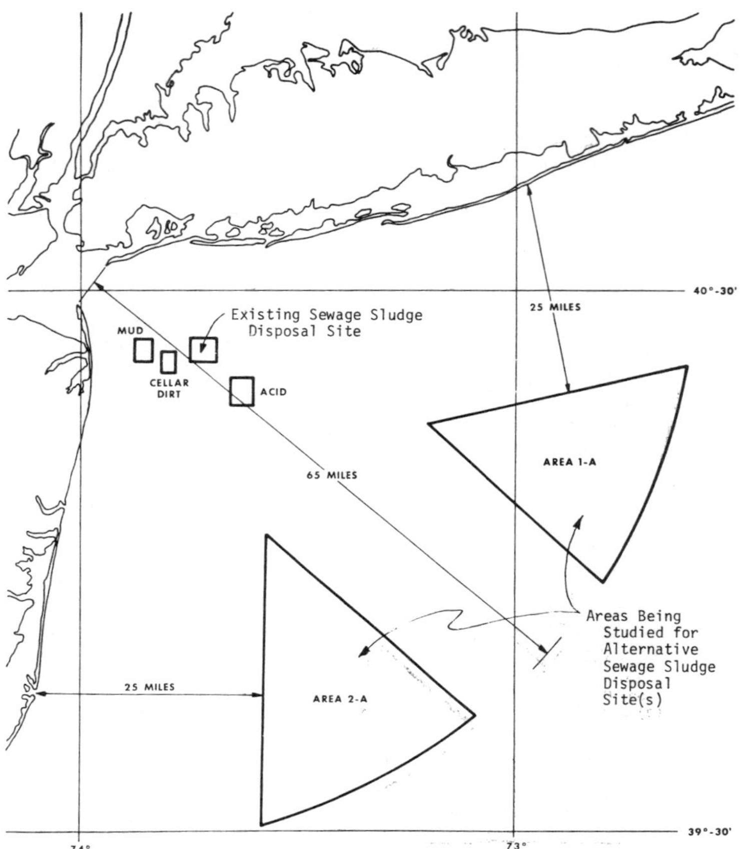

Source: EPA Report April 1975

ecosystem, with priority on ocean dumping, and identifying and describing the important ecological subsystems, processes, and driving forces operating in the New York Bight, and defining their interrelationships and rates of change.

The area encompassed by the MESA Project here is bounded on the offshore side by the edge of the continental shelf, and on the inshore side by the Long Island and New Jersey coasts to the Apex of the Bight, where the Lower New York Bay meets the coastal waters.

MESA's reason for being is that environmental scientists understand the interactions between human civilization and the marine environment in only a general way. It is obvious, for example, that some 20 million human neighbors, great cities, and large industries impact the coastal environment and ocean waters they border and use. But it is not at all obvious whether such impact is totally adverse, or how much use can be accommodated by an ecosystem such as the New York Bight without irreversibly damaging it. That it has been heavily stressed is known, what is not known are the recuperative powers of the Bight and its marine life.

Understanding of and the ability to predict events which control the condition of the Bight's marine life and environment is what is expected to be gained through this program.

The primary near-term emphasis of the research in the New York Bight Project is the delineation of the environmental effects of ocean dumping. NOAA, on behalf of the Secretary of Commerce, has research responsibilities for ocean dumping under the Marine Protection, Research and Sanctuaries Act of 1972, and the Bight, which is the site of most of the nation's ocean dumping is being used as a laboratory.

The characteristics of material disposed at individual dump sites in the Bight, dimensions and dispersal rates of dumped materials, long-and-short-term chemical, biological, and geophysical interactions which govern the fates and effects of such dumping, including the ability of dump sites to recover are being studied. Investigators are measuring currents at various depths and locations in the Bight, examining sediments and their transport rates and processes, and sampling bottom-dwelling and other organisms to determine the existence and extent of the biological and pathological impact of man's activities in the Bight.

The following technical information comes primarily from these ocean-dumping related studies. Despite the fact that this is a highly populated corner of the United States, insufficient historical data are available on these aspects of the New York Bight waters, sea floor, and marine life needed for understanding. In a few areas, most of the useful data available has been obtained this past year and a half.

To date, most of our field activities (the first of which began in July 1973) have been concentrated in the 625-square mile rectangular area with landward boundaries defined roughly by the New Jersey and Long Island coasts, from the mouth of the Hudson River. Activities there have accomplished the following:

(1) Ocean water depths and small-scale terrain features of the sea floor have been mapped with precision navigation and depth-measuring systems.

(2) The distribution and thickness of sea floor sediments has been mapped, using precision navigation and acoustic subbottom profiling equipment. In addition, more than 750 superficial sediment samples and 50 sediment cores have been collected to aid in sediment mapping and chemical analyses.

(3) Sea floor materials have been sampled on a quarterly schedule at selected sites to detect short-term changes in the sediments and associated microrelief. Sediment movement has also been studied using direct measurement techniques.

(4) Current meter arrays have been deployed at various locations for intervals as long as two and a half months.

(5) Twenty-six water-column stations have been sampled repeatedly at 18-day or 36-day intervals for water and suspended-material analyses.

(6) Intensive and continuous programs of biological sampling and analysis

have been conducted, including studies of bacterial roles in the ecosystem, oxygen-consumption by bottom-dwellers and sediments, distribution of fin-fish diseases in and around the Bight, and marine population distribution in Bight waters.

Sewage sludge is one of the products resulting from the treatment of waste-water, a mixture of human, animal, and industrial wastes. The material, although it varies with source and treatment plant, is about 5% solids and 95% liquid. In 1973, some 4.4 million cubic meters of this material, and 2.85 million cubic meters of industrial wastes, were dumped in ocean waters from 10 to more than 160 kilometers offshore.

Where does it go? When sludge is dumped, the mixture disperses, some of it swept laterally by Bight currents, and some of it sinks toward the sea floor. A small portion may float, and other waste constituents dissolve or remain suspended in the water column.

Based on core and grab samples, it is known that the portions of sewage sludge which sink to the bottom have not formed a massive lens of material.

In fact the bathymetry at the dump site is essentially the same as it was in 1936. On the other hand, roughly 8 meters of material have accumulated in the same period of time at the dredge spoil dump site. Generally, however, the sea floor near the sewage sludge dump site is an admixture of natural fine sands, natural muds of high organic material content, and sewage sludge materials at different stages of degradation. The surface layers of samples taken at the dump site have a consistency which ranges from sand to black organic material. An appreciable amount of sludge material extends beyond the sludge dumping site in all directions, but primarily to the north into Christiaensen Basin.

The concentration of sludge in the Christiaensen Basin may be caused by short dumping or by deposition of transported or resuspended sewage sludge materials in the water column. The bottom of the basin is essentially covered with a mixture of muds rich in organic matter (1) transported into the area from the Hudson River, (2) and produced in the area by marine organisms, primarily plankton, and (3) disposed as sewage sludge.

It is generally accepted by oceanographers that the overall movement of water on the continental shelf off southern New England and the Middle Atlantic States is to the west, southwest, and south, parallel to the continental margin. However, the overall pattern is quite thoroughly masked at particular locations and times by high variability. This variability in the New York-New Jersey area specifically the Bight Apex, has been the prime focus to date of the water current observation effort of the MESA Project. The major features of the spatial structure of circulation that relate to disposal of waste materials in shelf waters of the New York Bight are as follows:

In the immediate vicinity of the entrance to Lower New York Bay, the oceanographic regime is dominated by the influence of discharges from the Hudson and Raritan Rivers. There is a seaward flow of brackish water, from the estuary where the rivers meet, in surficial layers. At sea, this discharge turns to flow southward, paralleling the New Jersey shoreline. Lower in the water column, there is return flow of external water into the estuary. In spite of its importance to the oceanographic and ecological systems of the estuary and the Bight, this superposed flow system is recognized in measurements only as a slight imbalance of the much stronger ebb and flood tidal currents in the respective layers.

There is strong evidence from recent results of the MESA Project that outside the region of strongest influence of river dishcharges, there is a persistent, at least in a statistical sense, clockwise circulation or gyre. In its most inshore portion, off New Jersey and Long Island, flow in this eddy runs counter to the general flow over the continental shelf adjacent to this part of the coast. Although its existence in some sections of critical interest to the dumping question has been established, its horizontal extent and vertical structure, are imperfectly known at present. Evidence for the existence of this clockwise circulation or gyre comes from current meter measurements, studying where "drifters" dropped into the ocean at various locations go, and studying the red colored iron hydroxide particles introduced into the Bight waters from the acid waste dump site.

The drifter study, for example, shows that approximately 3% of those drifters released at the sewage sludge dump site were recovered within ten days following their release.

Material that becomes entrained in the Inner Bight circulation, and this refers to suspended, dissolved, and floatable portions of sewage sludge as well as naturally occurring organic matter, tends to move generally in a clockwise manner northward toward Long Island, then eastward out to sea, and then southward toward the open sea. In its travels, suspended material tends to be deposited in topographic lows, like the Christiaensen Basin and the area east of Cholera Bank, because of reduced wave and current action there.

The New York Bight is also marked by a characteristic nearshore flow, a result of which is that bottom water moves toward the beach. When this circulation enters the surf zone near the beach, natural hydraulic processes related to the breaking of waves permit suspended particles to be deposited on the bottom. The mud deposits formed in this way in depths of less than 20 meters are a common occurrence along the Atlantic coast, usually forming patches in small topographic lows.

In the nearshore zone, these organic rich mud patches tend to be small in extent, usually less than ten meters in diameter and less than 15 centimeters thick. SCUBA divers off Long Beach in July 1974 found that such patches covered less than 10% of the bottom. Most of these mud patches are probably modified by the strong wave action associated with winter storms.

Continental shelf floor muds are invariably rich in organic matter as a result of hydraulic sorting processes which tend to group finely divided inorganic particles and organic matter. Clays occur here, and characteristically contain about 5%, by weight, of iron oxide, which usually veneers a shelf mud with a thin reddish-brown layer. Depletion of oxygen by decaying organic matter several millimeters below the bottom surface has its effect on these muds, producing a black color.

This type of material can be found along both heavily populated and virtually deserted stretches of the Atlantic coast. This gross color and texture of mud patches near the Long Island coast are no guides as to whether or not they contain sewage sludge. Descriptive terms such as "sludge" or "black greasy stuff" or "black mayonnaise" do not indicate whether the organic mixture consists of degraded organic material of natural origin or of human waste.

Analysis of these muds for their natural and man-introduced constituents is not a straightforward procedure. There is no "litmus paper test" which differentiates sewage sludge from natural sea floor muds. Sewage sludge produced by one treatment plant differs in composition from that produced by another, and even material produced at one treatment plant varies greatly over time. Once dumped, the material is acted on by decomposition, interactions with various components of the marine biota, reaction with inorganic portions of the environment, and dilution due to mixing with water and sediment. A simple test or two will not differentiate transported or migrated portions of sewage sludge from organic material, raw sewage emanating from rivers, harbors, outfalls, and passing vessels, or from sewage sludge which is dumped short of the existing dump site.

At best, a series of tests on a sample containing "sludge-like material" may exclude one or more sources, but not all possible sources. Positive identification of sewage sludge, especially after it has been in the marine environment for any length of time, is not possible at this time.

MESA scientists have filtered and then microscopically analyzed bottom mud samples to sort out the fraction of the particles in the mud that could be identified as of human origin. Near the sewage sludge dump site about 15% of the mud grains were found to be processed cellulose fibers or soot-like particles considered to be of human origin. Near the Long Island beach roughly 1% of the grains were considered to be of human origin. Clearly, any dumped, sewage-sludge-related materials which move in suspension toward Long Island must be tremendously diluted by natural particles during transport.

Preliminary studies have been made to compare the chemistry of sewage sludge with that of shelf floor muds. These indicate that sewage sludge is rich in heavy metals and organic matter; however, so are Bight floor muds, including many natural muds. Thus, while certain inorganic analyses, including heavy metal ratios, may eventually prove to be indicators of sewage-sludge-derived material, they are not considered to be an adequate means of differentiating sewage sludge from other muds at this time. This was again verified by analysis of heavy metals and samples collected at a series of stations south of Long Beach and Atlantic Beach, including the sewage sludge dump site and inside Rockaway Inlet.

Similarly, total organic matter is not a suitable indicator of sludge-derived material because in an area such as the New York Bight, marine plants produce more than twice as much organic matter than is dumped in the form of sewage sludge.

However, some fractions of organic matter, such as carbohydrates (including cellulose), may provide a crude means of differentiating muds. When sewage sludge is dumped into sea water, the organic matter decreases with time because of microbial degradation. Most carbohydrates decompose less rapidly than many other organics; therefore, the ratio of carbohydrates to total organic carbon should increase with the age of the sludge.

Analysis of sediment samples shows that the highest total organic carbon values of about 5% by dry weight occurred in the Christiaensen Basin, a few miles north and west of the dump site. This is about one-tenth the total organic content that is found directly in sewage sludge. High total organic content and also high carbohydrates occur all along the Hudson Shelf Canyon. Mud samples from within one mile of Long Island beaches show relatively high carbohydrate content, indicating that some of this mud may be partly derived from sewage, raw sewage from the harbor, or from ships or from sewage sludge. But sediment samples from Rockaway Inlet and Jones Inlet showed rather low carbohydrate content, indicating mainly a natural origin.

Further, it can in general be concluded that the effect of dredge spoil and sewage sludge dump sites on nutrient concentration in the Bight Apex is rather localized, and secondary in importance to that of the nutrient-rich Lower Bay outflow.

MESA scientists have observed reduced abundance and diversity of benthic invertebrates in the vicinity of the sewage sludge and dredge spoil dump sites. Increased incidences of fin erosion have been observed in the Inner Bight, although positive identification of cause and effect has not been made. These findings, however, argue strongly against contamination of new areas by moving the existing sewage sludge dump site.

In sum, there is no evidence of a massive general movement of sewage sludge toward Long Island beaches. On the contrary, evidence suggests that milennia of natural discharges and 40 years of dumping sewage sludge have produced a well-established, rather stable distribution of organic-rich muds in the New York Bight. It is believed that the distribution patterns of these muds is likely to change appreciably with time except as the muds may be temporarily resuspended in the water column by the action of storm waves. Pockets of mud near beaches are a common natural occurrence and in this case appear to be mainly of natural origin with, perhaps, small admixtures of material derived from sewage. These patches have almost certainly existed for a long time.

NOAA does not believe that an immediate change in the New York Bight sewage sludge dump site is required; in fact, it is believed that more harm to the environment than good would come from such a precipitous act. However, in view of the increasing quanity of sewage sludge to be dumped in the Bight over the next few years, plans should be made now to change the dump site should future conditions warrant. Based on the evidence in hand, these are NOAA's recommendations:

(1) That the present sewage sludge dump site continue in use for the next year or two, unless a situation develops that truly indicates an endangerment to public health.

(2) That intensive ecological studies be conducted as a prelude to selecting an alternative dumping site for sewage sludge capable of assimilating the greatly increased volume anticipated for the near future. A three-week cruise, whose purpose is to continue the examination of these two suggested areas is being conducted.

(3) That a new dump site, if necessary, be selected upon completion and on the basis of these studies so that contamination of shoreline areas will not occur and that deleterious effects on fish and benthic communities will be minimized.

(4) That a new ocean dumping site be regarded only as an interim measure and that every effort be made to avoid long-term use of the New York Bight for disposal of sewage sludge.

(5) That it is understood that ocean disposal, while worthy of public concern, is very likely of secondary importance, when compared to other contaminant sources such as the Upper and Lower Bays and their associated waters.

Upon review of the data gathered in 1974, the sample collection frequency was changed to the following:

Type I	- Surf Zone	- biweekly (summer) - six-week intervals (Oct - May)
Type II	- Nearshore	- monthly (summer) - bimonthly for water column (Oct - May) - trimonthly for sediments (Oct - May)
Type III	- Transects	- quarterly

Based on data from the surf and nearshore waters along the Long Island and

FIGURE 6.21

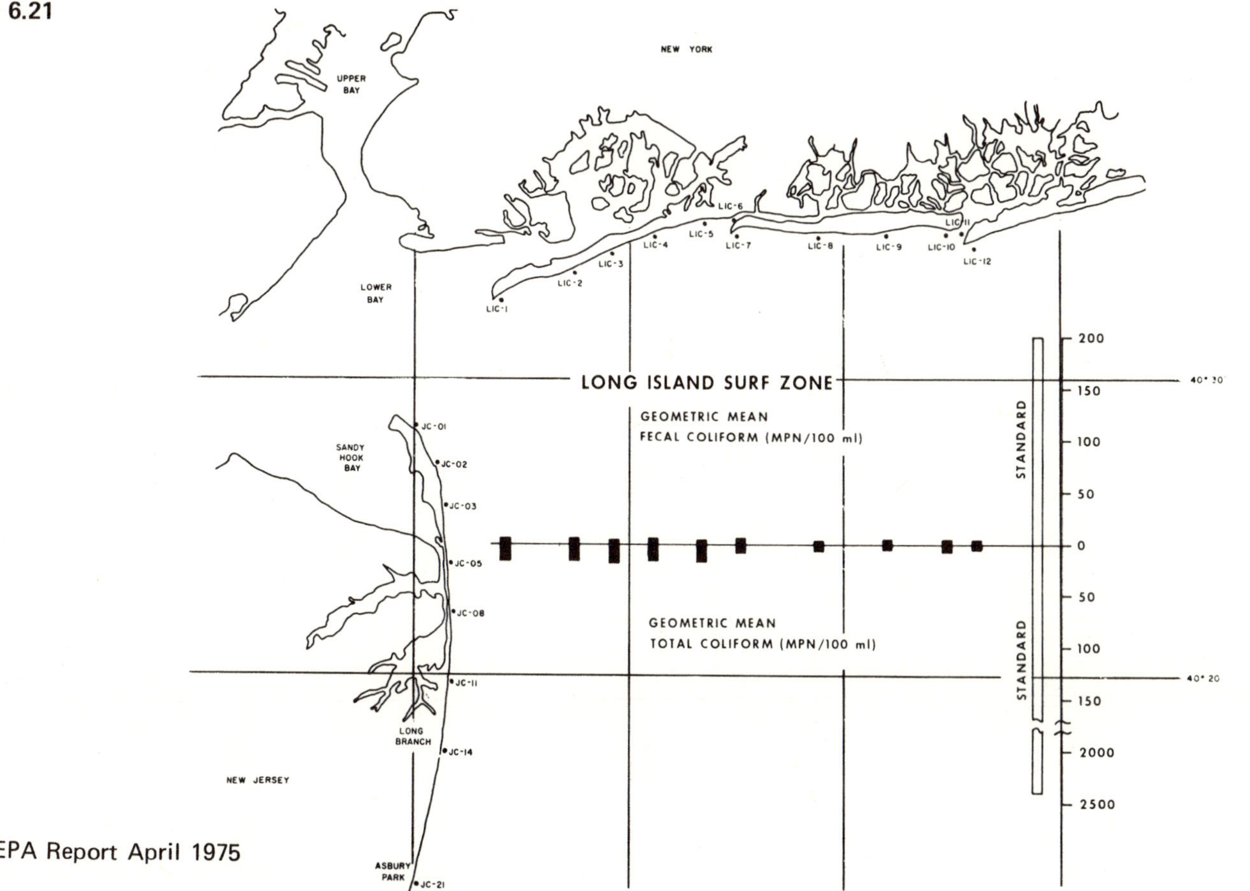

Source: EPA Report April 1975

FIGURE 6.22

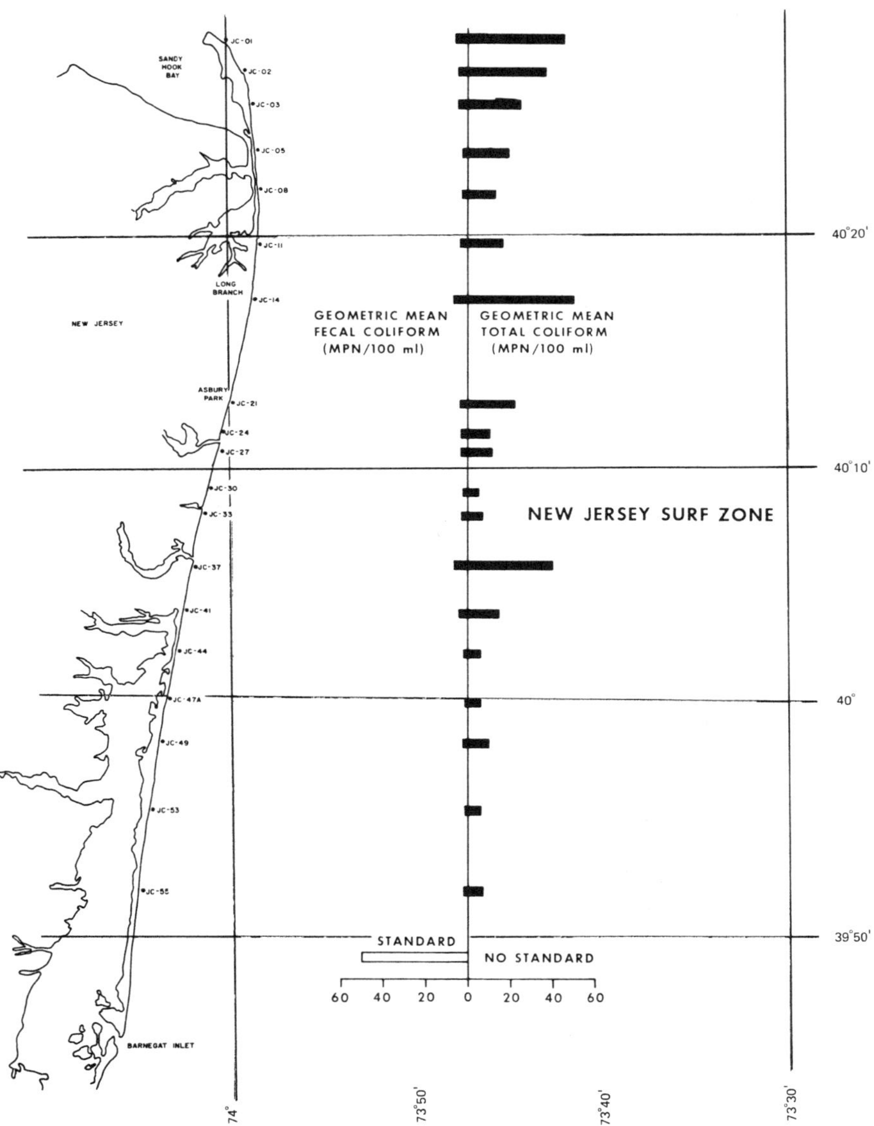

Source: EPA Report April 1975

New Jersey beaches, it appears that the water quality remains excellent with respect to coliform density and is acceptable for contact recreation (see Figures 6.21 and 6.22). Elevated coliform counts do appear at random, but this does not constitute any violation of standards nor does it indicate any systematic change or degradation of water quality.

A review of nearshore sediment data indicate slightly elevated bacterial counts at certain nearshore sampling stations. These elevated counts can be attributed to inland runoff or wastewater outfalls.

Two sampling runs were conducted on the Long Island Transect. The data obtained are consistent with data reported earlier. Specifically, a "clean water-sediment" zone of about 5½ to 6 miles separates the leading edge of the sludge mass from the Long Island coast. Based on this data there is no massive movement of sewage sludge towards the Long Island coast.

Sediment samples were collected and analyzed for Salmonella group organisms at 20 sampling locations. Members of this group were not detected in 100 g of sediment sample.

ASSESSING OCEAN POLLUTANTS

From May 1973 to August 9, 1973, EPA issued 42 permits in the Middle Atlantic Bight and off Puerto Rico. Five EPA-approved dump sites, referred to herein as the deep-sea sites, were considered as a group. Four of these sites are located near the outer edge of the continental shelf, from New Jersey to Virginia, the shallowest at approximately 1.8×10^3 meters. The fifth site is located over the Puerto Rico trench where the greatest depth (10^4 meters) in the Atlantic Ocean has been recorded.

The floor of the trench is flat, and the sediment consists of graded calcareous sand which contains shallow-water Foraminifera and the alga Halimeda. Thus, a transport of surface material to the hadal zone is indicated.

Dredge Spoil and Municipal Sewage Sludge

Although dredge spoil dumping is controlled by the U.S. Army Corps of Engineers, EPA formulates policies for choosing sites and regulations for the disposal of such wastes.

Dredge spoil accounts for the largest mass of wastes dumped in Atlantic coastal waters, amounting to 28.1×10^6 tons (all units contained in this report are metric unless indicated otherwise) annually in 1968 according to EPA estimates. Approximately 45% of the spoil is believed to be polluted. The average annual discharge of dredge spoil in the New York Bight between 1964 and 1968 ranged from 3.5×10^6 tons to 5.5×10^6 tons.

Municipal sewage sludge, a slurry of solid waste from primary sedimentation and secondary treatment by the activated sludge or trickling filter process, is second to dredge spoil in quantity dumped into the Middle Atlantic Bight. It is thickened by gravity, settling, centrifugation, or flotation to a solids content of 3 to 15%. Oxidizable organic matter is a major component of sewage sludge.

Sewage sludge contains an average of 23% oxidizable carbon in soluble and particulate form. The dissolved organics consist mainly of acids and sugars, and the particulate fraction contains proteins, carbohydrates, fats, esters, and unidentified organic matter. Unidentified organic matter in dredge spoil is considered to be of petroleum origin, but it sometimes contains sewage solids and industrial wastes.

In 1968, the Atlantic coastal waters of North America received 4.1 x 10^6 tons of sewage sludge averaging 4.5% solids by weight, or 0.18 x 10^6 tons/year of solids. The New York-New Jersey metropolitan area contributed about 90% of the total. It was estimated that an average of 0.15 x 10^6 tons per year of sewage sludge solids was dumped in the New York Bight. Current disposal operations account for 0.22 x 10^6 tons per year of solids.

The Philadelphia area discharges 0.077 x 10^6 tons per year of sewage sludge solids (assuming 13.8% solids) at its dump site near the mouth of Delaware Bay. EPA estimates that 0.05 x 10^6 tons per year of sludge solids were dumped in the Philadelphia dump site in 1972. These estimates are based on volumes stated in EPA dumping permits issued in April and May 1973, and on data on solids concentrations.

Analyzed sewage sludge contained 150 times more silver, 10 times more chromium, 50 times more copper, 50 times more lead, 30 times more tin, and 30 times more zinc than sedimentary rocks and soils. Sludges may also contain high concentrations of pesticides; concentrations 10^3 to 10^4 times as high as any other sludge sample were found in samples analyzed from Tallmans Island, New York which contained organochlorine pesticides, indicating that this treatment plant discharges 98 kg/yr of pesticides to the New York Bight.

Based on a 1969 estimate of the natural productivity of the waters in the disposal area of the New York Bight (100 g carbon/m^2 ocean surface/year), it was calculated that about 1.2 x 10^4 tons per year of carbon would be "fixed" by phytoplankton in the 125 km^2 disposal area, while current disposal operations are putting about 10^5 tons of carbon per year into the same area.

The distribution of carbon-rich sediments in the New York Bight was mapped in 1971. Uncontaminated sediments collected from the continental shelf more than 10 km from the dumping areas contained less than 0.2% total carbon. It was assumed that sediments containing more than 2% total carbon contained organic matter originating in wastes. According to this criterion, waste deposits cover an area in excess of 50 km^2 in the New York Bight. The highest total carbon concentration (6.4%) was in sediments near the dredge spoil disposal site.

The Sandy Hook Lab (1972) and the Franklin Institute Research Laboratories (1972) failed to find a buildup of organic matter in the sediments of the Cape Henlopen disposal area. They assumed that the wastes were dispersed in the water column and carried away from the dump site.

Aside from miscellaneous organic compounds, both of petroleum and synthetic origin, the bulk of the organic wastes dumped in the New York Bight and Cape Henlopen (Delaware) sites consists of biodegradable matter of natural origin. Petroleum and synthetic organics may be attacked by microorganisms. Secondary treatment of sewage sludge by aerobic or anaerobic microbial digestion decreases the content of oxidizable organic matter and, hence, the biochemical oxygen demand (BOD) and chemical oxygen demand (COD) of the waste. The oxidation process continues after the sludge is dumped. It has been suggested that the apparent absence of thick waste deposits in the New York Bight indicates either that the organic matter is rapidly degraded or that a transport mechanism is removing both organic and inorganic sediments.

Degradation of organic matter at the surface of waste deposits and in the water consumes oxygen at a rate between 16 and 330 g O_2/kg of volatile solids.

The difference in oxygen concentration between surface and bottom water of the Bight varies from 2 to 13 mg/l, with the most severe bottom water oxygen depletion occurring between July and October when the thermocline limits natural mixing. The oxygen content of bottom water over the waste deposits is, on the average, two to three mg/l lower than that at the same depth in areas outside of the dumps. In the summer, the oxygen content of bottom waters of the sludge dump reaches 2 mg/l, a level too low to support well-balanced marine life.

Another important component of the sewage sludge is its bacterial flora. Coliform bacteria were found in sewage sludge in numbers exceeding 2.4×10^9 per 100 ml and fecal coliforms numbering between 4.3×10^5 and 2.4×10^9 per 100 ml. The presence of coliform bacteria and fecal coliforms, in particular, are indicative of contamination with human wastes, and hence, possible presence of human pathogens.

While coliform and fecal coliform counts are useful as indicators of contamination with human wastes, they may not accurately reflect the presence or absence of pathogenic bacteria and viruses in estuarine and oceanic waters. Pathogenic species may survive in ocean water for different periods than the coliform group. Bacteria have been shown to survive in both particulate matter and marine muds.

The distribution of coliform and fecal coliform bacteria in the New York Bight was surveyed, and it was noted that the highest coliform and fecal coliform counts were found in sediments near the center of the dump sites. Moderate counts were found to extend down into the Hudson Channel, substantiating the contention found by the Sandy Hook Labs in 1972 that wastes are being transported from the dumps down into the Hudson Canyon. These results indicate that ocean dumping, rather than contaminated river water, is the source of the bacteria.

Clams harvested in the New York Bight for sale have been found to contain 50 to 80 times higher coliform counts than FDA standards (U.S. Council on Environmental Quality 1970). Excessive coliform counts were found in more than 80% of surf clams collected 8 kilometers from the center of this dump site. However, a similar study of the Cape Henlopen disposal area showed that clams were not contaminated with coliform at a distance of 1.6 km from the site. The FDA has closed both sites to fisheries within a 10-km radius as a margin of safety.

Dredge spoil from the area of the New York Bight and from other major metropolitan areas contains high concentrations of various heavy metals. High concentrations of silver, chromium, copper, tin, zinc, and lead were found in sediments composing dredge spoil from the New York area.

In Figure 6.23, the concentrations of chromium, copper, lead and silver in inner harbor sediments of the type constituting New York's dredge spoil, are given and these values are compared with their concentrations in sewage sludge, marine plants, and natural sediments.

FIGURE 6.23: SEDIMENT METAL CONCENTRATION, NEW YORK

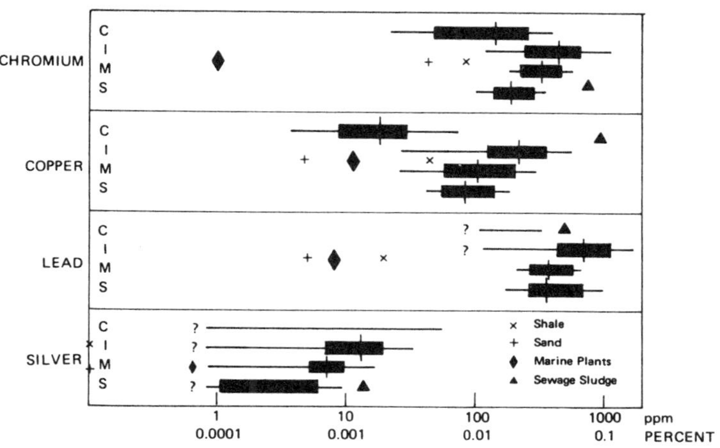

Median concentration value (indicated by vertical line) range, and limits for 70% of samples analyzed (shown by heavy bar) for surficial samples in New York Harbor and New York Bight. (C-continental shelf sediment, I-inner harbor deposits, M-deposits near "mud" disposal area, S-deposits near sewage sludge disposal area). Typical concentrations are shown for shales, sands, and marine plants (Bowen 1966) and sewage sludges. Question marks indicate the detection limits for the various elements.

Source: PB 240 917

The Sandy Hook Laboratory (1972) reported concentrations of heavy metals in spoil and sludge dump sites which were often as high as 100 times the background levels in apparently uncontaminated areas around the dump sites off Barnegat Bay Inlet and in Delaware Bay. The concentrations of heavy metals at the Hudson Channel stations indicate that contaminated sediments are being transported down the Hudson Canyon.

Lead was found to be the most useful indicator of wastes in the New York Bight. Its distribution correlates well with that of carbon, copper, and silver. Chromium was found in high concentrations throughout the New York Bight. It was particularly abundant in the dredge spoil deposits and in inner-harbor sediments. Optical emission spectroscopy does not detect mercury or other volatile metals.

The Franklin Institute Research Laboratory reported no significant increases in heavy metal concentrations in the sediments of the Cape Henlopen dump site. The EPA, on the basis of this report, concluded that "contrary to assertions in the report, the FIRL study has adduced evidence that a heavy metal-containing sediment fraction has been deposited on the bottom in the dumping area. For each of the seven metals investigated the mean for 'inside' is greater than that for 'outside'. The probability that this would occur purely by chance is less than 1%, a statistically significant result."

The contents of metals in sediments of the Cape Henlopen disposal area were rarely over 1 ppm, and in no case over 5 ppm. Concentrations of heavy metals in the New York Bight sludge disposal site are on the order of hundreds of parts per million. In view of the similarity in the quantities of these metals being discharged in the two locations, the question arises as to where the metals and other components of the waste dumped at Cape Henlopen are going.

Very little is known of the physical and chemical state of metals in dredge spoil and sewage sludge. It was found that only about 1% of the total metal in waste deposits could be extracted by hydrochloric acid with the exception of nickel, of which 5% was extractable. This phenomenon has been attributed to surface adsorption, chelation of metal ions by organic materials, and formation of insoluble heavy metal sulfides in anaerobic sediments with high BOD.

In reducing sediments, such as those of the New York Bight, cations of heavy metals are likely to exist as highly insoluble sulfides. Studies of fluxes of cadmium and zinc in Corpus Christi Bay demonstrated that, in the summer when the harbor is most stagnant, these metals are present in the oxidizing surface waters in dissolved states, but in the deeper waters, they react with sulfide ion to form precipitates.

In the winter, storms mix and aerate the water, and the oxygen-rich water redissolves some of the metals from the sediments. Adsorption on suspended sediment also plays an important part in transport of the metals through the water.

It had been stated elsewhere that marine bacteria are important in the deposition of heavy metals in estuarine sediments. It has also been suggested that heavy metals in sediments may inhibit microbial processes and thereby decrease the rate of waste degradation.

In the New York Bight and perhaps in other areas, ocean dumping of organic and metal-rich wastes creates anaerobic conditions in the sediments, resulting in generation of hydrogen sulfide which forms insoluble salts with heavy metal cations. If the same mechanism is acting as reported in Corpus Christi, dispersal of contaminated sediments by storms and deposition in areas where more oxidizing conditions prevail could cause release of the toxic metals to the waters.

It has been pointed out that collected data show a decrease in concentrations of heavy metals such as Cr, Pb, Ag, and Cu in the sediments of the dumping sites compared with those of the waters, suggesting that "it is possible that a transfer occurs by lateral loss due to sediment transport, mixing with shelf sediments, or flux to overlying water occurs."

The Sandy Hook Laboratory (1972) found that areas of the New York Bight which were "covered with sewage sludge and dredging spoils were devoid of normal, or are characterized by greatly reduced benthic macrofaunal populations." Meiofauna and macrofauna species diversity and the relative numbers of individuals were diminished, as was the total number of individuals. The meiofauna, especially the foraminifera, were the most ubiquitous organisms in the sediments, occurring even in the most polluted areas.

In other areas, such as between the dredge spoil and sludge dumps, heavy metal concentrations were greatly elevated, but benthic communities were not as severely impoverished as those actually in the dumps. The Sandy Hook Lab (1972) report gives three reasons for concern with infaunal meiofauna:

(1) they are near the base of the benthic food chain;

(2) their small size and general immobility would prevent them from extensive horizontal movement, so they should be indicative of the sediment conditions in which they were found; and

(3) excluding certain protozoan groups, they are generally the most common group of animals in marine sediments.

Stomach contents of fish collected from the dumps showed that yellow-tail flounder, winter flounder, and red hake mainly eat benthic organisms such as polychaete worms, amphipods, and bivalves, and that they also ingest debris including hair, band-aids, seeds, and cigarette filters. The report states that "the fact that they (fish) are ingesting some of these items opens up the question as to what, if any, of the harmful materials including heavy metals and pathogenic microorganisms are ingested by the fish."

Investigators of the Sandy Hook Laboratory found petroleum in dredge spoil dumped in the New York Bight. Sediments in New York Harbor, which accounts for a large portion of that dredge spoil, contain 8 to 10% organics and those in the lower Hudson River contain an average of 5.5%. Concentrations of hexane-extractable material in bottom samples from the dredge spoil dump site in the range 0.5 to 5.0% were reported.

Analytical results required by EPA permits show 0.002 to 4.00% oil in sludge samples from New Jersey and 0.065 to 0.282% in New York sludge. Philadelphia's sludge contains 11.2 to 22.0% hexane extractables from the Northeast Plant and 5.5 to 9.8% from the Southwest plant. Hexane extractables include other organics in addition to petroleum products. Organics of petroleum origin may affect marine organisms in three ways: by their toxicity, by their ability to concentrate pesticides, and by their ability to concentrate heavy metals in complexed form.

It has been suggested that the dredge spoil probably contains heavy tars and residues rather than lighter petroleum fractions. An oil spill which occurred near Woods Hole, Massachusetts in 1969 was studied and it was found that petroleum chemical contaminants persisted in sediments and marine life for at least two years and that the most persistent were highly toxic and carcinogenic aromatic compounds. They further observed that the contaminants spread within the sedimentary column.

Biocides are a major concern from the standpoint of ecological effects because of their high toxicity, their persistence in the environment, and their potential for biomagnification. The organohalogen compounds are important among the biocides. Polychlorinated biphenyls (PCB) are closely related to the organohalogen pesticides in terms of the properties listed above. These lipophilic compounds are absorbed by fatty tissues of marine animals, particularly benthic species such as oysters and clams where they interfere with certain processes of the metabolism of lipids.

The organophosphorus pesticides are of the same order of toxicity as the organohalogens, but they are much more quickly degraded in aqueous solution and are, therefore, less likely to accumulate in marine ecosystems.

Sediments from the dredge spoil dump site for pesticides were analyzed and 0.013 to 0.126 ppm DDT, and the DDT metabolites DDE and DDD, at concentrations of 0.013 to 0.019 ppm and 0.039 to 0.81 ppm respectively were found.

In view of the observations that pesticides are concentrated by oil slicks on the sea surface, monitoring of pesticides in the dumping grounds should include sampling of slicks produced by disposal operations.

It was reported that dredge spoil disposal has an appreciable effect on biota over an area of 18 km^2 in the New York Bight, where the average accumulation of dredge spoil amounted to 32 cm/yr between 1964 and 1968. Sludge disposal accounted for a sedimentation rate of about 4 mm/yr over an area of 36 km^2. Investigators feel that an accumulation as great as 30 cm/yr may be great enough to exterminate certain benthic organisms simply by burial.

A study of spoil dumping operations in Rhode Island Sound, where a total of 6.8×10^6 cubic meters of dredge spoil had been dumped, found direct effects of the dumping to be limited.

A study of shallow-water dredge spoil disposal in the upper Chesapeake Bay found significant loss of bottom animals due to burial, but there was rapid recolonization by certain species which returned to predredging levels within 18 months. The sediments covered an area at least five times larger than the designated disposal area. A study of other dredge spoil disposal areas in the lower Chesapeake Bay area indicated similar conclusions. Recolonization of infauna and epifauna was so rapid that only transitory effects were observed.

The same study included observations for sediment buildup in an oyster bed 1.4 to 3.2 km from the discharge point. No dredge spoil accumulation or abnormal mortality of the oysters could be detected.

On the whole, the sedimentation effects of dredge spoil dumping were confined to a relatively small area in the immediate vicinity of the dump site and, in the case of nonpolluted spoil, were quickly reversible by recolonization with the same bottom communities as previously present or by development of new bottom communities depending on the nature of the dredged sediment.

Therefore, no widespread harm would be likely to result from continued ocean dumping of nonpolluted dredge spoil if the material were thinly enough distributed and did not significantly alter a unique or rare habitat. Presently established dump sites could continue to be used, but future locations should be chosen so that indigenous bottom communities are of a common and widely distributed type to avoid the danger of burying and thus destroying a unique ecological niche.

The amount of turbidity produced by a load of dredge spoil or sewage sludge varies with the size and density of the sediment particles of which the waste is composed. Fine-grained sediments and those with low density will remain in suspension for a longer time than coarser and denser particulates. Dredge spoil from the New York metropolitan area is composed of particles of varying sizes with grain densities between 1.9 and 2.8 g/cm^3, while sewage sludge is composed primarily of fine particulate matter, with an average organic content of about 50% and a relatively low grain density of 1.3 to 1.7 g/cm^3.

It has been estimated, from studies of dredge spoil dumping in Rhode Island Sound, that up to 25% of the dumped material remains suspended for a time in the water while the remaining 75% or more settles to the bottom quickly. The formation of two turbidity slicks made up of fine particulate matter has been observed. One slick formed at the surface, and a second, which probably resulted from the density gradient at the thermocline formed near mid-depth. The slicks were observed to dissipate after four to five hours, but 1 to 5% of the sediment remained in suspension for a longer period.

A larger portion of sewage sludge solids than of dredge spoil will probably remain in suspension after a dump. Therefore, a significant portion of a load of waste may be carried outside of the specified dump by currents before settling out. The suspended matter is likely to be composed of low-density particulates such as organics and oil droplets. This may be an important mechanism for transportation of polluted matter to other areas.

The Franklin Institute Report on the Cape Henlopen sludge dump site states that "examination of the drift of digested sludge indicates that the particulates remain in suspension for a considerable time and that the material is well dispersed." The transport and dispersal of a turbidity cloud from a sludge discharge was observed for two hours. The sediment concentration decreased from 10^4 to 25 ppm in 76 minutes and to 10 ppm in 2 hours. A surface cloud was carried over 0.8 km downwind during the observations, and a second cloud at mid-depth was carried "several hundred feet" in the opposite direction.

Frequently, a slick of organic matter and debris is associated with sludge dumps in the New York Bight, and there is no reason to expect that Philadelphia's sludge would have very different characteristics.

In the New York Bight, resuspension of the fine sediments by wave action generates turbidity clouds which may carry contamination away from the disposal site. Storms in the New York Bight are capable of producing vertical homogeneity in the water column and of transporting contaminated sediments. The sludge-contaminated area appears to be spreading toward the northeast and east. It has been stated that turbidity currents probably play an important role in the removal of waste sediment in the New York Bight, and it was suggested that the Hudson Channel may conduct such sediments to deeper water in the form of density currents.

Aside from the potential for carrying contamination to areas outside of the dumping grounds, the generation of turbidity clouds may produce other environmental effects. High turbidity can be detrimental to aquatic life. Nine potential indirect effects of turbidity and siltation on aquatic organisms might be:

(1) reduction in light penetration and reduced photosynthesis;
(2) reduction of visibility to some feeding organisms;
(3) destruction of spawning areas;
(4) reduction of food supplies;
(5) reduction of vegetational cover;
(6) trapping of organic matter, resulting in anaerobic bottom conditions;
(7) flocculation of planktonic algae;
(8) absorption or adsorption of organic matter or inorganic ions;
(9) adsorption of oil.

Suspended sediment, if sufficiently concentrated, may clog gills and impair respiratory exchange in fish and invertebrates, and it may reduce the growth and survival of larval states of fish and shellfish. Fine suspended solid matter adversely, sometimes lethally, affects the gill epithelium of fish. Deleterious effects are to be expected in invertebrates, particularly filter feeders.

Studies on the effects of turbidity on eggs, larvae, and adults of pelecypod mollusks indicated that larval and adult forms will be adversely affected by fine suspended sediment concentrations of 3 to 4 ppt. The University of Rhode Island report on dredge-spoil dumping in Rhode Island Sound stated that fish and lobsters could withstand the high concentrations of suspended sediments for short periods and that lobstering was the least affected fishery in the area.

The Chesapeake Biological Lab dredging report stated that no gross effects were observed on microscopic plants and animals, on fish eggs or larvae, or on adult fish either held in cages near the disposal point or caught in the area. The Sandy Hook Laboratory's report on the New York Bight dumping grounds described crabs taken from the disposal site as being moribund and stated that their gills were fouled with granular materials.

The sediments in the sampling area are known to be highly polluted and the reported fouling may have resulted from the pollution load rather than from the sediment itself. Lobsters and crabs kept in aquaria with sewage sludge developed granular fouling and a brown coating which covered the exoskeleton. In a turbidity bioassay of lobster for the Rhode Island Sound study, one case of unusually high mortality was observed which was attributed to some unknown pollutant in the dredge spoil.

Industrial Wastes

In excess of 0.1×10^6 tons per year of industrial wastes are currently being dumped in the Middle Atlantic Bight under permits issued by EPA Region II. Over 90% of the industrial waste barged from New York City is acid-iron waste from the titanium dioxide pigment industry. EPA Region II has also authorized the dumping of more than 6×10^6 tons per year of industrial waste solids near Puerto Rico. In addition, 8.9×10^3 cubic meters of wastes of unknown composition are discharged in the Middle Atlantic Bight, and 3.2×10^6 cubic meters of solvents from pharmaceutical production are dumped in the Puerto Rico trench.

The composition of industrial wastes is as varied as the processes which produce them. The major industries currently using ocean disposal under EPA permits produce petroleum, chemicals, and pharmaceuticals. The chemical descriptions of these wastes are given in varying degrees of thoroughness.

Approximately 7 x 10^3 tons per day of spent sulfuric acid solution and mud produced in the extraction of titanium from ore are being dumped in the Mid-Atlantic Bight. The liquid waste consists of a solution of approximately 8 to 9% each of H_2SO_4 and $FeSO_4$.

Another waste acid dumped in this area is a liquid containing 27 to 31% HCl and 0.5 to 1.5% fluoride. It is generated in the manufacture of chlorinated and fluorinated methanes and tetrafluoroethylene and contains 200 ppb of the products.

Refinery wastes consist mainly of spent caustic solutions which contain sodium hydroxide, sodium sulfide, sodium naphthenates, sodium phenylates, sodium mercaptides, oil (2% from one refinery), and between 3 and 5% solids.

Approximately 3.2 x 10^6 cubic meters per year of solutions from pharmaceutical production are disposed of in deep water north of Puerto Rico, plus 6 x 10^6 tons per year of solids. The permits describe the materials dumped as primarily dissolved and suspended solids, with large quantities of iron present in the wastes (6.7 x 10^2 kg per month). The same barge is used to transport the combined wastes to the disposal area.

Other Wastes

One EPA permit provides for the dumping of construction rubble in the New York Bight at the "cellar dirt" site at a rate of 0.46 x 10^6 cubic meters per year. Two permits provide for the one-time disposal of confiscated "dangerous weapons and/or objects" in an "approved site." The Corps of Engineers holds a permit for the disposal of a derelict tugboat and "any additional sunken vessels which are deemed to be obstructions to navigation." upon EPA approval, in the New York Bight wreck disposal site.

The permit requires that "all crude oil, fuel oil, heavy diesel oil, lubricating oils, hydraulic fluids, and/or any fixtures containing the above shall be removed." In addition, loose floatable materials must be removed or secured prior to disposal.

ENVIRONMENTAL SURVEY OF A MID-ATLANTIC BIGHT DUMPSITE

Public Law 92-532, the "Marine Protection, Research, and Sanctuaries Act of 1972," was enacted in part to regulate ocean dumping practices by establishing a permit system which inventories quantity and quality of materials transported to sea for disposal as well as by a monitoring program to continuously assess the effects of these practices at specified sites.

Under this law, the designation of dumpsites and the quantities deposited thereon must be based on a scientific knowledge of the specific environments. While generic information was available for continental shelf environments, the background was sufficient only to design the preliminary steps in a monitoring program.

An oceanographic survey of a proposed interim dumpsite approximately 50 miles off the mouth of Delaware Bay, was conceived and executed in spring 1973 with the objectives: (1) attempt to establish ambient environmental parameter levels prior to active dumping at the site, (2) assist in developing a practical monitoring scheme for monitoring of ocean dumping practices, and (3) develop further insight into the continental shelf environment, which seems to be the next province to be assailed by man's wastes, but which, with foresight and current information can be managed intelligently.

The conclusions contained herein are based on four days of data collection within the area of concern. Bottom configurations in the interim dumpsite were found to be in accordance with published charts. No unexpected holes, trenches, or rises were found with detailed bathymetry. Bottom substrate was primarily quartz sand, with a gradation from coarse to fine as a function of depth and effective wave energy. The sorting values of sediment were evaluated to determine sites on the bottom where waste deposition may accumulate. Subsequent interpretation indicated that these regions occupy topographic "lows" within the dumpsite.

Shipboard current measurements indicated a flow of the water column at approximately 0.25 knot in a direction between 192° and 248° true, under the regime of observation. Seabed drifters released during the cruise indicate a net southwestward movement of near-bottom waters to the coastline.

Hydrographic conditions indicated the partial establishment of the thermocline and halocline. These conditions would be important in the settling and distribution of dumped materials. Measured water quality parameters indicated no abnormal concentrations of the compounds generally encountered in the marine environment. Ammonia determinations were not amenable to preservation for later analysis. Phytoplankton and zooplankton populations were typical of a normal temperate shelf environment. Vertebrates were observed by underwater television and by capture with an otter trawl. The demersal community examined appeared to be in sound physiological condition.

The benthic fauna was characteristic of a firm sand-shell-gravel community, dominated by sea stars, sand dollars and polychaetes. Suspension feeders and carnivores were extremely well represented. The sand bottom community was surprisingly diverse and abundant. Any significant changes, or more specifically reductions in populations of sand dollars, principal polychaetes and some of the more fleshy ectoprocts, would be indicative of degradation in bottom water quality. However, based on the benthic fauna, the site was unpolluted.

Coliform and fecal coliform concentrations in waters and sediments were negligible. The data indicated an environment relatively free of terrestrial influence. Heavy metals were not appreciable in water or sediments, but certain chemical species were evident in some biological components of the ecosystem. Copper and iron appeared to be accumulating in sand dollars in and southwest of this interim disposal site. However, as a result of these findings, we feel further investigation is warranted. Chlorinated hydrocarbon concentrations in sediments were found to be negligible. Dark flaky material observed by underwater television may be ferric hydroxide originating at an acid waste dumpsite approximately 10 miles northwest, suggesting interaction between the two sites.

In addition to the characterization of the physical, chemical, and biological aspects of the dumpsite, the objectives of this survey included (a) development of a practical scheme for monitoring ocean dumping practices and (b) gaining additional insight into the continental shelf environment. Both the sampling scheme and analytical techniques employed during this study have applicability to other sites on the continental shelf.

In conjunction with published data of regional scope, this brief yet intensive study has provided a detailed picture of the spring conditions at a mid-shelf site. The most troublesome aspects of any shelf study continue to be the difficulty of determining accurate position at sea (for reoccupation of stations) and the lack of current speed and direction data throughout the water column.

Final evaluation of the suitability of this site for waste disposal must await further study which will reveal seasonal changes in the biota and water-mass. The introduction of an unnatural perturbation (solids and fluid wastes) will require reassessment of its effect on the marine ecosystem. The concepts of accumulation and assimilation can then be considered.

WASTE DISPOSAL IN OTHER GEOGRAPHIC AREAS

The following sources were used for the information provided in this chapter.

PB 224 793
PB 230 944
PB 240 917

The disposal of industrial wastes by barge dumping is practiced mainly in the Gulf of Mexico and off the East Coast. The Southern California Bight provided the best available data on the discharge of municipal and industrial effluents through outfalls.

Wastes disposed by ocean outfalls in Southern California have had a harmful impact in the local ecosystem, but no widespread impact is evident. As concluded with regard to the New York Bight, the reduction of heavy metal and pesticide content in the disposal site areas will be achieved only as a result of eliminating the sources of these materials, since their removal from dredge spoil and sewage sludge is not expected to be feasible.

CHARLESTON STUDY

Maintenance dredging has been performed at Charleston since before 1875. Prior to 1940, the estuary had a drainage area of 1,400 square miles and a Cooper River flow of 72 cfs. Completion of the Santee-Cooper Hydroelectric project enlarged the drainage area to 16,000 square miles and increased the flow of the Cooper River to 15,000 cfs. Prior to completion of the hydroelectric complex, maintenance dredging averaged 120,000 yd^3/yr. It has since increased to the present 10,000,000 yd^3/yr. Ocean disposal is utilized for 17% of this material.

The majority of ocean disposal material is clean dredge spoil. One site is used, located close to the south jetty. Although limited environmental studies have been made, there is no reason to believe that present practices are dangerous to marine life. Expanded maintenance operations in the inner harbor pose a polluted dredge spoil disposal problem.

This problem is recognized by the Corps of Engineers and is under study by the Waterways Experiment Station.

History of Charleston Area Ocean Disposal

Dredge Spoil: Ocean disposal of dredge spoil has been done since 1875; however, the quantities were minor until the Santee-Cooper project was completed in 1942. The harbor maintenance dredging requirements increased rapidly to the current 10,000,000 yd^3/yr volume. It is estimated that this level will remain fairly constant until the proposed rediversion project is completed.

Material being placed in the disposal area within the cognizance of the Charleston District Corps of Engineers consists almost entirely of fine and coarse grain sand, interspersed with sparse amounts of silt. Materials deposited to date would come under the classification of unpolluted dredge spoil.

No major reports or studies have been located that assess the ecological impact of the disposal operations. Discussions with cognizant local scientific personnel indicate a high degree of confidence in the safety of the current disposal operation. Concern was voiced regarding possible future dredging operations in areas of the inner harbor where there are quantities of contaminated material. The consensus was that special processing or alternative disposal methods would have to be used for this material. The major problem seems to be the concentrations of heavy metals and toxic chemicals which have built up in the harbor area as a result of industrial discharges into the Cooper and Ashley Rivers.

Industrial Wastes: Currently, there is no ocean disposal of industrial wastes from the Charleston area. Disposal of waste in the Cooper and Ashley Rivers by industrial plants in and about the city of Charleston, and from municipal sewer outflows, appears to be coming under control. Major industrial facilities have installed extensive pollution control facilities and are in the process of upgrading these facilities to provide even greater purification processes.

The city of Charleston has installed a sewer treatment plant on Plum Island, which is designed to provide at least primary treatment of the effluents. Although these are not the source of the subject of the case study, it is necessary that they be mentioned here to provide an awareness that pollution occurring in the ocean in proximity to Charleston Harbor is most likely to occur from waste disposal into the Cooper and Ashley Rivers, rather than from dredge spoil being deposited at the disposal site.

Analysis of Disposal Activities

Dredging Operations: Dredging operations involving ocean disposal are primarily performed by government hopper dredges operated by and under the supervision of the Corps of Engineers. Disposal is normally carried out during a once-a-year maintenance period. Normally, 4 to 6 wk of intensive work are required for the dredging operation. The remainder of the dredge spoil is deposited on land from pipe dredges. Although the disposal site is in proximity to the recreational beaches along Folly Island, there have been no reported incidents of recreational deterioration due to the use of this site. This is significant, since Folly Island is one of three major areas of public recreational use in the Charleston area. With the exception of a carefully maintained navigation channel, the harbor area, including the nearby disposal grounds, is in relatively shallow water. Maintenance depth of the main ship channel is 35 ft. The dredge spoil disposal ground is located in proximity to the ship channel, the closest point being 1½ nautical miles from the center of the channel. The disposal ground is located in shallow water, with an average depth of approximately 34 ft. (Maximum depth is 50 ft, which occurs in only a few scattered locations.)

Boundaries of the disposal zone are currently not fixed. The two inland boundaries are defined as two sides of a parallelogram. There are no ocean limits set; therefore, no positive control of disposal operations is possible. Extensive analytical work has been performed on dredge spoil by the Corps of Engineers. In one Corps of Engineers study, a representative sample (177 analyses) was made of the dredge spoil.

Eighty-eight percent of the samples had a median diameter of less than 0.062 millimeter. This means that the majority of spoil is composed primarily of silt, clay, and colloidal-size particles. Division laboratory analysis revealed that there is no significant variation in the grain size distribution with respect to location or depth. This holds true for both the fine material (silt, clay, and colloid) and the coarse material (sand). This conclusion is further substantiated by comparison of the statistical constants determined from the cumulative curves.

Twenty-four samples were analyzed chemically for percentage of acid-soluble material, percentage of acid-insoluble material, percentage of loss of ignition, and percentage of organic matter. The results of these tests reveal no significant correlation between these properties and the depth or location of the sample.

Standard soil tests were made on 272 samples during the same study. Density, specific gravity, and plasticity measurements were made. In order to eliminate the effect of sand content on the results, only the material finer than 200 mesh (0.074 mm) was utilized for the specific gravity and plasticity tests.

The dry density measurements seem to be controlled by the sand content rather than by the depth, except where the finer material is found throughout. The fact that the average specific gravity of these samples is somewhat less than that normally associated with mineral-derived sediments can be accounted for by the organic material which is known to be present. The samples analyzed have a range of specific gravity from 2.32 to 2.52.

Values found for plasticity index are quite high. This situation is probably caused by the large quantity of colloidal material which is present. A lower value is obtained when the sample is dried out before determining its plasticity index, showing further evidence of colloidal character.

Alternative methods of waste disposal in the Charleston area comprise landfill projects, reclamation projects, and marsh building. Landfill projects generally consist of reclamation of the spartina marsh areas by filling with dredge spoil. These projects have come under opposition from many environmentalist groups. The value of the tidal marsh for shoreline protection and as a nursery ground for commercial and sports fisheries is highly recognized. In addition, acreage around the Charleston case study area is a prime hunting area of considerable economic significance. Therefore, use of this area for a disposal site for dredge spoil is quite limited.

Past experiences with landfill operations in this type of terrain indicate that unless they are scientifically carried out, the results can be very detrimental. The fine particle size encountered with the dredge spoil is such that it restricts the growth of plant life. Without the benefit of ground cover, the material is easily redistributed by the winds.

Much of the dredge spoil disposal in the Charleston survey area is accomplished by diked and undiked disposal in marsh. In the past, the marshland was considered of marginal economic value and was not closely scrutinized as to the total environmental impact of spoil disposal. The recent concern over the estuarine ecosystem has resulted in objections to the use of marshland for dredge spoil disposal.

For the current volume and makeup of ocean waste disposal materials, there are no immediate problems associated with disposal at the site now being used. Disposal of waste by the U.S. Navy has ceased to be a problem because of environmental controls that have been strictly enforced by the Navy. Reports from local fishing boat operators indicated that great progress has been made during the past two years. Prior to this period, significant quantities of garbage, wood, and oil were discharged by Navy vessels approaching Charleston Harbor. Navy disciplines now appear to have eliminated these problems.

The only potential problem arising from ocean waste disposal is the concern by scientists, within the Charleston area, that contaminated dredge spoil from the inner harbor may have adverse environmental effects if disposed of in the ocean.

The 1965 FWPCA study revealed that in the lower reaches of the Ashley River, pollution in the vicinity of a chemical company was extensive. Midchannel benthic environments in these areas lacked bottom-associated organisms. Deposits in the channel near industrial outfalls comprise dark colored muds and oily substances with odors similar to that of petroleum.

Bioassays conducted with these deposits using snails, shrimps, and fish as indicators demonstrated that the constituents of these muds were toxic to the organisms. Bottom deposits downstream toward the mouth of the area consist of black mud and organic matter from domestic sewage. Lower reaches of the Cooper River also contain similar deposits. Although the operational aspects of the Plum Island Sewer Plant will undoubtedly reduce future deposits of sewer sludges on the bottom of the river in these areas, the existing deposits will probably remain until dredged.

Secondary Disposal Area

During the conduct of the survey, information was obtained on four other disposal sites in use in proximity to the original survey area. These sites are Georgetown and Port Royal, South Carolina; Savannah Bar, and Brunswick Bay, Georgia.

South Carolina Sites: In addition to the Charleston disposal site, two other disposal areas within the Charleston district are presently being used by the Corps of Engineers. These areas are adjacent to Georgetown and Port Royal Harbors. As with the Charleston sites, these sites are used for disposal of dredge spoil. Material placed in these sites consists mainly of fine and coarse-grained sand interspersed with small amounts of silt. Material is from the annual maintenance dredging of the entrance channel and inner harbors dredging of Georgetown and Port Royal Harbors. Dredging operations are accomplished by use of government-owned hopper dredges operated and supervised by Corps of Engineer personnel. Sporadic site monitoring has been carried out by the South Carolina Wildlife and Marine Resources Departments. With the recent expansion of these agencies' laboratory facilities and capabilities, it is assumed that more comprehensive site surveys will be performed.

Each of the disposal sites is located on the continental shelf approximately 2 to 6 miles offshore. Typical water depths range from 30 to 50 ft. The disposal area at Port Royal Harbor has been in use since the project was constructed in 1957. The Georgetown disposal site has been used since prior to 1900. Quantities of dredge material placed in these disposal areas vary from year to year depending upon amount of shoaling and the amount of time the dredge is available for work.

Bottom profiling studies conducted by the Corps of Engineers and the South Carolina Wildlife Resources Department have detected no substantial buildup in the disposal areas. Dredging operations that require the use of ocean disposal are generally scheduled once a year at each harbor. Normal harbor maintenance periods of 4 to 6 wk are required for dredging. The Charleston District Corps of Engineers indicates that dredging operations requiring ocean disposal at these sites will continue at about the same frequency and quantities as has been experienced over the past 10 yr.

Since dredging operations are carried out by vessels operated and commanded by the Corps of Engineers, operational control is considered excellent. These vessels are adequately equipped with navigation equipment for accurate site location. In addition, crew standards are above the average of the private dredging industry. The vessels are fully equipped with radar, radio direction finders, gyrocompasses, recording fathometers, and other electronic navigation aids.

Georgia Sites: The Savannah Harbor Bar and Brunswick Harbor (Georgia) disposal sites are used by the Corps of Engineers for disposal of maintenance dredge material from Savannah and Brunswick Harbors. The sites have been in use for approximately 10 yr. Maintenance dredging usually amounts to one month each year, with a Corps of Engineers hopper dredge at each location. Quantities removed for ocean disposal are approximately 600,000 yd^3 annually; however, new channel dimensions are requiring more extensive dredging activities.

In Savannah Harbor, in 1972, approximately 2,000,000 yd^3 were dredged. Projections for 1973 are approximately 2,500,000 yd^3. This increased volume were due to deepening the channel from 36 to 40 ft, and widening it from 500 to 600 ft. To accomplish this, overall channel maintenance increased the 1974 disposal projections to 4,000,000 yd^3 in addition to the normal annual harbor maintenance.

As with the other Corps-commanded projects, extensive control of the dredging and disposal are imposed. Before and after surveys are performed, along with accurate control of the disposal. All dredging is coordinated with the state pollution control agencies. The hopper dredges used are under Corps of Engineers command, and are well equipped and professionally manned.

In addition, the Savannah Harbor disposal area is well marked by five buoys. Analyses of spoil material for Savannah and Brunswick are presented in Table 7.1.

TABLE 7.1: SEDIMENT SAMPLE ANALYSIS

Sample	Moisture	Dry Weight Basis Volatile Solids o/o	Zn	Pb	Per Weight Basis Hg ppm	COD Mg/gm	Oil-grease Mg/gm	Kjeldahl N ppm
Brunswick								
10	42.1%	3.8	30	--	0.06	23	1	500
Savannah Bar								
4	40.6	6.9	43	20	0.10	25	3	800
5	24.0	1	9	3	0.10	5	1	100

GULF COAST STUDY

The geographical area covered by this designation is the continental shelves and slopes bordering the Gulf of Mexico, from Port St. Joe, Florida to Port Isabel, Texas.

Fourteen interim disposal sites were approved in the Gulf of Mexiço (*Federal Register* 1973), with two classified as chemical disposal areas and the remainder as dredge spoil sites. As of July 1973, four companies with plants at seven locations were using the two areas referred to as A and B shown in Figure 7.1.

Disposal site A is situated on the upper part of the Texas-Louisiana continental shelf, approximately 125 miles southeast of Galveston, Texas. Disposal site B is located off the western side of the Mississippi Delta, approximately 60 miles south of the mouth of the Mississippi River.

The surface circulation pattern of the Gulf consists of a clockwise loop current flowing from the Yucatan to the Florida Straits in the eastern section and minor, semipermanent currents in the western section (the areas of dumping). Based on a comparison of six separate north-south transverses across the western Gulf, the winter pattern of east-west surface flow may be characterized by a broad westward flow through the southern section of the Gulf and a narrow zone of stronger eastward flow in the deepwater portion of the northern Gulf. The average direction of water movement over the Texas-Louisiana continental shelf is westward at approximately 1 knot.

The deepwater circulation pattern of the western Gulf of Mexico is unknown but has been shown to be greatly influenced by ring currents which break from the loop current and move westward. Within the Gulf of Mexico, the continental shelf ranges in width from 20 miles off the Mississippi delta to more than 100 miles off the Texas coast. Although the average slope of the area near Site **A** is slightly less than 1°, it has been described as "hummocky." Direct shear test results, coupled with the infinite slope theory, indicate that the marine sediments of this area would be stable to great thicknesses even on steep slopes.

FIGURE 7.1: EPA INTERIM-APPROVED DISPOSAL SITES, GULF OF MEXICO

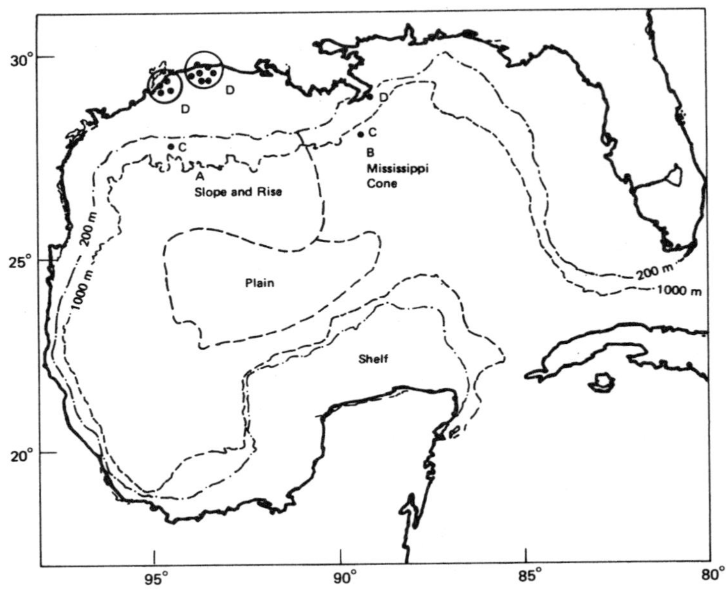

D = Dredge Spoil C = Chemical Wastes

Source: PB 240 917

In the area of disposal site **B**, there is a trough-like valley which does not approach the coast. The topography of this area is highly irregular and subject to landslides due to the dynamic instability of the sediments.

Three distinct sediment sources have been identified. These are the Mississippi, the Rio Grande, and the rivers of northeastern Mexico. In annual discharge, the Mississippi River (516.0×10^6 tons/yr) has been estimated to completely overshadow its nearest competitor, the Rio Grande (19×10^6 tons/yr). The sedimentation rate at dump site **B** has been estimated to be 30 cm/10^3 yr.

South and west of the Sabine River the influence of the Mississippi is reduced as a result of dilution by sediment derived from the major Texas rivers (Brazos, Colorado, Sabine, and Trinity).

Dump site A is located in this area, and the rate of sedimentation in this section has been estimated to be 3.4 cm/10^3 yr.

The fauna are semitropical, and the species of both invertebrates and fish are numerous and diverse. Fishing in the Gulf harvests over 60 species of fish and 20 species of invertebrates. Shrimp is the predominant commercial fishery product. The shrimp fishery operates across the entire continental shelf, but it is concentrated in the shallower nearshore waters.

During 1967 to 1968 the most valuable of the fish were menhaden; snappers; mullets; mixed groundfish species used for animal feed; weakfish; and groupers in decreasing order. Oysters and crabs are the important shellfish products. Sports fisheries are important in the area.

In the 1960s, the total Gulf fisheries production was approximately 28% of the United States' total production. Between Pascagoula, Mississippi and Port Arthur, Texas, 21% of the country's total fishery production was landed. This region has been called the Fertile Fisheries Crescent (Figure 7.2) and reflects the proximity of the largest estuarine region in North America. Dump site B lies on the periphery of the Crescent.

FIGURE 7.2: GULF OF MEXICO

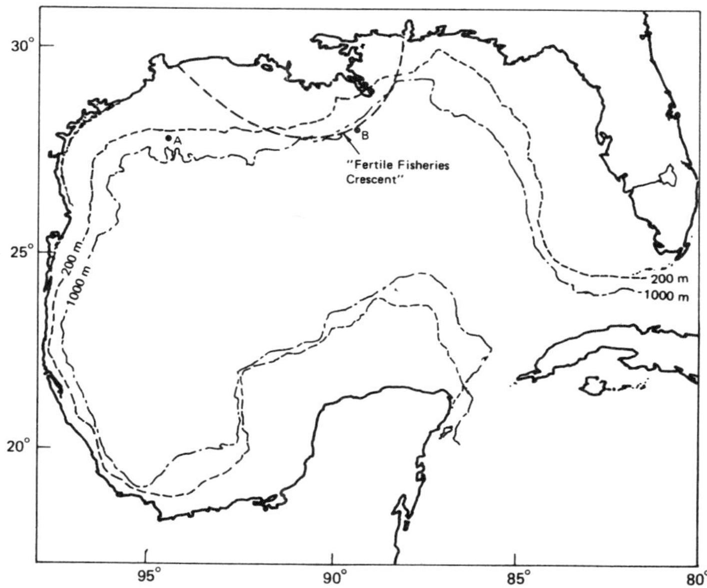

Source: PB 240 917

In the Gulf of Mexico, two disposal methods are used. The first method involves the discharge of wastes through a submerged pipe into the turbulent wake of a barge. This technique is employed at dump site A. Site B, off the Louisiana coast, is used exclusively for materials which are placed in barrels before discharge.

Waste disposal within the area is primarily of two types: dredge spoils and chemicals. Currently, 40 million tons of dredge spoil are dumped annually from three hopper dredges, all of which are operated by the Corps of Engineers. There are 33 active nearshore sites, located as shown in Figures 7.3 and 7.4.

Currently, 1,161,000 tons of chemicals are disposed of annually by bulk dumping, with an additional 6,650 tons dumped in 55 gal drums. Only two deepwater areas (A and B on Figure 7.3 and Figure 7.4) are authorized for the disposal of these industrial wastes; however, it is known that some of this material is dumped short of the target areas.

Disposal of each of these two types of wastes presents a short-term problem. Each may also have adverse long-range effects which are yet to be determined. Because the locations, depths, environmental conditions, users, amount and character of wastes, effects, problems, and potential solutions are so different between the two types, yet similar within them, these factors will be discussed separately.

History of Gulf Coast Dumping

Dredge Spoils: The ocean disposal of dredge spoils in the survey area has been continuous since 1926. All of this dumping is performed by hopper dredges, because the other types of dredges used in ports and inland waterways are not considered seaworthy. The volumes vary widely from year to year in each of the three districts because of project funding, priorities, and natural variations in streamflow and sedimentation. Annual averages over five year periods are indicative of the general trends.

In the Mobile District, an annual average of 5 million tons was dumped in the early 1960s, but this has since been reduced to an average of 3 million tons. This is primarily the result of increased usage of dredge spoils for beach nourishment in the Florida areas. This trend is expected to continue, with expansion into Alabama and Mississippi.

It is estimated that the annual amount dumped offshore will be less than 1 million tons in the late 1970s. Most of the dredged material is brought into the seaward portions of the channels by littoral drift and consists of silt to fine grained sand, with minor amounts of shell, gravel, and clay.

The hopper dredge activity in both the New Orleans and Galveston Districts has tripled during the past 10 yr, primarily the result of providing new channels and deepening and lengthening the old channels. New Orleans now dumps almost 24 million tons per year, compared to an estimated 8 million tons as recently as 1968. The equivalent Galveston figures are 13 million tons in fiscal 1973 and 5 million tons in 1968. The material from the outer channels dredged by the Galveston District is similar to that handled by the Mobile District. Although the silt content is probably greater, more could be used for beach replenishment

FIGURE 7.3: WESTERN GULF OF MEXICO WASTE DISPOSAL AREAS

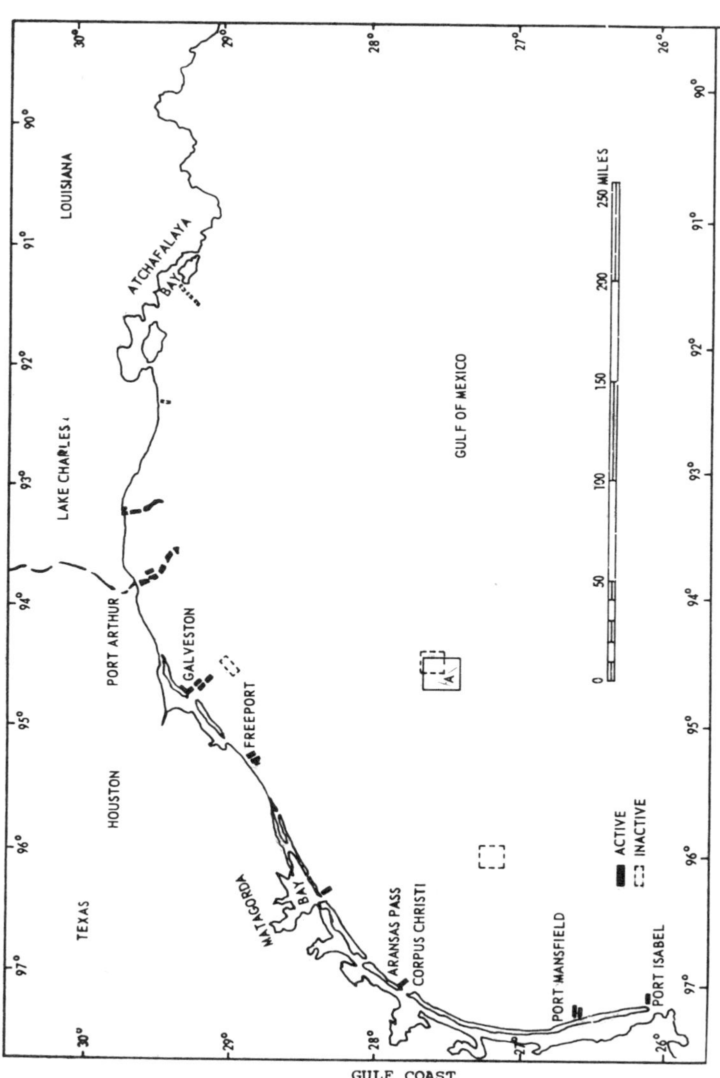

Source: PB 224 793

FIGURE 7.4: EASTERN GULF OF MEXICO WASTE DISPOSAL AREAS

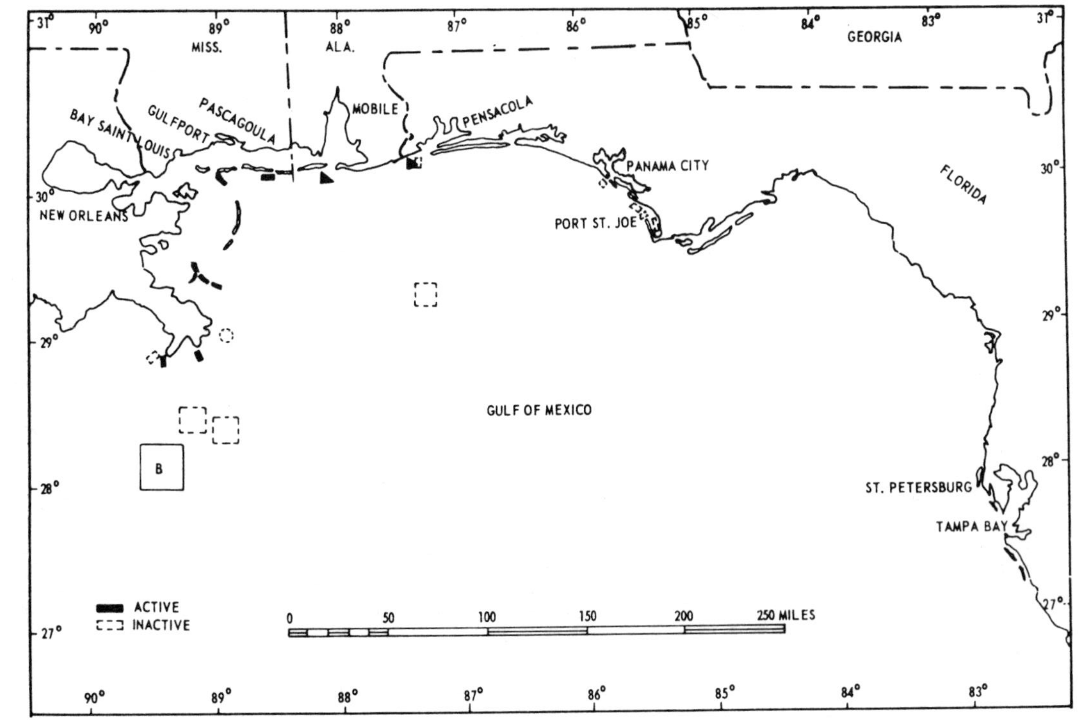

Source: PB 224 793

than is now the case. The costs would be appreciably greater because of the need for installing direct pumpout capability on the dredges and probable use of floating pipeline discharges in some areas.

The material in the channels dredged by the New Orleans District is mud or silty clays, with very little sand. It is not suitable for beach replenishment, even if there were beaches to nourish. It could, however, be used for marsh accretion because the existing marshes were developed of this same material, transported by distributaries of the Mississippi River. In the past 100 years, the river has been channelized by levees so that these sediments no longer are distributed, but are deposited at the ends of the ship channels. Thus, discharging the material to the edges of the marsh is essentially a return to nature. The added cost of this redistribution would be even greater than for the Galveston District because of longer hauls and/or pipelines.

In all three districts, the normal operating methods are the same because the same dredges are used between districts. Mobile uses the Langfitt from New Orleans for most of its hopper work. The Mackenzie usually stays in Texas, but the McFarland may be working anywhere on the Gulf Coast.

All operations are performed with the ship underway and are conducted around the clock, seven days a week. The daily performance ranges from 10,000 to 50,000 cubic yards. Spoil volumes are measured in cubic yards by the Corps of Engineers, but for comparison with other types of wastes and to be consistent with previous reports on ocean dumping, all amounts have been converted to tons using the factor: 1 cubic yard = 1.2 tons. Except for measurements to detect shoaling, there has been no environmental monitoring at any of the ocean spoil disposal sites.

Industrial Wastes: The dumping of chemical wastes in the survey area began about 1952 and, in theory, has been confined to the large, deepwater areas shown as **A** and **B** on Figures 7.3 and 7.4. The present annual amount is 1,168,000 tons, which is double the 1964-1968 average of 520,000 tons. Most of this material comprises carbonaceous solids and halogenated organic residues from petrochemical processes; spent acids and caustics; ammonium sulfates; filter aids; and sump cleanings, including dirt and shell.

Appreciable quantities of cyanides, herbicides, fungicides, and insecticides, with more than trace quantities of copper and other heavy metals, are present. All of the material is toxic and/or persistent in varying degrees, dependent on its concentration in soil or inert brine solution. The disposal of sodium hydroxide and sulfite and sulfate liquors has been reduced because many plants are now reclaiming this material through treatment processes.

Area **B**, south of New Orleans, was probably designated by a branch of the Department of Interior in March 1955, when a permit was issued to the Ethyl Corporation to dispose of drums containing sodium-calcium sludge. The New Orleans District Engineer honored the permit and formalized the bounding coordinates prior to June 1969, when Amoco Chemicals Corporation was granted a Letter of No Objection for bulk dumping of material of the same description. The Dallas Regional Office of EPA accepted these boundaries in April 1973. Neither of the charted explosive dumping areas northeast of area **B** is now active, although the easterly one was considered still active as recently as 1968.

The general vicinity of area A, south of Galveston, was recognized by the Galveston Corps of Engineers in Letters of No Objection, but no formal bounding coordinates were established. Instead, the sites were designated variously as "100 (110) miles South of Galveston" and/or "in 100 (200,400) fathoms or more." Verber, in the 1971 Dillingham Report lists seven sites with overlapping areas in this general vicinity which are known to have been in use since 1955. In April 1973, the Dallas Regional Office of EPA designated the boundaries of area B as shown in Figure 7.3.

The annual tonnages, therefore, were not concentrated entirely within the recently restricted dimensions, but were scattered over the larger area. One industrial user has expressed concern that the ship operators may not be able to navigate with sufficient accuracy to ensure dumping in the designated area.

The annual amount of waste disposal in area B is now only 13% of the 1969 tonnage. This reduction from 169,000 to 22,000 tons is entirely the result of the installation of treatment facilities at three refineries for the annual recovery of 158,000 tons of spent sodium hydroxide and sodium phenate.

For the same period, annual dumping in or around area B has increased five-fold, from 235,000 to 1,146,000 tons. Part of this growth is the result of plant expansions and part is from the shutting down of discharges into the estuaries. Although large enough for navigational purposes, area A may be too small to provide adequate dilution, dispersion, and exchange for the amount of bulk dumping that can be extrapolated for the next five years.

The bulk dumping is performed from tank barges with capacities from 1,600 to 4,800 tons and discharge rates up to 18 tons per minute. Drums are sometimes carried as a deck load, but usually they are transported on smaller ships in lots of 1,000 or less (approximately 250 tons at 1 drum = 500 pounds). There has been no environmental monitoring at either dump area. Some short-term studies have been performed on toxicities and near-surface dispersion by the industrial users.

Analysis of Dumping Activities

Effects of Dredging and Spoil Disposal: During dredging, most of the microorganisms and all of the larger animals on and in the removed sediments are killed by passage through the drags, suction pumps, and discharge lines to the hoppers. These remains add to the existing organic material in the sediments and increase the biochemical oxygen demand (BOD) loading. The aeration of the material during filling of the hoppers partially makes up for this increase by raising the dissolved oxygen content of the accompanying fluids.

Turbidity, which may be objectionable aesthetically, is also temporarily increased in the operating area; however, most of the shallow areas of the Gulf Coast have high normal turbidities during windy periods, and several studies have shown that sessile fauna such as oysters are unaffected by the turbidity alongside a working dredge.

The initial effect during hopper discharge is to increase turbidity and bury any existing benthic fauna, killing the sessile forms. Several studies have shown that the biomass returns to normal in six months to two years, but species diversity is reduced. Productivity is usually increased the second year because of the greater nutrient availability, similar to the effect of plowing a field. None of the studies was of sufficient duration (maximum of three years) to determine the length of time required to redevelop a full assemblage or normal community.

A secondary effect can be a change of substrate, such as covering a mud bottom with sand, resulting in a semipermanent change of habitat and faunal assemblage. The covering of a sand or rock bottom with mud would be temporary, because the existing energy level would quickly transport the finer sediments to a natural mud bottom area.

Long-term effects of dredging are in changes to estuarine circulation and salinity patterns caused by channel deepening, which usually increases saline intrusion and linear spoil banking. This usually restricts circulation and can cause low DO areas or retention of pollutants from adjacent discharges. These effects are strictly an estuarine or lagoonal problem and do not concern ocean or offshore dumping along the Gulf Coast.

The most obvious alternative to offshore disposal of dredge spoils is to use this material for beach or shoreline replenishment, because it is generally compatible.

Another alternative is to transport this material inland for enhancement of clayey soils, mixing with sewage sludge for agricultural lands. The major problems, besides the economics of transportation, are desalting and dewatering. Upland sites with abundant fresh water for pumpout would be required as interim dewatering enclosures.

A Corps of Engineers research group at the Waterways Experiment Station, Vicksburg, Mississippi, has started a five-year program of study on the effects, alternatives, and beneficial uses of dredge spoils and disposal methods.

Effects of Industrial Ocean Waste Disposal: At the present time, most disposal of toxic chemical wastes in the Gulf of Mexico is performed by discharge of liquids or slurries from tank barges in area A, 110 miles south of Galveston. Some of the companies have commissioned studies to determine surface toxicity gradients, dispersion rates, etc. during dumping operations. Their conclusions have been that adverse biological effects are minimal with mortality of organisms only at the discharge point, and no detectable surface effects after 2 to 8 hours.

There have been no studies of effects on benthic or demersal life, although one study showed reconcentration in the bottom muds and stratification at intermediate depths. There is no monitoring to determine seasonal and annual trends in chemical or biological parameters. Even the subsurface current characteristics are unknown. Surface currents tend northwest during the summer and southwest during the winter, as shown by a drift bottle study which covered a one-year period at approximately bimonthly intervals. The diluted wastes remaining in surface waters are transported through the Texas shrimping grounds to the coast. Barrelled wastes are no longer dumped at area A.

Most of the industrial waste dumped in area B, 35 miles south of Southwest Pass, is containerized in 55-gallon drums which are pierced by pickaxe or rifle fire during jettisoning. This practice does not take advantage of a unique attribute of this site. It is located in the mouth of a submarine canyon at the base of the foreslope of the Mississippi River delta. The rate of sediment deposition is probably the greatest of any comparable deepwater area in the world. It has been estimated that drums would be completely buried in three to five years. Drums which are completely filled to prevent rupture, adequately weighted, and sealed with a preservative coating could, therefore, be permanently buried, removing the contained toxics and their possible effects from the environment.

There is no available information on the effects of the present dumping operations. As in the case of area A, the surface currents are northwest towards shore and pass through the most economically important shrimping area in the country. Seasonal subsurface and bottom current data are lacking, and there is no monitoring program of any kind in the area.

The estuarine, littoral, and shelf areas of the Gulf Coast have been, and are being, studied and monitored to a considerable degree by state and federal water quality and fisheries agencies, the universities, and even the oil companies in their offshore lease blocks. Except for routine hydrographic surveys and a few isolated studies such as coring by the Glomar Challenger at Sigsbee Knolls, there has been little work in the deeper areas of the Gulf of Mexico.

Until this deficiency in baseline data is rectified by an integrated, multidisciplinary program of monitoring the critical physical, chemical, biologic, and geologic parameters, there can be no scientific foundation on which to base valid criteria and regulations for the control of industrial ocean dumping.

Many other disposal schemes have been proposed for this type of waste. Reclaiming and recycling of caustics has already proven profitable and could be extended to other chemical wastes if barge disposal becomes more expensive. Up to now, the relative economics of recovery versus dumping has been the sole governing factor. The same is true for treatment or conversion to less objectionable compounds or even marketable by-products.

Waste disposal in the Gulf of Mexico has caused only one major problem up to the present time. This is the collecting of drums in the nets of shrimp trawlers. If the nets are not torn or completely lost, the recovery of a leaking drum of toxic chemicals means that particular catch is completely dead and, further, that all previous catches on board are killed or contaminated, resulting in loss of the whole load.

There are reports of explosion injuries and chemical burns to crew members when they bring these drums aboard, or when attempting to jettison them after hauling them to deeper water. Some drums have floated ashore, where they are a similar hazard to bathers, fishermen, and other users of the beaches. These incidents could not have occurred if the drums had been adequately weighted and not prematurely dropped in the shallow fishing grounds which extend out to about the 60-fathom contour. Only one instance of a premature dump has been reported. In April 1971, drums were offloaded short of the 400-fathom line because their ship was in danger of capsizing in 15-foot waves. The number of drums picked up in nets and ashore indicates that short-dumping was very common.

Drums are no longer dumped at the Galveston site, only off the delta. By using the burial technique and a simple but rigorous surveillance system which ensures that dumps are made only at the approved site, future incidents of this nature should diminish. It is assumed that weather and sea-state prediction is presently adequate to prevent any repetition of a premature dump caused by danger of capsizing.

Recommended Procedures

The EPA Ocean Disposal Program Office should maintain close liaison with the Study Program for Disposal of Dredge Spoil, U.S. Army Engineer Waterways Experiment Station. In this manner, EPA can be kept abreast of research findings and indications where criteria or regulations may need to be modified. In turn, the Corps of Engineers may be alerted to any specific problems which may come to the attention of EPA.

No reason has been discovered to discourage use of the present spoil disposal sites pending interim surveys by the Corps of Engineers and/or EPA as to their suitability on the basis of hydrography, geology, and biology.

The deepening and lengthening of ship channels on the Gulf Coast to accommodate super tankers would vastly increase the amount of spoil that would be dumped offshore. These vessels should remain in uncongested deep water at all times, using buoys or platforms for loading and unloading. The Corps of Engineers is presently studying the development and location of such deepwater terminals. Economics and maritime safety, as well as environmental protection, would seem to dictate the use of such facilities for handling cargoes that can be transported via pipeline.

Sites A and B should be surveyed and should continue to be monitored on at least a quarterly basis to determine seasonal variations in current, salinity, temperature, dissolved oxygen, and pH profiles and to determine the seasonal pelagic and benthic biota. Annual measurement of bottom sediment characteristics is considered sufficient. A third area should be selected as a comparative control for measurement of water quality parameters if the present sites continue to be used.

As an interim surveillance procedure, the EPA should develop an affidavit form to be completed and signed by the captain of the disposal vessel, which stipulates the client's name; load amount; time and date of departure; time, location, and water depth at beginning and end of dumping; intervening courses, speeds, and discharge rates; and time of return to port. The original with a supporting, continuously recorded fathogram should remain with the USCG Captain of the Port with a copy to the EPA Regional Office.

Additional copies may be desired by the USCG District Office, the client, and the barge company. An affidavit should also be required from the waste producer, stipulating the load amount and a bioassay and/or chemical analysis of each load. A more sophisticated system,should be developed to record this information automatically instead of manual entry by the vessel operator. A feasibility study should also be made on the concept of permanent burial of drummed wastes in area **A.**

SOUTHERN CALIFORNIA STUDY

Southern California Bight

The Southern California Bight is essentially the continental margin off one of the largest population centers of the United States. It is an open embayment extending roughly from Point Conception, California to Cabo Colnett, Mexico (Figure 7.5), with its oceanward boundary at approximately the 10^3 meter contour. The geography, sources of contaminants, and the physical and chemical characteristics of the bight have been extensively reviewed by the Southern California Coastal Water Research Project in a three-year study.

The topography of the Southern California Bight consists of a smooth, narrow continental shelf, a narrow slope, a wide continental borderland, and a true continental slope. The area has many faults and as many as 32 submarine canyons. The borderland area is a rough region of basins and troughs interrupted by ridges and eight offshore islands. The submarine canyons tend to guide material from land into the offshore sediments. They do not penetrate the true continental slope.

The circulation pattern, in general, is southward along the edges of the Bight and northward through the offshore islands (Figure 7.6). There should be a net loss of water to the south. The patterns vary seasonally, with eddies common in summer and autumn. Circulation patterns near land are greatly affected by winds and bottom topography. Little information is found on bottom currents in this area. It is an area of upwelling with large volumes of the deepwater coming to the surface, often at the heads of the submarine canyons.

The principal water characteristics affecting the organisms of the Southern California Bight are temperature, salinity, and concentrations of dissolved oxygen. Upwelling is especially significant in controlling the plankton base of the marine food chain in the Bight, and nitrogen is the limiting nutrient. Fluctuations in temperature, salinity, dissolved oxygen, or nutrient concentrations are rapidly reflected in the plankton.

There are three principal benthic faunas in the Southern California Bight: the soft-bottom shelf community, the basin fauna, and those organisms living in the submarine canyons. The basin fauna is less diverse and lower in species density and total biomass than the shelf community. Fauna in the canyons is generally intermediate betweeen that of the shelf and the basins, it is subject to large fluctuations, and the fauna of each canyon is relatively isolated.

Each year approximately 5.2×10^{11} kg of solid material enters the Southern California Bight from municipal waste treatment outflows (1.4×10^9 m^3/yr), discrete industrial sources (0.25×10^9 m^3/yr) and surface runoff (0.6×10^9 m^3/yr). In addition, 7.7×10^9 m^3 of returned cooling water and 2×10^{13} m^3 of advected ocean water enter the Bight area annually.

Approximately 85% of the municipal wastewaters of the southern California coastal basin are discharged to the ocean through outfalls. The remainder is reused or released at inland locations. The discharge of effluent and sludge by this mechanism results in a buildup of high levels of organic matter in the sediments in the area of the outfall.

FIGURE 7.5: THE SEA FLOOR OF THE SOUTHERN CALIFORNIA BIGHT

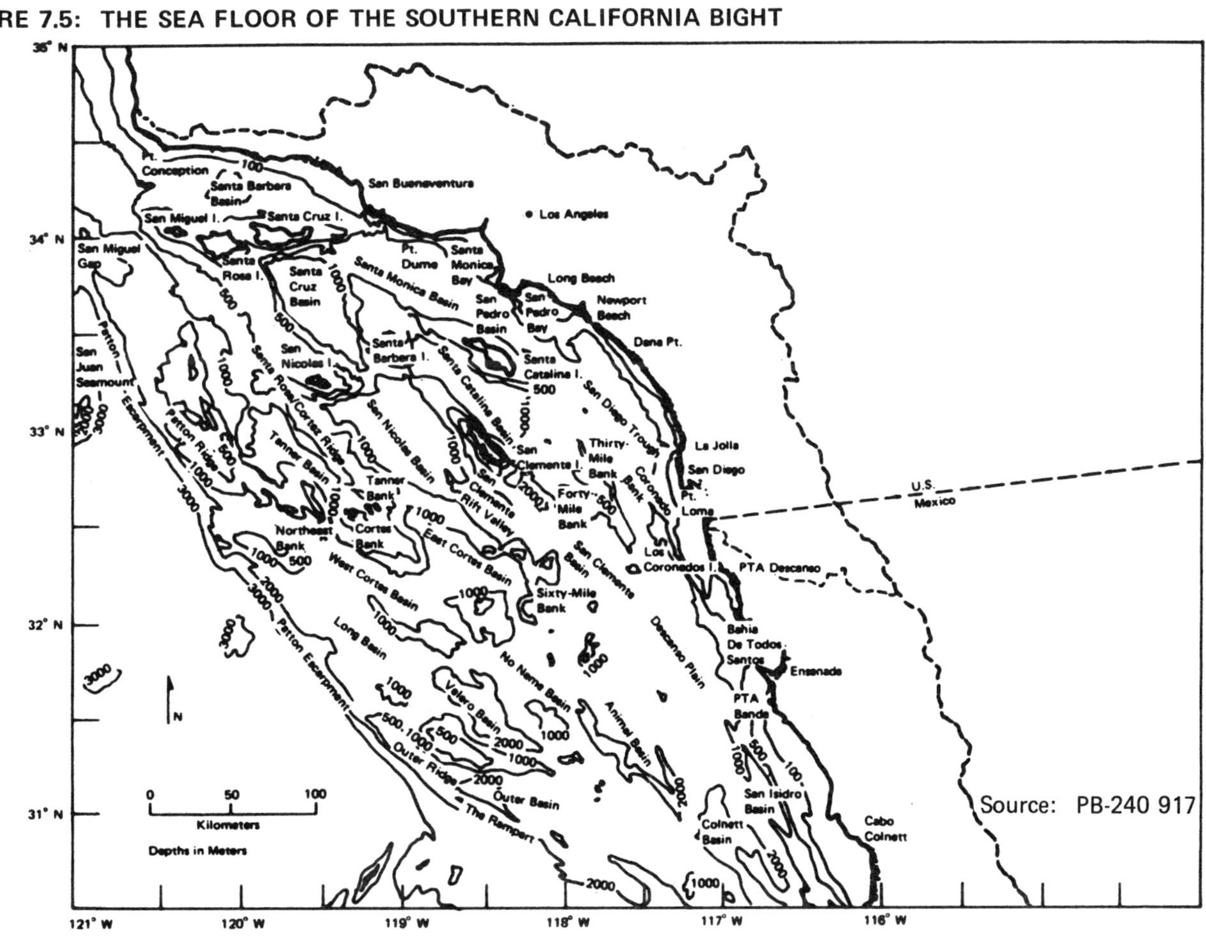

Source: PB-240 917

FIGURE 7.6: MEAN GEOSTROPHIC FLOW AT 200 m DEPTH IN THE SOUTHERN CALIFORNIA BIGHT

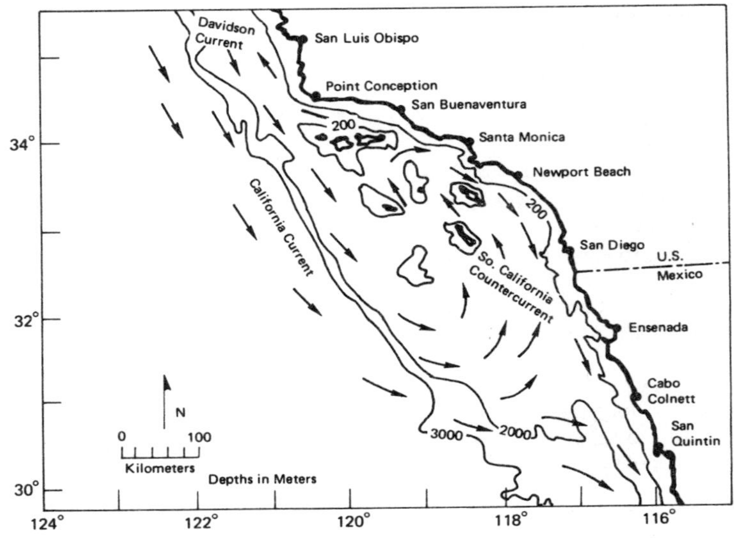

(Arrows Show Direction and Magnitude of Flow)

Source: PB-240 917

The area subject to organic loading varies with the dynamics of the surrounding water system and with the degree of waste treatment employed. Although the use of secondary treatment as required by NPDES regulations will reduce the organic loading, these procedures would not be expected to reduce the quantities of nonbiodegradable pesticides and heavy metals introduced into the coastal waters by sludge disposal. One hundred fifty-six industries discharge wastes directly or by means of storm drains into the Los Angeles and Long Beach Harbors.

History of Dumping in the Southern California Area: Materials in eight categories have been disposed of in the ocean areas of Southern California: (1) industrial wastes, (2) refinery wastes, (3) trash and garbage, (4) filter cake, (5) oil drilling wastes, (6) explosives, (7) radioactive wastes, and (8) miscellaneous wastes.

Industrial Wastes – The main sources of industrial wastes are: (1) industrial, medical, institutional, and academic laboratories; (2) heat treating, plating, and film processing plants; (3) chemical and petroleum processing plants, and (4) electronic and aerospace manufacturers. Typically, these materials have been handled either as containerized or bulk wastes. Of the containerized wastes, between 1965 and 1971, 510 tons per year were dumped in the San Pedro Channel. Between 1960 and 1967, 155 tons of bulk sodium cyanide were dumped west of San Diego. These two activities have been curtailed. Bulk disposal activity is mostly located at a site north of Santa Catalina Island, with amounts of approximately 195 tons per year. Dumping at this site has been active since 1960.

Refinery Wastes – It has been estimated that 530,000 tons of refinery wastes were dumped into the Southern California Area from 1946-1971. Annual rates for 1968 to 1971 were approximately 2,000 tons. The majority of this material was dumped in the San Pedro Channel. Information on the composition of this material is not available. These wastes can contain petrochemical solvents and chemicals, catalyst waste, cleaning wastes and oil and grease. Refinery wastes are not presently dumped in the Southern California Area.

Refuse and Garbage – This category of waste is one of the two that are still actively being disposed of in the waters of the Southern California Area. The U.S. Navy, between 1944 and 1970, dumped about 107,000 tons of garbage and trash from vessels in San Diego and Long Beach in two areas. This dumping has been terminated and the material is now incinerated and disposed of at landfill sites. Since 1931, approximately 51,000 tons of trash and garbage collected from commercial vessels in Long Beach and San Pedro Harbors have been dumped in a site southeast of Santa Catalina Island. This enterprise is presently disposing of approximately 623 tons of waste annually in the same site.

Filter Cake – A residue of algin extraction from locally harvested kelp, filter cake consists of half perlite and half cellulose. Approximately 350,000 tons of filter cake were disposed of during 1969 and 1970 in an area west of San Diego, before the dumping was prohibited in 1970.

Oil Drilling Wastes – Muds and cuttings derived from oil well drilling were disposed of at sea during the period 1966 through 1970. Approximately 3.3 million tons were disposed of in the San Pedro Channel until the practice was prohibited in 1970. The muds can contain oils, chemicals, heavy amounts of clays in suspension, and barite. This method of disposal has apparently been discontinued.

Explosives – In 1971, the disposal of military explosives at sea was interrupted by the U.S. Navy and it is present policy not to allow such disposal. It has not been possible to determine any amounts for materials dumped between 1944 and 1971. Small amounts of ammunition have been disposed of on an irregular basis by various police agencies.

Radioactive Wastes – Amounts of radioactive wastes disposed of in designated areas at sea off the Southern California Area were not determined. This material was disposed of during a period of 1946 through 1968 until the prohibition of such methods of disposal in 1968.

Miscellaneous Wastes – Disposal of slugs found in telephones, weapons collected by police agencies, and similar materials has, by one estimate, been placed at 275 tons per year. Some of the material is disposed of at the site used for garbage and trash disposal.

Problems generated by the present and past ocean dumping in the Southern California Area are varied in complexity because of the lack of knowledge about the effects of some materials. The industrial wastes that are presently disposed of include some very toxic (cyanide wastes) and some highly reactive (phosphorus) materials. The composition of the material to be disposed of leads to problems in personnel safety and detrimental effects upon the site environment. Present methods of disposal in this area are very crude and can, therefore, accentuate these problems. The composition of the garbage and trash, with some nonbiodegradable (plastics) and some floatable materials (wood), causes problems with esthetics and the introduction of undesirable materials into the ocean environment.

A lack of knowledge of munitions disposal amounts, types, and locations requires a larger area to be quarantined for disposal of this type of waste than may actually be required. Also, the future effect upon the ecology of the site is not determinable because of the lack of information on degradation of the containers of this waste.

Any future or continuing disposal must be closely analyzed for appropriate site location, allowable amounts and concentrations of materials, and disposal methods in order to comply with the present public health, public safety, and the environmental objectives for the local area, after all alternatives are rejected in order to preclude adverse reaction from any source. California State Law requires that these decisions be made in a public forum and that all information be made available to the local public audience.

Present dumping in the Southern California Area is concentrated in two areas west of Los Angeles. The materials, generally described, are industrial wastes and garbage and trash from commercial vessels. A site east of San Diego may be used for the disposal of dredge spoil in future dredging of San Diego Bay, but is not an active site at this time.

The industrial waste disposal site is centered over the sill between the Santa Monica and San Pedro Basins. The garbage and trash disposal is centered on the lower slope into the southern end of the San Pedro basin. Basins in Southern California are characterized as impoverished areas, as are the slopes below the sills of the basins. Both the Santa Monica and San Pedro Basins have large areas of impoverishment below their combined sill depths of 2,418 feet (737 meters). These basins are connected in such a way that the same sill controls both. The bottoms of both basins are barren of living benthic animals.

Impoverishment is a result of the low oxygen content of the bottom water and interstitial water. The low oxygen content of the bottom water is a reflection of the position of the sill within the oxygen minimum in the open sea. Because of the low oxygen content of the overlying water, organic debris reaches the bottom in an incomplete stage of decomposition which depletes the interstitial water of oxygen and causes the condition found at the bottom of the basins, that of an anoxic environment.

Above sill depth, animals dominating are siliceous sponges and ampharetid worms, but include ophiuroids, crustaceans, and mollusks. The benthic forms are very well represented in the waters on the slopes of basins above sill depth.

State and Federal Regulations: The disposal of garbage from vessels is subject to control by the California State Departments of Public Health and Agriculture, and by the United States Department of Agriculture. Among the control provisions of significance are the following:

(a) The dumping of garbage in or upon the navigable waters of the state, or at any point in the ocean within 20 miles of its coastline, is a misdeameanor.

(b) The removal of garbage from any vessel for dumping into any territorial waters or onto land is prohibited except for immediate incineration, approved treatment or disposal under supervision of the Director of the State Department of Agriculture, or delivery to a garbage collector licensed by the director or the Federal Government.

(c) Section 774 of the California Administrative Code permits, as an approved method of garbage disposal, reduction to a liquid state by grinding and discharge into seawater; location and distance from shore of such approved disposal are not specified.

These and equivalent restrictions enforced by the United States Department of Agriculture are intended to prevent hoof-and-mouth disease, rinderpest, and other highly contagious diseases which can infect cattle, sheep, or swine.

Los Angeles Regional Board Regulations: The Los Angeles Regional Water Quality Control Board under provisions of the California water code has set the following waste discharge requirements on disposal of these materials at sea.

(a) Disposal of wastes shall be prohibited into or onto any waters of the state (within three miles of shore).

(b) Disposal of wastes shall be prohibited into, onto, or within any place or waters from where they can affect the waters of the state.

(c) Disposal of wastes into the ocean, as proposed, shall be limited to the locations herein before described; that is, within a radius of one nautical mile of either latitude 33°17'N, longitude 118°10'W (prime site) or latitude 32°33'30"N, longitude 119°05'48"W (alternate site).

(d) The disposal of wastes shall not impair any of the beneficial uses of the offshore coastal waters enunciated for protection by this board in its Pacific Ocean Water Quality Control Plan.

(e) This waste discharge shall not cause the appearance of visible oil or oily slick or any other floating or suspended matter in waters of the state or along beaches or shores.

(f) The disposal operation shall be conducted in such a manner that no trash, garbage, or other waste material will be deposited along beaches or be present in waters of the state.

(g) Wastes discharged shall not cause concentrations of toxic materials at any point within coastal waters which would be detrimental to human, animal, fish, plant, or bird life.

(h) Wastes discharged shall cause no odors at any point within the coastal waters, or along the shore.

(i) Wastes discharged shall not cause any increase in turbidity within waters of the state.

(j) Wastes discharged shall not cause dissolved oxygen concentrations in waters of the state to fall below 7.0 mg/l as an annual average, and shall not cause such concentrations to fall below 5.0 mg/l at any time.

(k) Waste handling and loading operations shall be conducted in a manner that will prevent spillage or dropping of any waste material into waters of the state.

(l) Handling, storage, or disposal of wastes shall not cause pollution or nuisance.

(m) All Federal, State, County and City rules, regulations, laws, and ordinances pertinent to this disposal of wastes shall be complied with.

(n) Adequate facilities shall be provided to handle and store wastes without spillage or leakage, pending disposal.

(o) Upon verbal request by board staff, the discharger shall notify the board at least 24 hours prior to the departure of the disposal vessel so that an observer may accompany the vessel.

(p) Complete logs of all trips for purposes of waste disposal shall be kept and shall be available for inspection during regular business hours by the board staff. These shall include names of vessel and captain, times of departure and return, and latitude and longitude of disposal point, and vessel speed.

(q) A copy of these waste discharge requirements shall be made available at all times to personnel conducting waste disposal operations.

(r) In accordance with Section 13260 of the Water Code, the discharger shall file a report of any material change or proposed change in the character, location, or volume of the discharge.

(s) A report of any change or proposed change in ownership or name of this facility shall be filed with the board.

(t) The board shall be notified immediately by telephone of the presence of adverse conditions resulting from this discharge; written confirmation shall follow.

(u) In accordance with Section 13267 of the Water Code, furnish, under penalty of perjury, technical reports shall be submitted in accordance with specifications prepared by the Executive Officer, which specifications are subject to periodic revisions as may be warranted.

The technical reports required are submitted quarterly and contain quantities and types of materials, dates of discharge, sums of quantities discharged during period covered by report, a separate accounting of industrial wastes hauled (if any), a copy of the log on each trip made during the report period, and any deviations from the discharge requirements noted. Also, an annual report is submitted covering all the above information.

The Regional Board has sent an observer on at least one such trip to assess the disposal operation. Several complaints, one by the board and one by the California Department of Fish and Game, have been voiced. Neither complaint led to any legal action, but administrative inquiries were initiated. In each case, the dumper was able to guarantee that no further problems would arise and that settled both inquiries to the satisfaction of the board.

The San Francisco Area

The San Francisco area is located in Central Northern California and is the only major port in the northern part of the state. San Francisco has a history of ocean dumping that has in recent years been under close scrutiny by local and state regulatory agencies. As a result of the local activities, most ocean dumping has stopped, either by direct regulation or because of monetary pressures and environmental concern expressed by the regulatory agencies.

Dumping of materials at sea in the San Francisco area is at present limited to the disposal of dredge spoil materials. These materials are generally those that are deemed polluted and, thus, not suitable for disposal within the Bay itself, except for the material removed from the main ship channel on an annual basis and deposited just south of the channel. One dumping activity that had been active in the past few years, a cannery waste disposal, has been terminated recently. This was the only remaining operation, other than dredge spoil disposal, that entailed ocean disposal.

History of Dumping in the San Francisco Area: Dumping of wastes in the study area offshore of the Golden Gate have been in six categories of materials: (1) refinery wastes, (2) acid wastes, (3) cannery wastes, (4) radioactive wastes, (5) munitions, and (6) dredge spoil. A summary of the wastes disposed of between 1931 and 1972 is presented in Table 7.2.

TABLE 7.2: SUMMARY OF WASTES DUMPED INTO THE OCEAN AREAS OFFSHORE OF SAN FRANCISCO

Type of Waste	Period	1931-72 Estimated Total	Present Estimated Total
Refinery wastes	1966-72	315 M gal	-
Acid wastes	1948-71	240 M gal	-
Cannery wastes	1960-72	246 K tons	-
Radioactive wastes	1946-68	44,563 containers	-
Munitions	1968-69	746 tons	-
Dredge spoil	1935-72	-	1 M yards

Refinery Wastes — The estimate of 315 million gallons of refinery wastes is from one of two known generators of this waste.

Acid Waste — About 10 million gallons per year of spent steel pickling acid were discharged into the ocean about 14 miles southwest from the Golden Gate and about 9 miles offshore (at approximately 37°72'N, 122°40'W in about 120 feet of water). This operation began in 1948 and was terminated April 30, 1971.

Cannery Wastes — Beginning in 1960, a commercial scavenger company was contracted by six East Bay fruit and vegetable canneries to dispose of their wastes. The volume of the discharge was about 22,000 tons per year. The disposal site was in water depth of about 260 feet, at a location approximately 20 miles offshore of San Francisco. The latitude and longitude has been determined to be 34°35'N, 122°50'W. The operation has been terminated because of increasing costs associated with the monitoring requirements.

Radioactive Wastes – Licensees regulated by the U.S. Atomic Energy Commission dumped, in the period 1946 to 1968, about 44,563 containers of radioactive wastes into the ocean about 16 miles southwest of the Farallon Islands. This location would be approximately centered at 37°31'N, 123°14'W. Dumping was terminated by changes in Federal policies in the late 1960s. Positive determination as to what is called a container is not possible, although the normal disposal method is to encase an amount of the waste in concrete within a 55-gallon drum. The actual amount of waste per barrel can vary from a few grams to a few pounds, depending on the storage capability of the licensee.

Munitions – During 1968 and 1969 the U.S. Navy disposed of some 510 tons of conventional munitions in a designated disposal site some 18 miles west of the Farallon Islands. The disposal site is a trapezoid with an approximate center at 37°40'N, 123°25'W, and averaging 6,600 feet in depth. In addition, a ship was sunk on this site in 1964 under the "Chase" disposal operation. It contained approximately 236 tons of waste ammunition and explosives. These methods of disposal have been discontinued.

Dredge Spoil – The U.S. Army Corps of Engineers discharges materials from maintenance and improvement activities both near the main channel entrance and within San Francisco Bay. Also, activities under Corps of Engineer permits have discharged materials outside the bay. Main channel entrance maintenance programs have placed about 600,000 cubic yards annually in a location just south of the channel. This amount is expected to nearly double, with planned deepening to 55 feet of the entrance channel in the near future.

Also planned is the disposal of about 900,000 cubic yards of material that exceeds heavy metals criteria, offshore of the 100 fathom isobath some 30 miles out from the Golden Gate and 8 miles south of the Farallon Islands. The Corps of Engineers and permittees to the Corps are the major users of ocean dumping areas in the San Francisco area. Approved dumping areas for use inside San Francisco Bay are now limited to five specific sites.

Ocean disposal in the San Francisco area is presently limited to the disposal of materials derived from dredging the main channel at the entrance to San Francisco Bay, and to the disposal of materials considered undesirable to dump in the Bay proper.

Except for temporary turbidity and possible burial of some bottom dwellers, the main channel project probably does not cause any serious problems in the local dump site biota and it allows the retention of possible beach replenishment materials within the local littoral regime. The disposal of unwanted materials from within the bay is another matter. The materials could contain concentrations of heavy metals above present criteria and chlorinated hydrocarbons above present standards, and could be toxic because of high oxygen demanding constituents.

Present Dumping Activities in the San Francisco Area: Dumping activity in the study area is, at present, limited to two dredge spoil activities, both under U.S. Army Corps of Engineers administration. One site is used for the disposal of material from the main ship channel into San Francisco Bay. The other site (not yet determined positionally) is for the disposal of contaminated material from future projects within the Bay. Also, a site used, until recently, for disposal of cannery wastes will be discussed as it is an example of this type of operation, and because an extensive study of the effects of the materials upon the biota has been made, which should receive recognition.

One site is the location where dredge spoil obtained from the maintenance of the entrance channel to San Francisco Bay is deposited. On an annual basis, the U.S. Army Corps of Engineers deposits approximately 600,000 cubic yards of spoil in the dump area.

The site is on a submerged sand bar (San Francisco Bar) which extends in an arc outside the Golden Gate at a distance of about five miles. It is 3,000 feet south of the main ship channel and runs the length of the dredged portion, or about 5,000 yards at the present project depth of 50 feet and is 1,000 yards wide. Water depth in the disposal site ranges from 50 feet at the ends of the site to less than 36 feet near the center of the site. If the new 55-foot project depth is used, the disposal site would increase to 3½ miles in length and an annual maintenance quantity of 900,000 yards is projected. The center of the disposal area is at 37°45'06"N, 122°35'45"W.

The shape and position of the bar is maintained by a dynamic balance between tidal flows, prevailing wave actions, and coastal currents. The materials of the bar are 90 to 95% fine sands, with the remaining materials falling into the ranges of silts and clays. It is believed that this material is predominantly derived from erosion of north coastal cliffs and outwash of streams. It is transported to the bar by longshore coastal currents.

No measurable part of the material on the bar is believed derived from the bay system itself. Apparently, the outflow of fines from the bay system by tidal flows is carried out over the bar into deep water. The result is submerged beach, maintained by longshore sand transport, and held offshore by the repulsion of an ebb tidal prism. The central bar has approximated its position and depth since 1855, when a study recorded its position and shape. This position has varied less than 1,500 feet and the depth by less than 10 feet over the central portion of the bar, exclusive of dredged areas since that date.

In the past, materials from the main ship channel were deposited in deep water about one mile southwest of the seaward channel entrance. After a comparative assessment of the use of the deep site, land disposal, or deposition on the bar itself, the Corps of Engineers has determined that redeposition is the best alternative, hence the use of this present site.

The only user of this site is the Army Corps of Engineers, San Francisco District. They deposit materials dredged from the main ship channel annual maintenance program. Maintenance of the 50-foot project depth achieved in 1959 has required removal of approximately 580,000 cubic yards annually from the channel. The new project depth of 55 feet will increase the annual maintenance amount to about 940,000 cubic yards. During the period required to achieve the 5-foot increase in depth (about four years) a projected annual dredging volume of 1.1 million cubic yards will be removed from the channel and deposited at the disposal site. This project started in 1971 and should have been achieved in 1974.

The composition of the dredged materials was the subject of a study performed by the U.S. Army Corps of Engineers, San Francisco District, during the period from December 1970 to June 1971. Three sets of samples were collected on a hopper dredge, the second by a bottom dredge. A summary of the results of these samplings is presented in Table 7.3.

TABLE 7.3: SUMMARY OF SEDIMENT ANALYSIS MAIN SHIP CHANNEL U.S. ARMY CORPS OF ENGINEERS, SAN FRANCISCO

	Concentrations in Parts/Million by Weight				
	12/28/70	4/5/71			6/8/71
Parameter	Sample	Sample 1	Sample 2	Sample 3	Sample
COD	174	6,800	26,000	5,000	1,650
Oil-grease	356	1,000	1,000	1,000	NA
Kjeldahl-N(T)	13	220	620	120	170
TVS	NA	21,000	39,000	14,000	11,900
Lead	4.79	14.20	27.20	12.20	13.00
Mercury	0.08	0.02	0.03	0.01	0.03
Zinc	37.81	48.60	79.90	34.50	55.00
Arsenic	0.01				
Chromium	1.99				
Cadmium	2.08				
Copper	1.67				
Nickel	37.81				
Phosphate(T)	32.90				
Sulfides(T)	0.28				

In addition to the analyses shown in Table 7.3, other tests were performed on some samples. Tests for pesticides were run and, with a composite of samples, a 96-hour static bioassay was performed on three spined stickleback fish, both by the EPA Regional laboratory in Alameda. The results of the bioassay were inconclusive because of the high survival rate of the test fish. The pesticide tests (maximum concentration achieved from 3 samples) are shown in Table 7.4.

TABLE 7.4: PESTICIDE CONCENTRATIONS (MAXIMUM 3 SAMPLES) SAN FRANCISCO MAIN SHIP CHANNEL APRIL 5, 1973

Pesticide	Concentration (ppb)
OP' DDE	1.64
PP' DDE	1.64
PP' DDD	1.34
OP' DDT	0.54
PP' DDT	3.74
Arochlor 1254	19.60

The second site, approximately 11 miles southeast of the Farallon Islands and 16 miles west of the mainland shore, is the location where, until recently, cannery wastes were discharged.

The site is an area defined only as centered at 37°35'N, 122°50'W with no shape or dimensions. It is located just shoreward of the 50-fathom isobath, 254 ft deep. It is outside the Gulf of the Farallones and near the top of a slope outlining the coastal plain of the San Francisco area. The ocan bottom at the disposal site ranges from mud to fine, hardpacked sand; currents are generally associated with the California current and countercurrent, and are not as large as found inshore near the Golden Gate. A marked misdummer upwelling and winter downwelling are associated with SW winds. Surface water normally can vary 9°C during the year. Salinity increases with depth, the surface about 33 parts per thousand; variations are marked and coincide with upwelling periods.

As previously mentioned, a private scavenger was contracted by food processing plants to dispose of canning wastes. The Ocean Disposal Program started in 1960 and terminated in 1972. The dumping operation entailed disposing of about 22,000 tons of waste per year during the months between July and October. The disposal vessel has a single-load capacity of 1,040 tons of, which 20 to 25% is water used to improve waste pumping quality. This left a single load with 780 tons of waste, for which 30 trips during the 4-month period would be required. This would be about one every four days.

Actually, the disposal frequency was predicated upon the accumulation of the waste at the land holding site, and it is expected the frequency would gradually peak then taper off near the end as the canning season does.

The dumped materials consist of solid residuals produced from canning of fruits and vegetables. Approximately 90% of the wastes are peach and pear residuals. Of the peach residuals, 90% is obtained from overripe and underripe fruit, culls or undersize fruit, and normal peaches diverted from production. The pear residues are composed mainly of peel, core, stem, and blossom end cuts. This accounts for about 80% of the residue. The rest consists of rejected fruits considered unfit for canning.

Analysis of Dumping Operations: The maintenance of the main ship channel is the only ongoing ocean disposal operation in the San Francisco area. Problems associated with the disposal of dredge materials from the main ship channel at San Francisco Bay are:

(a) Direct burial of bottom-dwelling marine life
(b) Toxic materials in the sediments
(c) Increased turbidity inherent in disposal of dredge spoil
(d) Depression of dissolved oxygen by oxygen demanding chemical and biological materials
(e) Increases in productivity above normal levels because of nutrients released from the sediments

Direct burial of marine organisms is a factor that was studied at the main ship channel disposal site. The study determined that the present community could survive, to a great extent, the addition of materials that were found attributable to the disposal operation. A diving survey conducted at the time of dredging determined no noticeable addition to the bar by the disposal runs. A layer, in fluid state, exists under most normal sea conditions and seems to absorb the new material quite readily. The natural erosion and deposition is believed to exceed the capacity of present dredging equipment, even under sustained operations.

Toxic materials, primarily heavy metals, are present in the main ship channel sediments. The only material which exceeds present criteria is zinc, which in two samples was 10% and 60% above the criteria. Although not exceeding the criteria, lead was fairly high in one sample. These high zinc and lead counts may reflect a naturally high content of the sediments.

There is some evidence that zinc ions are strongly absorbed in a permanent manner in silt, with resultant inactivation of the zinc. Lead is at its minimum toxic capability in seawater as compared with fresh water. A concentration of 70×10^{-4} parts lead in soft tapwater was not toxic to minnows and stickleback fish over a three-week test period.

Increased turbidity is an expected problem associated with any method of disposing of dredge spoil. The toxic effects of increased turbidity have been the subject of many studies, the results of which have not shown great problems except with very high levels of turbidity not expected on this disposal operation.

Depression of the oxygen available for supporting life is a problem when dredge spoil is contained in a small area for long time periods. When spoil is allowed to disperse over a large area this depression rarely exceeds levels that would be dangerous to marine life and is short term in its effects.

Nutrient release causes rapid population surges in algae communities. The nutrients in the sediments of the main ship channel are not too high to cause large blooms, although at times of low currents and calm weather blooms could occur. Blooms caused by artificial nutrient inputs can cause immediate and drastic depressions of oxygen levels that can be fatal to active types of fish. Several alternatives are available to the dredging of San Francisco Bar. They are as follows.

No Dredging Alternative — If the San Francisco Bar main ship channel were allowed to return to its natural state, the effect upon the San Francisco Bay area would be an economic disaster. Ships now trading the world's ocean would no be able to enter the bay because of the decreased water depths to be expected at the bar.

Land Disposal Alternative — Material could be deposited on land, but the benefit gained would be doubtful. Unless it was deposited on the beach, the sand would be lost from the littoral regime and erosion of beaches would be accelerated.

Disposal in Deep water — This method would have all the problems of the present method, plus the sediment would be lost to the littoral regime. Under present conditions the best available method for disposal of these materials is the existing method. Until more is learned concerning the problem areas, it is considered to be the rational approach to disposal of this material.

PUGET SOUND STUDY

The geographical area covered is the waters of the Strait of Juan De Fuca, Puget Sound, and the interconnecting inlets and adjacent bays south and east of the Canada boundary. Figure 7.7 outlines the area and shows both active and inactive dump sites.

FIGURE 7.7: LOCATION MAP (PUGET SOUND DISPOSAL SITES)

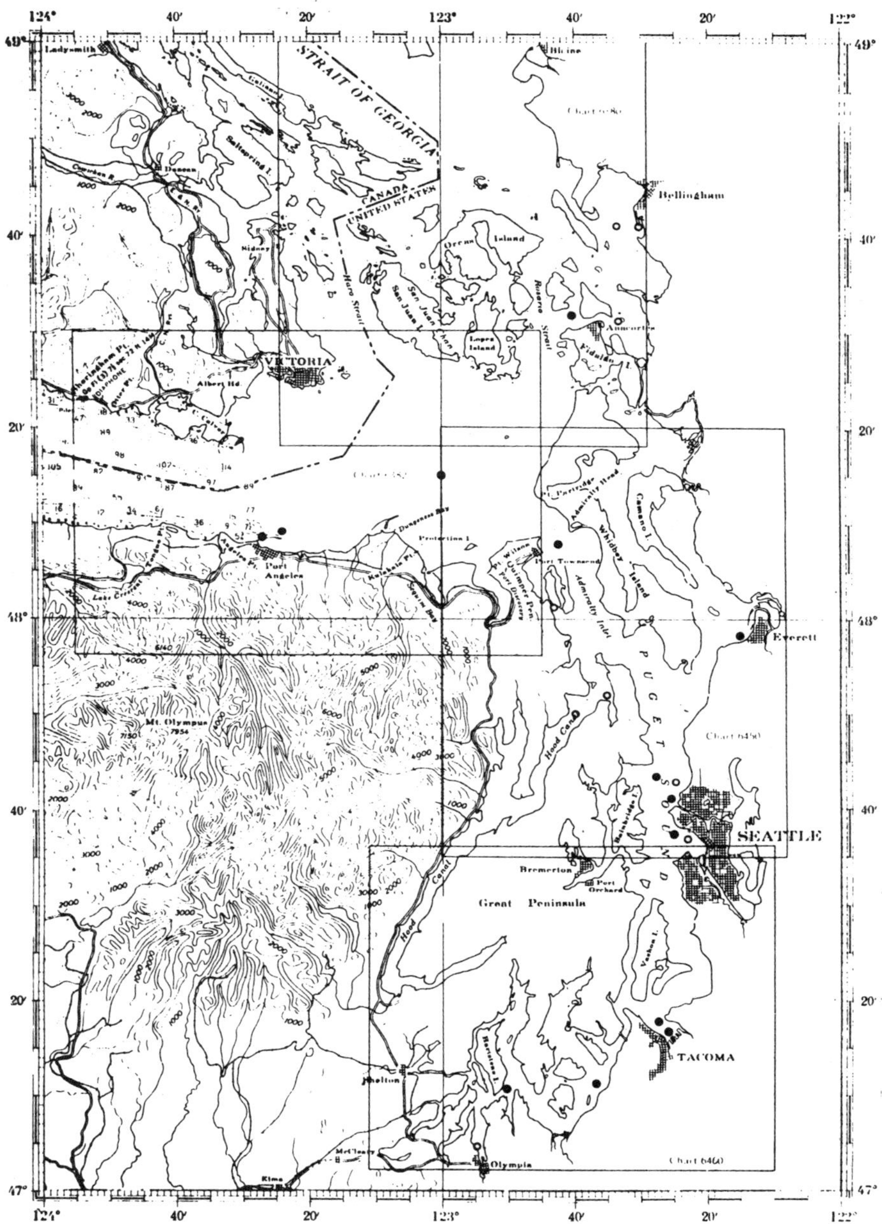

Source: PB 240 917

Although there are many outfalls discharging municipal and industrial wastes in the waters of Puget Sound, only dredge spoils are transported and dumped by vessels at nearshore sites. There is a barging operation which takes refuse from Seattle to a shoreside landfill on the Tulalip Indian Reservation north of Everett. 225,000 tons per year of vanillin black liquor from a chemical company are disposed of in the Strait of Juan De Fuca. There is an explosives disposal site located 80 miles west of Cape Flattery, but the last dump was in 1970.

Since 1970, in-water disposal of unpolluted spoils has been restricted to twelve sites which were designated by the State of Washington Department of Natural Resources (WDNR). Permit and lease applications for use of these sites are routed as shown in Figure 7.8. Spoils which do not meet the *Criteria for Determining Acceptability of Drege Spoil Disposal to the Nation's Waters* are placed in diked landfills. Approximately 250,000 cubic yards or 300,000 tons of dredge spoils are dumped in the waters of Puget Sound and the Strait of Juan De Fuca.

History of Puget Sound Dumping

Dredge Spoils: The disposal of dredge spoils in the waters of Puget Sound started at least as early as 1920 when Port Gambel Harbor entrance was first dredged. Although records of amounts dumped prior to 1939 are unavailable, it is reasonable to assume that most of the sediments removed in harbor construction were deposited as backfill behind adjacent bulkheads rather than dumped in deepwater.

From 1939 until 1970, material was needed for fill and, consequently, most of the spoil from maintenance dredging was taken out by dump scows to the nearest convenient spot with a depth in excess of the project depth. In 1970, the WDNR selected 12 sites to be used for future in-water dumping of dredge spoils which met the EPA criteria as unpolluted. However, a major dredging operation at Olympia Harbor was completed in October 1972 and most spoils (196,400 cubic yards) were dumped at older sites in Budd Inlet; only 21,000 cubic yards were hauled to the approved Dana Passage site. This operation was the subject of cooperative studies on the effects of dredging and spoil disposal by the Corps of Engineers and the Washington Departments of Natural Resources, Ecology (WDE) and Fisheries (WDF).

Dredging and disposal techniques vary in the Portland and Seattle Corps of Engineers districts. The hopper dredges Biddle, Harding, and Pacific are owned and operated by the Portland District and do all of the dredging of harbor inlet channels along the exposed Pacific Ocean coast of California, Oregon, and Washington. The Biddle and Harding also work within San Francisco Bay and the Columbia and Lower Willamette Rivers.

Most of the areas require maintenance dredging only once every ten years because of the dynamic tidal action. Bellingham and Tacoma Harbors have needed dredging at about five-year intervals because of their location at the mouths of rivers with high sediment loads. Major projects are contemplated in the next two or three years at these two sites, as well as one in the Bremerton area.

Much of the harbor sediments have high BOD ratings and will require land disposal. Dredging at Bellingham has reportedly been postponed because of this factor. The Seattle District Corps of Engineers office routinely takes core samples

FIGURE 7.8: WASHINGTON STATE SPOIL DISPOSAL PERMIT SYSTEM

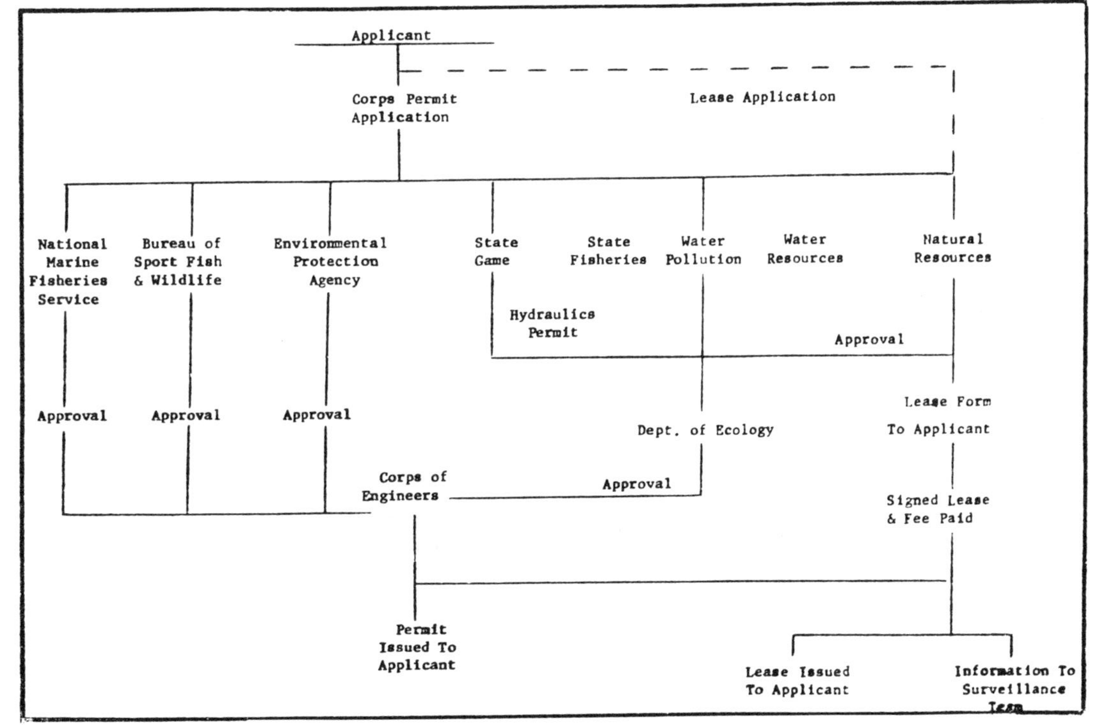

Source: PB 240 917

before dredging to determine compliance with the EPA criteria. Prior to the Olympia Harbor studies, a study was made of the 1969 disposal operation from dredging in the Whatcom Waterway at Bellingham, and an attempt was made to study the effects on water quality during a disposal operation at or near the Steilacoom site.

Surveillance of disposal operations at the present time is performed by the WDNR. They require 12-hour notice prior to dumping and they use both aircraft and reports from cooperative property owners near the sites to detect irregularities. The Marine Environmental Protection Branch of the Thirteenth Coast Guard District has recently received a list of the Corps of Engineers sites along the Oregon-Washington coast from Coast Guard Headquarters. Shore station personnel have been instructed to report any waste loading activities, and air station flight crews are directed to report any dumping.

Industrial Wastes: Disposal of pulp mill and other industrial wastes has been primarily through private outfalls or into municipal sewer systems. At the present time, the only reported barging and deepwater dumping is done by a chemical company which barges to Seattle from Bellingham. At the Georgia Pacific a calcium base sulfite pulp mill, waste sulfite liquor is processed to obtain alcohol and the spent liquor is then clarified and air-oxygenated in settling ponds.

After the vanillin is extracted, the remaining alkaline aqueous solution of lignin fragments (vanillin black liquor) is sold to Kraft paper mills for further reclamation of chemicals and the unsold excess is dumped in the vicinity of latitude 48°15'N, longitude 123°00'W as the barges return to Bellingham. The present barge has a capacity of 336,000 gallons and makes two to three trips per week.

A new one-million gallon barge will soon be used to make one trip per week. The present barge discharges at the rate of 980 gallons per minute while underway at three to six knots. The new barge will average eight knots and the discharge rate will be 2,800 gallons per minute from a deeper discharge nozzle. The liquor weighs 9.55 pounds per gallon and annual dumping is about 225,000 tons. About 35% of the output, or 64,000 tons, was sold in 1972. The permit from the Washington State Department of Ecology limits them to 350,000 tons per year. Black liquor has been dumped at this site in the Strait of Juan De Fuca since 1956.

Analysis of Dumping Activities

Effects of Puget Sound Dredging and Spoil Disposal: There have been two studies in Puget Sound on the effects of dredging and spoil disposal. Both were conducted at relatively shallow sites compared with the depths of the presently approved dump areas. The WDNR hopes to obtain funding for observations of these deeper sites from a manned submersible. WDNR located and approved the sites primarily on the basis of avoidance of unique habitats. That is, they selected areas on the channel side slopes so that fauna, which is restricted to either the shallow edges or the deep basins, is not disturbed.

The study at Bellingham showed that spoils from a submerged discharge of a pipeline dredge caused no surface turbidity plume, but formed a mudflow down a bottom slope of less than 0.5%. There was no measurable dissolved oxygen in this spoil flow, which was seven feet thick at a distance of 400 feet from the discharge, four feet thick at 750 feet, two feet thick at 1,000 feet, and still one foot thick at a distance of 1,400 feet. All benthic organisms covered by this flow were considered to have been smothered.

It is highly unlikely that spoils dumped at the WDNR sites will remain in place because the tidal currents are faster at all of them than at Bellingham, and bottom slopes range from 0.7 to 12%. Thus, it is concluded that spoils dumped at such places will soon be moved down into the basins, and it is conceivable that massive slides may be triggered down the steeper gradients.

As in other areas of the country, the obvious alternative to in-water disposal is land or shoreline filling, creation of mudflats, or addition to existing shallow-water habitats. At the present time only unpolluted spoils are dumped in the water, while polluted spoils are used for fill. Rather than dumping the clean material, it has been suggested that it be mixed to dilute the poor quality material to acceptable levels.

Even better is the concept of using clean spoils as the containing dikes and topping layer for an area to be filled with substandard spoils. In this way, the Sound would be separated by a buffer zone from the polluted spoil fluids. Sufficient time for biodegradation of the contaminants or, at least, a very slow leaching and percolation rate could be obtained by proper soil engineering in the design of the dikes.

While not truly an alternative to in-water dumping, the concept of wider spread dispersion is seldom considered although it more closely emulates nature. Until it stops raining and the law of gravity is repealed, sediments will erode from hills and be dumped in holes. Man's dredging work is simply an acceleration of this natural transport process at selected spots. Floods and tides, slowly but inexorably, are going to fill and level the basins of Puget Sound.

In the process, nutrients are supplied to the living organisms and the dead are buried. If man spread his nontoxic dredge spoils over the basin at annual or more frequent intervals, the benthic fauna could accept the thin veneer of added sediments as readily as they have accepted the spring runoff for millions of years. The logic of establishing small sites for concentrated dumping of spoils in order to reduce the area of adverse effects to a minimum is laudable in theory, but loses its credibility if the mound is flattened and spread by gravity and currents within a week.

WDNR regulations provide for a fine of fifty cents for each yard of spoil that is dumped outside the approved site boundaries. At the same time, the terminal sections of sewer outfalls must be equipped with diffuser ports to spread the discharge over a wide area. Even considering that dredge spoils are mostly solids in a fluid medium, and outfall discharges are mostly liquid with only a small suspended solid component, this differentiation in dispersion concepts is difficult to rationalize.

A joint Federal-State study program on the effects of pulp mill wastes on water quality and marine life was conducted from April 1962 until June 1966. The studies which concerned sulfite and similar wastes concentrated on the effects of seven mills in four separate areas: Bellingham, Anacortes, Everett and Port Angeles.

The studies determined the character and amount of sulfite waste liquor produced at each mill, the concentration and dispersion in the receiving waters, and the toxic effects on selected biota.

In addition to the effects of sulfite waste liquor, whenever sludge beds at these mills were disturbed, release of hydrogen sulfide and depression of dissolved oxygen caused almost instantaneous fish kills. Benthic biota was almost nil where the sludge was more than one inch thick, and both population and species diversity increased with distance from the mills.

Normal dredging techniques and disposal of spoils from harbors at Bellingham, Anacortes, Port Angeles, Everett and Tacoma present a special problem because of the past practices of pulp mills which dumped waste sulfite liquors and settleable suspended solids into the harbor waters forming bottom sludges with high toxicity and BOD. Even if spoil sites for landfill or diked enclosures can be located and constructed to prevent return of polluted fluids to groundwater or Puget Sound, there is still the problem of minimizing deleterious effects during the dredging.

As previously noted, any disturbance of these sludge beds results in the release of hydrogen sulfide and an immediate oxygen demand, with resulting drop in the DO level. The timing of the operation should be such to avoid fish migrations and critical hatching or spawning periods for as much of the local fauna as practical. The use of an enclosing plastic sheet barrier should be seriously considered, to keep organisms out of the dredging zone and to confine the turbidity and pollutants to the smallest possible area.

The disposal enclosures should be designed with internal baffles or diked compartments to allow maximum sedimentation prior to returning excess water to the sound. Planning of the disposal site in coordination with local or regional sewage disposal authorities could be of mutual benefit. For example, part of the site could be used for building sewage treatment facilities with incompletely filled compartments used as aeration ponds. The same could be true if a disposal site next to an existing treatment plant is feasible. Some treatment may be given to the return water if there is any excess hydraulic capacity in the facilities.

REFERENCES

PB 195 225 — *Economic Aspects of Solid Waste Disposal at Sea,* Vol. 3, Massachusetts Institute of Technology, Cambridge, Mass., Sept., 1970.

PB 198 032 — *Background Papers on Coastal Wastes Management,* Vol. 1, National Academy of Engineering, Washington, D.C., 1969.

PB 204 868 — *The Barged Ocean Disposal of Wastes—A Review of Current Practice and Methods of Evaluation,* B.D. Clark, et al, Pacific Northwest Water Laboratory, Corvallis, Oregon, July, 1971.

PB 210 710 — *Eutrophication in Coastal Waters: Nitrogen as a Controlling Factor,* Institute of Marine Resources, Scripps Institution of Oceanography, University of California, San Diego, Calif., December, 1971.

PB 213 473 — *Ocean Disposal of Barge Delivered Liquid and Solid Wastes from U.S. Coastal Cities,* D.D. Smith and R.P. Brown, Applied Oceanography Division, Dillingham Corp., La Jolla, Calif., 1971.

PB 221 684 — *Municipal Waste Disposal by Shipborne Incineration and Sea Disposal of Residues,* M.W. First, et al, Harvard School of Public Health, July, 1973.

PB 224 793 — *Ocean Waste Disposal in Selected Geographic Areas,* Interstate Electronics Corp., July, 1973.

PB 229 761 — *Environmental Survey of an Interim Ocean Dumpsite,* Middle Atlantic Bight, Cruise Report, May 1–5, 1973, H.D. Palmer, et al, Westinghouse Ocean Research Laboratory, Sept., 1973.

PB 230 944 — *Ocean Waste Disposal Practices in Metropolitan Areas of California,* C.F. McFarlane, Interstate Electronics Corp., Anaheim, Calif., Feb., 1974.

PB 230 945 — *Navigation Aids for Ocean Waste Disposal Control,* K.W.

Herkimer, Interstate Electronics Corp., Feb., 1974.

PB 233 019 *Guidelines for Development of Criteria for Control of Ocean Waste Disposal,* Interstate Electronics Corporation, Feb., 1974.

PB 240 917 *Assessing Potential Ocean Pollutants.* A report of the Study Panel on Assessing Potential Ocean Pollutants to the Ocean Affairs Board Commission on Natural Resources, National Research Council/National Academy of Sciences, Washington, D.C., 1975.

AD 735 378 *System Study for Surveillance of Ocean Dumping Operations,* Sperry Systems Management, Division Sperry Rand Corp., Great Neck, N.Y., Sept., 1971.

AD 757 599 *Disposal of Dredge Spoil—Problem Identification and Assessment and Research Program Development.* Army Engineer Waterways Experiment Station, Nov., 1972.

AD 775 826 *Dredged Material Research Program Miscellaneous Paper D74-14.* "Discussion of Regulatory Criteria for Ocean Disposal of Dredged Materials: Elutriate Test Rationale and Implementation Guidelines," J.W. Keeley, R.M. Engler, U.S. Army Engineer Waterways Experiment Station, Office of Dredged Material Research, Vicksburg, Miss., March 1974.

COM 73 11292 *Biological Effects of Ocean Disposal of Solid Waste,* S.D. Pratt, S.B. Saila, A.G. Gaines, Jr. and J.E. Krout, Rhode Island University, Kingston, R.I., 1973.

COM 74 50632 *Report to the Congress on Ocean Dumping and Other Man-Induced Changes to Ocean Ecosystems,* October 1972 through December 1973, Public Law 92-532, Title II. National Oceanic and Atmospheric Administration, Washington, D.C., March, 1974.

Reports by the United States Environmental Protection Agency, Region II, Surveillance and Analysis Division.

Ocean Dumping in the New York Bight—Facts and Figures, July, 1973.

Briefing Report—Ocean Dumping in the New York Bight Since 1973, April, 1974.

Ocean Disposal in the New York Bight—Technical Briefing Report Number 1, July, 1974.

Ocean Disposal in the New York Bight—Technical Briefing Report Number 2, April, 1975.

OCEAN FLOOR MINING 1975

by John S. Pearson

Ocean Technology Review No. 2

This book represents a sincere attempt to gather together and to collate the available information concerning the state of the art of deep sea mining. It is hoped that it may prove helpful to those interested in tapping these valuable ocean resources.

Although manganese nodules now occupy center stage on the undersea minerals scene, there are other promising prospects. Natural barium sulfate, for instance, which is practically insoluble in water, is needed in ever increasing quantities as an additive to drilling muds for oil wells. Promising deposits of this substance have been found off Alaska. Tin-bearing sands from the ocean floor are likely to prove commercially exploitable.

Ocean floor mining offers many advantages which are not possible with traditional land mining. As a whole new concept it lends itself readily to automation. The minerals are available without removing the so-called "overburden," without the use of explosives, and with very little drilling. Transportation by boat, so very desirable for mined ores, is already available without the now prohibitively expensive building of canals connecting to inland waterways.

A partial and condensed table of contents follows. Chapter headings and some of the more important subtitles are given here.

ISBN 0-8155-0569-8 **205 pages**

OIL SPILL PREVENTION AND REMOVAL HANDBOOK 1974

by Marshall Sittig

Pollution Technology Review No. 11

Energy Technology Review No. 2

Ocean Technology Review No. 1

This handbook describes and expounds the most feasible and sophisticated technologies available at present for dealing with a menace of worldwide proportions, i.e. marine pollution caused by accidental contamination with crude or (less often) refined petroleum oils. To make matters worse the crude oil has a specific gravity very close to that of seawater.

Oil slicks, tar balls, and dispersed hydrocarbons are found on the surface as well as on the bottom of the ocean. Very little has been accomplished so far about relieving the latter calamity. Once the spilled oil has found something solid to cling to—it may be the sand of a beach, underwater rocks, wooden structures, the feathers of a gull, or the hair and skin of a bather—it does not readily let go.

Aside from most rigorous safety regulations and preventive measures now in force for tankers, offshore drilling, and seabed mining (still in its infancy), the worldwide effort towards combating oil spills is best illustrated by the myriad, albeit rudimentary, devices being designed by companies engaged in the development of containment booms and other mechanical containment apparatus for dealing with oil slicks on the surface of the sea.

Research continues to obtain improved performance, but, because of natural turbulences and gulfstream-type currents, the effectiveness of floating booms and other recovery systems is severely limited.

Once a spill has been contained and the sea is calm, cleanup measures can be instituted. The two major techniques used today are mechanical gathering and recovery with adsorbents. Mechanical recovery or skimming may be used without restriction in carrying out the contingency plan, while adsorbents may be used only if these materials do not by themselves or in combination with the oil increase the pollution hazard. Straw, manufactured fibers, and natural clays may be mixed with the oil slick, and collected. Straw is considered the most cost-effective agent, because it floats readily even when wet, holds five times its weight in oil and costs little.

Other methods include use of dispersing agents, sinking agents, biological agents and burning. Approval by the appropriate governments is required for each of these measures and is practically never granted for U.S. Coastal Waters. It has been suggested that relatively small beaches can be cleaned by feeding the contaminated sand through the municipal incinerator.

Even a relatively small oil spill may have effects that are catastrophic and long-lasting, but prompt application of the methods described in this handbook will at least help to prevent successive biological and degradative changes of the marine biota and the ecological entity.

In this handbook are condensed vital data from numerous government and other sources that are unusually difficult to find and to pull together. Important oil containment and removal processes are interpreted and explained by examples from recent U.S. patents.

A partial and condensed table of contents follows here. Numbers in parentheses indicate the number of processes or designs per topic. Chapter headings are given, followed by examples of important subtitles.

EFFECTS OF OIL SPILLS ON THE ENVIRONMENT
- Effects on Man
- Effects on Agriculture
- Effects on Industry
- Effects on Wildlife
  - Waterfowl
  - Shellfish
  - Fish
  - Marine Mammals
  - The Marine Food Chain
- Effects on Marshes

PREVENTION AND CONTROL OF SPILLS FROM OFFSHORE OIL PRODUCTION
- Mud Cleaning
- Leakage Control at Well Head (23 processes)
- Wastewater Purification (3)

PREVENTION AND CONTROL OF SPILLS FROM OIL TRANSPORT
- Oil Transfer Operations
  - Pre-Transfer Conference
  - Transfer Procedures
  - Vessel-to-Shore Transfer
  - Shore-to-Vessel Transfer
  - Miscellaneous Transfer Precautions
  - Terminal Facilities
  - MacLean Process
  - Wittgenstein Process
- Tank Cleaning (3)
- Ballast Disposal (3)
- Oil Removal from Tanker Accidents (4)
- Drydock Operations

PREVENTION AND CONTROL OF SPILLS FROM OIL REFINING
- Treatment of Oily Wastewaters (3)

PREVENTION AND CONTROL OF SPILLS OF INDUSTRIAL OILS
- Treatment of Oily Wastewaters (2)

OIL SOURCE DETECTION, IDENTIFICATION AND MONITORING
- Detection and Monitoring (2)
- Spill Quantity Estimation
- Sampling
- Tagging
- Identification

ORGANIZATION FOR OIL SPILL CONTROL
- The On-Scene Commander
- Industry Spill Cleanup Cooperatives
- Relations with Government Organizations
- Financing
- Use of Independent Contractors

OIL CONTAINMENT
- Air Barriers (7 designs)
- Chemical Barriers (4 processes)
- Floating Booms (103 designs)
- Combined Containment and Removal Devices (18 designs)

OIL REMOVAL FROM OPEN WATER SURFACES
- Weir Devices (16)
- Floating Suction Devices (14)
- Sorbent Surface Devices (32)
- Vortex Devices (6)
- Screw Device
- Jet Devices (2)
- Other Devices (4)
- Skimming Vehicles (49 designs)

OIL REMOVAL FROM STREAMS, LAKES, HARBORS
- Proprietary Processes (6)

OIL REMOVAL FROM UNDERGROUND WATERS
- Proprietary Processes (2)

OIL/WATER SEPARATOR DEVICES
- Proprietary Devices and Processes (18)

PHYSICAL AND CHEMICAL TREATMENTS OF OIL SPILLS
- Dispersion Treatments (3)
- Sorption Processes (33)
- Sinking Processes (4)
- Gelling/Coagulation Processes (8)
- Magnetic Separation (3)
- Combustion Processes (14)

OIL REMOVAL FROM BEACHES
- Spill Control Measures
- Chemical Control Agents
- Combustion Methods
- Culpepper Process
- Froth Flotation Treatment

DISPOSAL OF RECOVERED SPILL MATERIAL

ECONOMICS OF OIL SPILL PREVENTION AND CONTROL

FUTURE TRENDS

ISBN 0-8155-0543-4 **466 pages**

INCINERATION OF SOLID WASTES 1974

by Fred N. Rubel

Pollution Technology Review No. 13

This book gives advanced technical and economic information to promote efficient incineration of municipal refuse, and will help to erase the bad neighbor image of incinerator installations within the community. Special incinerator applications, largely outside the scope of municipal waste disposal, have not been neglected.

Incineration offers the most significant volume reduction of solid organic wastes when compared with all other disposal methods.

The most undesirable side effect is, or rather was, air pollution, and here the primary concern is with emissions of particulates, rather than with gases and odors. Many sophisticated devices for nearly complete retention of these solid particles have been developed in recent years, and special emphasis has been placed upon them in the descriptions of apparatus.

Most of the data and other information presented in this book are based upon detailed reports on government contracts fulfilled by industrial companies under the auspices of the Environmental Protection Agency.

A partial and condensed table of contents follows here. Chapter headings are given, followed by examples of important subtitles.

ISBN 0-8155-0551-5 **246 pages**

OFFSHORE DRILLING TECHNOLOGY 1975

by Frank R. Carmichael

Energy Technology Review No. 5
Ocean Technology Review No. 3

The pecuniary interests, impacts, and high stakes involved in offshore drilling for petroleum are clearly revealed in the patent applications of the last few years.

Indeed, many of the concepts and design parameters now being implemented by the major construction companies around the world for use in the North Sea and other offshore locations are fully described and illustrated in the U.S. patent literature.

This book presents over 190 different processes and equipment designs for all phases of modern offshore drilling techniques including descriptions and uses of drilling ships, platforms, wellhead completions and a considerable number of very complex subsea facilities. Many of the processes in this area are described as complete systems, thus consolidating multiple functions in this emerging technology. However, for continuity, these processes are presented in the sections deemed most appropriate for their overall contributions to offshore exploration and exploitation.

A partial and condensed table of contents follows here. Numbers in parentheses indicate a plurality of processes per topic. Chapter headings and some of the more important subheadings and subtitles are given here.

ISBN 0-8155-0566-3 **392 pages**